Yehua Dennis Wei (Ed.)

Urban Land and Sustainable Development

MDPI

This book is a reprint of the Special Issue that appeared in the online, open access journal, *Sustainability* (ISSN 2071-1050) from 2015–2016, available at:

http://www.mdpi.com/journal/sustainability/special_issues/UrbanLand_SustainableDevelopment

Guest Editor
Yehua Dennis Wei
Department of Geography, University of Utah
USA

Editorial Office
MDPI AG
St. Alban-Anlage 66
Basel, Switzerland

Publisher
Shu-Kun Lin

Managing Editor
Guoshui Liu

1. Edition 2017

MDPI • Basel • Beijing • Wuhan • Barcelona • Belgrade

ISBN 978-3-03842-260-0 (Hbk)
ISBN 978-3-03842-261-7 (PDF)

Table of Contents

Section II: New Dimensions of Urban Land Use and Urban Space

Section III: Effects of Urban Development and Land Use Change

List of Contributors

Troy D. Abel Huxley College of the Environment on the Peninsulas, Western Washington University, P.O. Box 1699, Poulsbo, WA 98370, USA.

Wen Chen Key Laboratory of Watershed Geographic Sciences, Nanjing Institute of Geography and Limnology, Chinese Academy of Sciences, Nanjing 210008, China.

Stacy Clauson Department of Environmental Studies, Western Washington University, MS 9085, Bellingham, WA 98225, USA.

Qingyun Du Key Laboratory of GIS, Ministry of Education; School of Resource and Environmental Science, Wuhan University, 129 Luoyu Road, Wuhan 430079, China.

Xianwei Fan School of Economics and Management, University of Chinese Academy of Sciences, Beijing 100190, China.

Jinlong Gao Key Laboratory of Watershed Geographic Sciences, Nanjing Institute of Geography and Limnology, Chinese Academy of Sciences, Nanjing 210008, China.

Chengcheng Gu Center for Modern Chinese City Studies, East China Normal University, Shanghai 200062, China.

Jiong Gu School of Economics and Management, Nanjing University of Information Science & Technology, Nanjing 210044, China.

Ji Guo School of Economics and Management; Collaborative Innovation Center on Forecast and Evaluation of Meteorological Disasters, Nanjing University of Information Science & Technology, Nanjing 10044, China.

Huiran Han School of Geography, Beijing Normal University, No. 19, Xinjiekouwai Street, Haidian district, Beijing 100875, China.

Jianqin Hang School of Economics and Management, Jiangsu Maritime Institute, Nanjing 211170, China.

Chunyang He Center for Human-Environment System Sustainability (CHESS), State Key Laboratory of Earth Surface Processes and Resource Ecology, Beijing Normal University, Beijing 100875, China.

Daquan Huang School of Geography, Beijing Normal University, No. 19, XinJieKouWai Str., HaiDian District, Beijing 100875, China.

Li Huang Department of Geography, University of Idaho, Moscow, ID 83844, USA.

Felix Haifeng Liao Department of Geography, University of Idaho, Moscow, ID 83844, USA.

Juan Lin Center for Modern Chinese City Studies, East China Normal University, Shanghai 200062, China.

Hui Liu School of Economics and Management, Southeast University, Nanjing 211189, China.

Xiaoxing Liu School of Economics and Management, Nanjing University of Information Science & Technology, Nanjing 210044, China.

Zhen Liu School of Geography, Beijing Normal University, No. 19, XinJieKouWai Str., HaiDian District, Beijing 100875, China.

Zhifeng Liu Center for Human-Environment System Sustainability (CHESS), State Key Laboratory of Earth Surface Processes and Resource Ecology, Beijing Normal University, Beijing 100875, China.

Cheng Luo College of Public Administration, Huazhong Agricultural University, 1 Shizishan Street, Wuhan 430070, China.

Corrin McCarn Department of Geosciences, Mississippi State University, Starkville, MS 39762, USA.

Qingmin Meng Department of Geosciences, Mississippi State University, Starkville, MS 39762, USA.

Mário Monteiro Department of Earth Sciences, Faculty of Science and Technology, University of Coimbra, Coimbra 3030-790, Portugal.

Yunfei Peng Shenzhen Urban Planning and Land Resource Research Center, 8009 Hongli Road, Shenzhen 518040, China | Key Laboratory of GIS, Ministry of Education; School of Resource and Environmental Science, Wuhan University, 129 Luoyu Road, Wuhan 430079, China.

Jing Qian Shenzhen Urban Planning and Land Resource Research Center, 8009 Hongli Road, Shenzhen 518040, China | Key Laboratory of GIS, Ministry of Education; School of Resource and Environmental Science, Wuhan University, 129 Luoyu Road, Wuhan 430079, China.

Guangping Shi School of Economics and Management, Southeast University, Nanjing 211189, China.

Minjun Shi School of Economics, Renmin University of China, Beijing 100872, China.

Jinping Song School of Geography, Beijing Normal University, No. 19, Xinjiekouwai Street, Haidian district, Beijing 100875, China.

Shunfeng Song School of Economics and Management, Tianjin Chengjian University, Tianjin 300384, China | Department of Economics, College of Business, University of Nevada, Reno 89557, NV, USA.

Yinshan Tang Henley Business School, University of Reading, Reading RG6 6UD, UK.

Alexandre Oliveira Tavares Centre for Social Studies; Department of Earth Sciences, Faculty of Science and Technology, University of Coimbra, Coimbra 3030-790, Portugal.

Mingfeng Wang Center for Modern Chinese City Studies, East China Normal University, Shanghai 200062, China.

Ran Wei Department of Geography, University of Utah, Salt Lake City, UT 84112, USA.

Yehua Dennis Wei Department of Geography, University of Utah, Salt Lake City, UT 84112, USA.

Ming Wen Department of Sociology, University of Utah, Salt Lake City, UT 84112, USA.

Jonah White Department of Geography, Michigan State University, Geography Building, 673 Auditorium Rd., Room 116, East Lansing, MI 48824, USA.

Chao Wu Key Laboratory of GIS, Ministry of Education; School of Resource and Environmental Science, Wuhan University, 129 Luoyu Road, Wuhan 430079, China.

Jianguo Wu School of Life Sciences and School of Sustainability, Arizona State University, Tempe, AZ 85287, USA | Center for Human-Environment System Sustainability (CHESS), State Key Laboratory of Earth Surface Processes and Resource Ecology, Beijing Normal University, Beijing 100875, China.

Xianhua Wu School of Economics and Management; Collaborative Innovation Center on Forecast and Evaluation of Meteorological Disasters, Nanjing University of Information Science & Technology, Nanjing 10044, China.

Weiye Xiao Department of Geography, University of Utah, Salt Lake City, UT 84112, USA.

Chengfeng Yang School of Geography, Beijing Normal University, No. 19, Xinjiekouwai Street, Haidian district, Beijing 100875, China.

Yongchun Yang College of Urban and Environmental Science, Northwest University, Xi'an 710127, China | Key Laboratory of Western China's Environmental Systems, School of Resource and Environmental Science, Lanzhou University, Lanzhou 730000, China.

Dengbao Yao School of Economics and Management, Southeast University, Nanjing 211189, China.

Komali Yenneti Key Laboratory of Watershed Geographic Sciences, Nanjing Institute of Geography and Limnology, Chinese Academy of Sciences, Nanjing 210008, China.

Deli Zhang School of Geodesy & Geomatics Engineering, Huaihai Institute of Technology, Lianyungang 222005, China | Key Laboratory of Western China's Environmental Systems, School of Resource and Environmental Science, Lanzhou University, Lanzhou 730000, China.

Mengzhu Zhang Centre for Urban Planning and Transport Studies, School of Urban and Environmental Sciences, Peking University, Yiheyuan Road 5, Beijing 100871, China.

Xu Zhang School of Economics and Management, Southeast University, Nanjing 211189, China.

Pengjun Zhao Centre for Urban Planning and Transport Studies, School of Urban and Environmental Sciences, Peking University, Yiheyuan Road 5, Beijing 100871, China.

Xingshuo Zhao Urban-Rural Planning Administration Center, Ministry of Housing and Urban-Rural Development of China, No.9 San Li He Road, Beijing 100835, China.

Dan Zheng School of Economics and Management, University of Chinese Academy of Sciences, Beijing 100190, China.

About the Guest Editor

Yehua Dennis Wei (Ph.D., UCLA) is a professor in the Department of Geography and Institute of Public and International Affairs. Wei is the author of Regional Development in China and more than 100 referenced journal articles. He has many publications in leading journals in geography, urban studies/planning, and regional studies/science, such as *Applied Geography* (AG), *Economic Geography, Environment and Planning A, Journal of Economic Geography, Landscape and Urban Planning, Progress in Human Geography, Urban Geography* and *Urban Studies*. He has also edited about a dozen journal special issues including China's Restless Urban Landscapes I, II (*Environment and Planning A,* 2002) and Urbanization, Land Use and Sustainable Development in China (*Stochastic Environmental Research and Risk Assessment,* 2014). He has received excellence in research awards from the University of Wisconsin-Milwaukee, the National Natural Science Foundation of China, and the Association of American Geographer's (AAG) Regional Development and Planning Specialty Group. His professional services include: advisor/panelist for the U.S. National Science Foundation, consultant to the World Bank, Chair of AAG's China Geography, Asian Geography, and Regional Development and Planning specialty groups, and is vice president of the Chinese Professionals in Geographical Information Science.

Preface to "Urban Land and Sustainable Development"

According to the 2014 revision of the World Urbanization Prospects by UN DESA, urbanization could add another 2.5 billion people to urban population by 2050, with close to 90 percent of the increase concentrated in Asia and Africa. This unprecedented increase in urban population not only poses challenges to providing jobs, housing, and infrastructure, but also exerts an increased pressure on urban land and sustainability. As land is a vital yet limited resource, sustainable management of urban land to cater the needs of this growing urban population is seen as one of the key challenges for achieving an economically efficient, socially equitable, and environmentally safe society. A key tenet for sustainable economic development and smart growth is promoting equitable sustainable urban land development and mitigating land use conflicts.

While a large body of literature has dealt with both land use and sustainable development, the study of the interactive effects of these two remains limited. We need more sophisticated empirical studies examining processes, mechanisms, institutions, equity, and sustainability of urban land use. This book is based on a Special Issue published in *Sustainability*, and examines patterns, structure, and dynamics of urban land development and sustainability from multiple perspectives, in various contexts and at multiple dimensions (economic, social, political, developmental, and environmental, etc.). The research articles examine urban land development and sustainability arising from globalization, urbanization and institutional change.

We have drawn papers from both developed and developing countries, with a focus on China due to the rapid rate of urbanization and change there. This book includes 14 articles, with various research emphases, methodologies and study areas, reflecting the inter-disciplinarily that characterizes urban and land use studies. The book has examined patterns and processes of urban land expansion and sprawl, and also studies new dimensions of urban space, including social media. These papers can provide us insights into the underlying structure and mechanisms of urban land expansion, and open new frontiers on the effects of urban land development on both the natural and human environments. The papers can be grouped under the following three aspects: (1) investigating urban land space by employing varied measurements and methods; (2) addressing the structure and mechanisms of urban land expansion and urbanization; (3) exploring ecological, environmental and socioeconomic effects and footprints of urban development and land use change. We hope that this book will contribute to the understanding of urban land use patterns and processes, and their implications for sustainable development.

Lastly, we would like to acknowledge the constructive comments of Robert Argenbright, for his research assistance to Han Li and Komali Yenneti, and the funding of the Ford foundation (0155-0883) and the National Natural Science Foundation of China (41329001). We would also like to express our gratitude to all the authors, reviewers and the MDPI team.

Yehua Dennis Wei
Guest Editor

Towards Equitable and Sustainable Urban Space: Introduction to Special Issue on "Urban Land and Sustainable Development"

Yehua Dennis Wei

Abstract: The unprecedented wave of global urbanization has exerted increased pressure on urban land and made land-use sustainability an urgent concern. This Special Issue examines patterns, structures, and dynamics of urban land use from the economic, social, and, to a lesser extent, environmental standpoints, in light of the goal of equitable and sustainable development. This introduction discusses the background and design of the Special Issue and highlights the contribution of the selected papers.

Reprinted from *Sustainability*. Cite as: Wei, Y.D. Towards Equitable and Sustainable Urban Space: Introduction to Special Issue on "Urban Land and Sustainable Development". *Sustainability* **2016**, *8*, 804.

1. Introduction

According to "World Urbanization Prospects" [1], urbanization could add another 2.5 billion people to the global urban population by 2050, with close to 90 percent of the increase concentrated in Asia and Africa. The largest urban growth will take place in India and China. This unprecedented increase in urban population not only poses the challenges of providing jobs, housing, and infrastructure, but also exerts an increased pressure on urban land and makes sustainability a matter of overriding concern.

Sustainability is a broad topic with multiple dimensions, including economic development, health, energy, land use, water consumption, and air pollution, among others. A key component of sustainable development is equity. The recent financial crisis exacerbated income inequality in most countries, although it has had an uneven impact on cities, regions and people. Social and environmental disparities have also been intensified in many countries. Consequently, equity has become a top sustainable development goal of the UN's post-2015 development agenda. Indeed, equality and sustainability are fundamental issues for human society and major concerns of governments worldwide [2,3].

Given the importance of these issues, there are relatively limited studies that explore the global characteristics and mechanisms of urban land change and their consequences in regard to equitable and sustainable development. Land is a vital yet limited resource, and how to cater the needs of the growing urban population and achieve an economically efficient, socially equitable, and environmentally sustainable society are challenging issues. A key tenet is promoting equitable and sustainable land development and mitigating land use conflicts [4]. We need more studies to examine the processes, mechanisms, and effects of urban land use change globally and comparatively. We also need more efforts to better understand the myriad interconnections between urban land use and sustainable development.

This Special Issue examines patterns, structures, and dynamics of urban land use and their implications for equitable and sustainable development from multiple perspectives (inter alia, economic, social, and environmental). We examine urban land use and, more broadly, urban space in the contexts of urbanization, institutional change, and governance, as well as the general socio-economic context. During the review process, a high percentage of the manuscripts submitted were rejected, and what remaining all went through 2–3 rounds of review and revision. Papers selected for this Special Issue therefore have been subject to a rigorous peer review procedure with the aim of achieving high quality research and wider dissemination of research results.

2. Research Background and Progress

Urban land has been a subject of inquiry for decades. Earlier studies of urban land mainly focused on spatial patterns and internal structure of urban land use, which could be roughly divided into two groups: evolution of different types of urban land use and description of urban land use structure. While land use studies rely on land use data, studies of urban structure look at broader patterns, such as industrial location and population distribution.

Studies of the evolution of urban spatial structure, i.e., the patterns of land use in urban areas and the distribution of different functional zones in the city, were closely associated with the process of industrialization and advances in transportation and communication [5]. Major theoretical contributions of this sort were made by German location theorists and the Chicago School, including the development of traditional models of urban spatial structure, such as the concentric zone model, the sector model, and the multiple nuclei model.

The post-World War II era from the 1950s to the 1960s witnessed the rise of neoclassical economics in social sciences, including the use of factorial ecology in urban social analysis and the examination of urban land value and its linkage with urban form by means of the bid-rent theory [6]. Mounting social problems and the limitations of neoclassical economics prompted the development of alternative thinking in the 1970s and the 1980s, including neo-Marxist urban theories, behavioral approaches and Third World perspectives. Meanwhile, the neoclassical perspective rebounded with the "quantitative revolution", which provided new methods of analyzing urban land use and spatial structure. In this regard, point pattern analysis and fractal dimension were used in the 1980s and the 1990s to quantitatively define urban land use and spatial structure.

With the intensification of globalization and urbanization in the late 1980s, scholars turned their attention to patterns and dynamics of urban expansion firstly in developed countries, followed by studies in the developing countries [7]. Rapid urban transformation prompted a new wave of theorization, which was led by the "Los Angeles School" in the 1980s and 1990s [8]. Post-World War II suburbanization and the resulting urban sprawl in the United States became key topics of academic inquiry. Such studies are often linked to studies of urban form and spatial structure of cities. Scholars have also sought to describe the regularities of urban space based on the spatial concentration of modern human activities, such as traffic flow, imbalance of housing and jobs, mobile communication frequency, and even Wi-Fi access.

GIS and remote sensing have greatly aided empirical studies of land use, especially as regards urban land expansion, sprawl, and land use change. We now have more accurate data on patterns of global urban expansion and sprawl, not just in the West but also in large developing countries undergoing rapid urbanization such as China and India [9,10]. In addition, various modeling approaches such as agent-based modeling, geographically weighted regression (GWR), spatial regressions and spatial regime models have been applied to the analysis of the underlying determinants of rural-urban land conversion. Results suggest that multiple natural and socio-economic factors, such as access to central business districts (CBDs), transportation improvement, development of mega-projects and industrial districts, and availability of land in suburban areas, are the primary determining factors of urban land expansion [11–14].

Scholars continue to try to understand the process of urban land expansion and sprawl. Following the notions of Asian development states, scholars have analyzed the role of the Chinese government and growth coalitions in urbanization, land development and urban expansion [15–17]. It is argued that even urban sprawl in the United States has been supported by federal housing and transportation policies. Scholars have also investigated issues of power and motivations [18–20].

Urban expansion in developing countries has also been viewed as a process closely linked to the globalization of capital and flow of FDI to developing countries. Adopting the concept of teleconnection, Seto et al. [21] linked global urban land use change to globalization, which broadened the conventional land-use science focus on local factors. Wei and his associates have attempted to apply the notion of China's triple transition of globalization, decentralization and marketization to the understanding of rapid urban expansion in China [22,23].

Great attention has been paid to the interactive effects among urban land use, human society and the physical environment. A rich body of literature has examined the cost of urban expansion and sprawl in developed countries, which tend to see more costs than benefits. Scholars have also linked a variety of urban problems to urban expansion and sprawl, such as spatial inequality in general, residential segregation, job-housing imbalance, spatial mismatch, and food deserts. In addition, a recent study has found that urban sprawl holds down upward intergenerational mobility [24]. As regards health, adults living in sprawling metropolitan areas tend to walk less, are more obese, and suffer more from high blood pressure [25]. Urban sprawl may provide more housing choices for residents, but more centralized metropolitan areas actually exhibit lower median home prices and lower proportions of homes in upper-end price categories than sprawling regions [26].

Scholars have linked urban expansion and sprawl with sustainability and transportation issues [27]. "Compact city" and transit-oriented development have been promoted for more effective use of public transportation systems [28]. Sprawl is associated with significantly higher fatal crash rates [29], and higher costs of infrastructure [30]. Environmentally, air pollution tends to a more serious problem with urban sprawl. Large sprawling metropolitan regions experience a greater number of excessive ozone concentrations than more spatially compact ones [31], and the geographical scope of the urban heat-island effect also tends to expand [32].

As Ewing et al. [24] summarize, there have been reported benefits of urban sprawl as well. Suburban sprawl is seen as a natural manifestation of the American Dream of a big house in the suburbs and tends to improve resident's quality of life in some respects [33,34]. While urban sprawl has decentralized jobs, the rise of suburban population and employment can lead to shorter average commute time [35].

Overall, a large body of literature has dealt with urban land and space due to the consequences of intensified human activity upon resources, social relations and the natural environment [4]. However, more studies are still needed to provide comprehensive analysis of patterns, mechanisms, and effects of urban land use change. In the context of globalization and institutional change, more studies are also needed to better understand urbanization and urban expansion, and their influences on urban space, structure and environment.

3. Research Themes, Topics and Questions

While there is a rich literature on urban land and impacts of urban sprawl, there is little understanding of how urban land and space may affect so called "topical outcomes" such as life expectancy, spatial inequality, housing, and transportation costs. Rising income inequality, and associated lack of upward mobility, have become two of the most important issues of our time [2,36]. While income inequality often makes headlines of news, urban spatial inequality remains less studied and upward mobility is barely on the radar of geographers and planners [3]. The effects of urban space on rising inequality, which are of urgent concern both for developed and developing countries, is largely missing in the literature.

While the discourses on sustainability emphasize overall interactions among the three pillars of sustainable development—society, economy, and environment—most of the research has been focused on preserving the ecological capital against economic growth and development or vice versa. There is little research on the intertwinement of the economic, social (e.g., health, inclusion), and environmental dimensions of sustainable development, which constitutes the essence of sustainability of urban societies.

This Special Issue examines patterns, structure, and dynamics of urban land use change from multiple perspectives, in various contexts and with multiple foci (economic, social, political, developmental, environmental etc.). The research articles examine urban land development and sustainability arising from globalization, urbanization and institutional change. We have identified the following key areas as relevant to the theme of this Special Issue:

- Patterns and trajectories of urban land use change
- Drivers of urban land expansion/sprawl
- Urban land use, metropolitan development and global change
- Urban land and economic/social/environmental sustainability
- Sustainable land use policies and practices

We have also proposed the following potential topics for study:

- Urban land use: patterns, restructuring, efficiency and equity
- Urban land expansion/sprawl: directions, structures, dynamics/trajectories, drivers
- Land development: globalization, development zones, mega projects, global(izing) cities, and "land grabs", space/place/geography as agents
- Urban land use policy and governance; role of institutions
- Global-local: effects of globalization, land development and global change
- Urban land and economic/social/environmental equity and sustainability
- Sustainable land use and urban development policies and practices
- Methodology: ESDA, spatial regression, GWR, multi-agent based modeling, spatiotemporal dynamics, joint decision-making scenario analysis, etc.

We have further raised the following questions to address in research:

- What are the nature, structure and trajectory of urban land expansion, sprawl and development in different contexts?

- What are the underlying mechanisms and the main drivers of urban land use change? What is sustainable urban land use in the context of increasing global urban population and climate change and how can it be achieved?
- What implications do the practices of urban land use have on sustainable development at the local, regional, and global scales?
- What are the governance and political processes?
- How to measure the roles of geography, contexts and institutions?
- How are urban land use and sustainability displayed in terms of convergence/divergence, dimension, scale and time?
- How to use developments in GIS and big data to advance our knowledge?
- What are the theoretical and policy implications of the research?

4. Content and Contribution of the Special Issue

This Special Issue includes 14 articles, with various research emphases, methodologies and study areas, reflecting the inter-disciplinarily that characterizes urban and land use studies. These papers can provide us insights into the underlying structure and mechanisms of urban land expansion, and open new frontiers on the effects of urban land development on both the natural and human environments. For the purpose of this editorial introduction, the papers can be grouped under the following three aspects: (1) investigating urban land space by employing varied measurements and methods; (2) addressing the structure and mechanisms of urban land expansion and urbanization; (3) exploring ecological, environmental and socioeconomic effects and footprints of urban land development.

4.1. Multi-Scalar Patterns and Mechanisms of Urban Land Expansion

This group includes six papers, ranging from the global to the national and local scales, all with a focus on China. At the global scale, Liu et al. test the four prominent hypotheses about patterns of urbanization for sixteen world cities over the period from 1800 to 2000 [37]. The result shows that all these cities have shared three common patterns in urban landscape change: urban landscapes become compositionally more diverse, structurally more fragmented and geometrically more complex as urbanization progresses. Moreover, the dominant patterns of urban land expansion are mainly shifting among infilling, edge expanding and leapfrogging.

On the national scale, using panel data of 253 prefecture-level Chinese cities from 1999 to 2012 [38], Fan et al. investigate the mediating effect of public infrastructure between urban land development and economic growth, and find that public infrastructure is a partial but significant mediator. Land development has a positive impact on public infrastructure, which further facilitates urban economic development.

The next three articles examine urban land use in specific cities and regions in China, including the Yangtze River Delta (YRD), Shenzhen and Beijing. In the first, Gao et al. apply the economic transition framework of globalization, marketization, decentralization, and urbanization, and find that urban land expansion in the YRD is mainly characterized by the growth of residential and industrial land [39]. They identify six factors—foreign direct investment, labor force, government competition, institutions, population, and job-housing relations—as the major underlying driving forces underlying urban expansion.

Qian et al. describe and evaluate the evolution of urban land expansion and land use policy in Shenzhen, China since 2005 [40]. They find that the current mandatory, top-down mode of control, which relies on the central government, has very limited effects on controlling urban expansion. However, such policies as governmental self-restraint, governmental identity change, and policy innovation are effective on guiding China's urbanization.

Relying on CLUE-S model, Markov model and remote sensing data, Han et al. develop two scenarios of urban land expansion (development scenario and protection scenario) to explore the characteristics of land use change and to simulate the future land use demand in Beijing [41]. They find that the conversion from cultivated land to built-up land is still the primary feature of land use change in Beijing. Based on their prediction, the conversion of cultivated land to built-up land in mountainous areas of Beijing will be more prevalent by 2020, and the city will still face a great need for ecological and cultivated-land protection.

Moving to the local scale, Zhao et al. explore the formation process and the determinants of gated informal housing communities in Beijing [42]. This study addresses the role of villages and townships in informal land development and their conflicts with state regulation, and suggests that decentralization and marketization have significantly changed the roles of different levels of government. Instead of following the goals settled by upper-level administration, a township government seems to be more concerned about the quality of local livelihood.

In addition to Chinese cities, this Special Issue also includes a study focusing on urban land use change in Western cities. Monterio and Tavares analyze the spatial and temporal dynamics of land use change in a medium/small city in Portugal between 1958 and 2011, with an explicit emphasis of the planning framework [43]. They find that urban spatial structure changes in Viseu were mainly characterized by a decrease in annual crops and contraction of arable land, as well as an increase of the continuous and discontinuous urban fabric.

4.2. New Dimensions of Urban Land Use and Urban Space

The second group of articles focuses on new dimensions of urban land use and urban space, and how space changes in global/globalizing cities. Based on enterprise registered data in 2010, Huang et al. investigate the employment density distribution as an indicator of urban spatial structure change in Beijing [44]. Their result shows that spatial structure of Beijing is still characterized by mono-centric development. To evaluate the process of polycentric development, they identify five sub-centers of Beijing, most of which are located in suburban areas.

Wang et al. examine Shanghai's adoption of wireless technologies in urban and infrastructure development and the spatial distribution of public Wi-Fi access in Shanghai [45]. By mapping both government-sponsored hotspots through the "i-Shanghai" project and the hotspots established by China Mobile Communications Corporation (CMCC), they find that Shanghai's telecommunication providers have been proactive in arranging WLAN hotspots (a proxy of public Wi-Fi or wireless access). Moreover, for the old neighborhoods in the city center and remote sub-districts, the public Wi-Fi infrastructure development of both government and CMCC facilities is still relatively backward. The authors further address the policies in terms of how to promote Shanghai as a more sustainable wireless city.

4.3. *Effects of Urban Land Use Change*

The six papers in the third group deal with how urban land development affects environmental and socioeconomic structures at both the national and city scales. At the national scale, Guo et al. analyze the panel data of 577 recorded disasters in 30 provinces of China from 1985–2011 to identify their links with economic growth [46]. Based on an integrated indicator system with entropy weighting, they find that human capital can turn meteorological hazards into economic growth, while geological disasters cannot trigger local economic growth.

Zhang et al. test how urban rail transit affects housing prices in China [47]. Relying on panel data of 35 Chinese cities from 2002 to 2013, and by controlling other determinants, such as GDP per capita, land price, investment in real estate and population density, they find that a 1% increase in rail transit mileage will advance housing price by 0.027%. Moreover, they point out that the new rail transit lines have increasingly positive influences on housing price of China.

Yang et al. find that there is a linkage in Chengdu, China between residential land use change and the massive migration and relocation of employees in government departments (EGD), on the one hand, and displacement of urban households by demolition (UDDH), on the other [48]. By tracking the employees hired by state government departments and related branches, they compare residential relocations of EGD and UDDH groups and find that most of the UDDH have migrated from urban centers to fringes, while a large number of EGD still agglomerate close to urban centers. Government interventions differentiate residential relocations of EGD and UDDH, and market mechanisms reinforce this relocation process.

Besides the research focused on China, two studies focus on cities in the United States: Salt Lake City and Seattle. Relying on Utah Household Travel Survey data, Wei et al. analyze the linkages between physical activity (PA), neighborhood land use, accessibility to transportation, and socio-demographic status in Salt Lake County, Utah [49]. By using four-component walkability indices at various geographic scales, they find that PA is associated with neighborhood land use and social demographic status, including the compact design of the neighborhood. Specifically, the current neighborhood design in Salt Lake County only supports people's 20-min walk.

Abel et al.'s paper analyzes the intersections of urban industrial and nonindustrial land use planning, gentrification, and environmental injustice in Seattle, WA [50]. Based on geographic cluster analysis and longitudinal air toxic emission comparisons, the authors quantitatively investigate socioeconomic changes in Seattle Census block-groups between 1990, 2000, and 2009. They find that there is a growing tension between industrial and non-industrial land use, and that Seattle's pattern of development is highly stratified by occupation, income and property values.

5. Conclusions

This Special Issue provides comprehensive studies of urban land use change and its implications for equitable and sustainable development. We have drawn papers from both developed and developing countries, with a focus on China due to the rapid rate of urbanization and change there. We hope that this Special Issue will contribute to the understanding of urban land use patterns and processes, and their implications for sustainable development. The Special Issue has examined patterns and processes of urban land expansion and sprawl, especially in terms of the conversion of rural land to urban use. It also studies new dimensions of urban space, especially social media. Such studies on the effects of urban land use change are particularly valuable, and more efforts are still needed to advance our knowledge.

Globalization and urbanization are continuous processes, and we understand that many questions have not been fully addressed by papers in this special issue. The papers included could not represent many countries in the developing world, despite our efforts trying to broaden the geographical scope of research. The large number of papers on China is evidence of rising scholarship and the rapid advance of science and education in China. We feel the social consequences of urban expansion and sprawl remain under-investigated, and consequently, we have proposed a special issue on urban expansion, sprawl and inequality in *Landscape and Urban Planning*. We are looking forward to coordinate more efforts to better understand the processes and consequences of urbanization and urban land use change.

Acknowledgments: I would like to acknowledge the constructive comments of Robert Argenbright, research assistance of Han Li and Komali Yenneti, and the funding of the Ford foundation (0155-0883) and the National Natural Science Foundation of China (41329001). I would also like to express my gratitude to the authors, reviewers and the MDPI team.

Conflicts of Interest: The authors declare no conflict of interest.

References

1. United Nations. *World Urbanization Prospects: The 2014 Revision*; United Nations: New York, NY, USA, 2014.
2. Stiglitz, J.E. *The Price of Inequality: How Today's Divided Society Endangers Our Future*; W.W. Norton & Company Publishing: New York, NY, USA, 2012.
3. Wei, Y.H.D. Spatiality of regional inequality. *Appl. Geogr.* **2015**, *61*, 1–10.
4. Wei, Y.H.D.; Ye, X. Urbanization, urban land expansion and environmental change in China. *Stoch. Environ. Res. Risk Assess.* **2014**, *28*, 757–765.
5. Anas, A.; Arnott, R.; Small, K. Urban spatial structure. *J. Econ. Lit.* **1998**, *36*, 1426–1464.
6. Alonso, W. *Location and Land Use*; Harvard University Press: Cambridge, MA, USA, 1964.
7. Turner, B.L.; Lambin, E.F.; Reenberg, A. The emergence of land change science for global environmental change and sustainability. *Proc. Natl. Acad. Sci. USA* **2007**, *104*, 20666–20671.
8. Dear, M. Los Angeles and the Chicago school: Invitation to a debate. *City Community* **2002**, *1*, 5–32.
9. Deng, X.; Huang, J.; Rozelle, S.; Uchida, E. Cultivated land conversion and potential agricultural productivity in China. *Land Use Policy* **2006**, *23*, 372–384.
10. Mertes, C.M.; Schneider, A.; Sulla-Menashe, D.; Tatem, A.J.; Tan, B. Detecting change in urban areas at continental scales with MODIS data. *Remote Sens. Environ.* **2015**, *158*, 331–347.
11. Seto, K.C.; Kaufman, R.K. Modeling the drivers of urban land use change in the Pearl River Delta, China. *Land Econ.* **2003**, *79*, 106–121.
12. Luo, J.; Wei, Y.H.D. Modeling spatial variations of urban growth patterns in Chinese cities: The case of Nanjing. *Landsc. Urban Plan.* **2009**, *91*, 51–64.
13. Schneider, A.; Mertes, C.M. Expansion and growth in Chinese cities, 1978–2010. *Environ. Res. Lett.* **2014**, *9*, 024008.
14. Li, H.; Wei, Y.H.D.; Liao, F.H.F.; Huang, Z. Administrative hierarchy and urban land expansion in transitional China. *Appl. Geogr.* **2015**, *56*, 177–186.
15. Wei, Y.H.D. Restructuring for growth in urban China: Transitional institutions, urban development, and spatial transformation. *Habitat Int.* **2012**, *36*, 396–405.
16. Liu, T.; Lin, G.C.S. New geography of land commodification in Chinese cities. *Appl. Geogr.* **2014**, *51*, 118–130.
17. Wei, Y.H.D. Zone fever, project fever: Development policy, economic transition, and urban expansion in China. *Geogr. Rev.* **2015**, *105*, 156–177.
18. Healey, P.; Barrett, S.M. Structure and agency in land and property development processes: Some ideas for research. *Urban Stud.* **1990**, *27*, 89–103.

19. Gore, T.; Nicholson, D. Models of the land-development process: A critical review. *Environ. Plan. A* **1991**, *23*, 705–730.
20. Ganderton, P. Modeling the land conversation process: A realist perspective. *Environ. Plan. A* **1994**, *26*, 803–819.
21. Seto, K.C.; Reenberg, A.; Boone, C.; Fragkias, M.; Haase, D.; Langanke, T.; Marcotullio, P.; Munroe, D.K.; Olah, B.; Simon, D. Urban land teleconnections and sustainability. *Proc. Natl. Acad. Sci. USA* **2012**, *109*, 7687–7692.
22. Gao, J.; Wei, Y.H.D.; Chen, W.; Chen, J. Economic transition and urban land expansion in Provincial China. *Habitat Int.* **2014**, *44*, 461–473.
23. Huang, Z.; Wei, Y.H.D.; He, C.; Li, H. Urban land expansion under economic transition in China. *Habitat Int.* **2015**, *47*, 69–82.
24. Ewing, R.; Hamidi, S.; Grace, J.B.; Wei, Y.H.D. Does urban sprawl hold down upward mobility? *Landsc. Urban Plan* **2016**, *148*, 80–88.
25. Ewing, R.; Meakins, G.; Hamidi, S.; Nelson, A.C. Relationship between urban sprawl and physical activity, obesity, and morbidity—Update and refinement. *Health Place* **2014**, *26*, 118–126.
26. Wassmer, R.W.; Baass, M.C. Does a more centralized urban form raise housing prices? *J. Policy Anal. Manag.* **2006**, *25*, 439–462.
27. Ewing, R.; Cervero, R. Travel and the built environment: A synthesis. *Transp. Res. Rec.* **2001**, *1780*, 87–114.
28. Cervero, R. Built environments and mode choice: Toward a normative framework. *Transp. Res. D* **2002**, *7*, 265–284.
29. Ewing, R.; Schieber, R.A.; Zegeer, C.V. Urban sprawl as a risk factor in motor vehicle occupant and pedestrian fatalities. *Am. J. Public Health* **2003**, *93*, 1541–1545.
30. Lietske, S.N. Determining the relationship between urban form and the costs of public services. *Environ. Plan. B* **2012**, *39*, 155–173.
31. Stone, B., Jr. Urban sprawl and air quality in large US cities. *J. Environ. Manag.* **2007**, *86*, 688–698.
32. Yue, W.Z.; Fan, P.L.; Wei, Y.H.D.; Qi, J.G. Economic development, urban expansion, and sustainable development in Shanghai. *Stoch. Environ. Res. Risk Assess.* **2014**, *28*, 783–799.
33. Bruegmann, R. *Sprawl: A Compact History*; University of Chicago Press: Chicago, IL, USA, 2006.
34. Brueckner, J.K.; Largey, A.G. Social interaction and urban sprawl. *J. Urban Econ.* **2008**, *64*, 18–34.
35. Gordon, P.; Lee, B. *Spatial Structure and Travel Trends in Commuting and Non-Commuting Travel in US Metropolitan Areas in International Handbook on Transport and Development*; Edward Elgar: Cheltenham, UK, 2013.
36. Piketty, T. *Capital in the Twenty-First Century*; Harvard University Press: Cambridge, MA, USA, 2014.
37. Liu, Z.; He, C.; Wu, J. General spatiotemporal patterns of urbanization: An examination of 16 world cities. *Sustainability* **2016**, *8*, 41.

38. Fan, X.; Zheng, D.; Shi, M. How does land development promote China's urban economic growth? The mediating effect of public infrastructure. *Sustainability* **2016**, *8*, 279.
39. Gao, J.; Wei, Y.H.D.; Chen, W.; Yenneti, K. Urban land expansion and structural change in the Yangtze River Delta, China. *Sustainability* **2015**, *7*, 10281–10370.
40. Qian, J.; Peng, Y.; Luo, C.; Wu, C.; Du, Q. Urban land expansion and sustainable land use policy in Shenzhen: A case study of China's rapid urbanization. *Sustainability* **2016**, *8*, 16.
41. Han, H.; Yang, C.; Song, J. Scenario simulation and the prediction of land use and land cover change in Beijing, China. *Sustainability* **2015**, *7*, 4260–4279.
42. Zhao, P.; Zhang, M. The role of villages and townships in informal land development in China: An investigation on the city fringe of Beijing. *Sustainability* **2016**, *8*, 255.
43. Monteiro, M.; Tavares, A.O. What is the influence of the planning framework on the land use change trajectories? Photointerpretation analysis in the 1958–2011 period for a medium/small sized city. *Sustainability* **2015**, *7*, 11727–11755.
44. Huang, D.; Liu, Z.; Zhao, X. Monocentric or polycentric? The urban spatial structure of employment in Beijing. *Sustainability* **2015**, *7*, 11632–11656.
45. Wang, M.; Liao, F.H.F.; Lin, J.; Huang, L.; Gu, C.; Wei, Y.H.D. The making of a sustainable wireless city? Mapping public Wi-Fi access in Shanghai. *Sustainability* **2016**, *8*, 111.
46. Guo, J.; Liu, H.; Wu, X.; Gu, J.; Song, S.; Tang, Y. Natural disasters, economic growth and sustainable development in China: An empirical study using provincial panel data. *Sustainability* **2015**, *7*, 16783–16800.
47. Zhang, X.; Liu, X.; Hang, J.; Yao, D.; Shi, G. Do urban rail transit facilities affect housing prices? Evidence from China. *Sustainability* **2016**, *8*, 380.
48. Yang, Y.; Zhang, D.; Meng, Q.; McCarn, C. Urban residential land use reconstruction under dual-track mechanism of market socialism in China: A case study of Chengdu. *Sustainability* **2015**, *7*, 16849–16865.
49. Wei, Y.H.D.; Xiao, W.; Wen, M.; Wei, R. Walkability, land use and physical activity. *Sustainability* **2016**, *8*, 65.
50. Abel, T.D.; White, J.; Clauson, S. Risky business: Sustainability and industrial land use across Seattle's gentrifying riskscape. *Sustainability* **2015**, *7*, 15718–15753.

Section I:
Patterns and Mechanisms of Urban Land Expansion

General Spatiotemporal Patterns of Urbanization: An Examination of 16 World Cities

Zhifeng Liu, Chunyang He and Jianguo Wu

Abstract: Urbanization is the most dramatic form of land use change that has profoundly influenced environmental and socioeconomic conditions around the world. To assess these impacts and promote urban sustainability, a better understanding of urbanization patterns is needed. Recent studies have suggested several spatiotemporal patterns of urbanization, but their generality is yet to be adequately tested with long-term data. Thus, the main goal of our study was two-fold: (1) to examine the spatiotemporal patterns of urbanization of 16 world cities over a period of 200 years (1800–2000); and (2) to test four prominent hypotheses of urbanization patterns. Using a set of landscape metrics, we quantified temporal changes in the urban landscape pattern of the 16 cities and examined the four hypotheses individually. Our results show that these cities exhibit several common urbanization patterns: the urban landscape becomes compositionally more diverse, structurally more fragmented and geometrically more complex as urbanization progresses. Our study also suggests that urbanization is a process of shifting dominance among three urban growth modes: infilling, edge expanding and leapfrogging. However, idiosyncrasies do exist for individual cities, as detailed attributes of urbanization patterns often depend on the environmental and socioeconomic settings of cities. In addition, the choice of specific landscape metrics and the scales of analysis both influence the urbanization patterns revealed. Our study examined the urbanization patterns, for the first time, on long-term and global scales. The findings shed new light on the patterns and processes of urbanization, with implications for future studies of the ecology, planning and sustainability of cities.

Reprinted from *Sustainability*. Cite as: Liu, Z.; He, C.; Wu, J. General Spatiotemporal Patterns of Urbanization: An Examination of 16 World Cities. *Sustainability* **2016**, *8*, 41.

1. Introduction

Urbanization across the world has resulted in dramatic changes in landscape patterns and profound effects on biodiversity, ecological processes and regional sustainability [1–6]. From 2010–2050, the percentage of the global urban population is projected to increase from 51.6%–67.2% [7], while the global built-up area will increase three times [8]. To assess the impacts of urbanization on ecosystems and regional sustainability, understanding the spatiotemporal patterns of urbanization is an important and necessary step [9,10].

The importance of understanding urbanization patterns is evidenced by the increasing number of studies around the world during the past few decades [5,6,11–17]. For example, Dietzel *et al.* analyzed the spatiotemporal dynamics of the built-up area in the Central Valley of California from 1940–2040 and in the Houston Metropolitan Area from 1974–2002, concluding that urbanization is a two-step alternating process of diffusion and coalescence [18,19]. Li *et al.* quantified the spatiotemporal patterns of urbanization in the central Yangtze River Delta region, China, from 1979–2008, describing urbanization "as a spiraling process of shifting dominance among three urban growth modes: the wax and wane of infilling, edge-expansion, and leapfrog" [20]. Based on their long-term ecological research in the Phoenix metropolitan region, Wu *et al.* stated that "the general pattern of urbanization was that the increasingly urbanized landscape became compositionally more diverse, geometrically more complex, and ecologically more fragmented" [10,21–23]. Since 2000, numerous studies have reported similar findings concerning urbanization patterns [24–26].

These findings have provided much insight into urbanization patterns and their underlying processes, showing both common and idiosyncratic features among cities, which depend on city size, socioeconomic drivers, land use policies and the scale of analysis. However, a comprehensive examination of these findings with cities in different parts of the world is lacking. It remains unclear, therefore, whether these observed patterns of urbanization are general and whether they change with scales in time and space. To address these questions requires comparing and contrasting a number of cities together, using consistent methods and long-term data (at least several decades or longer). Such studies have been hindered by the lack of data and other related logistical reasons.

In this study, we took advantage of the recently available long-term data of global urbanization (see the Materials and Methods for detail) and attempted to address the questions of generality and idiosyncrasy in urbanization patterns. Specifically, our main objectives are two-fold: (1) to identify the generalities and idiosyncrasies in the spatiotemporal patterns of the urbanization of the world's major cities over a period of 200 years (1800–2000); and (2) to test several hypotheses of urbanization patterns. These hypotheses include:

(1) the "diffusion and coalescence" hypothesis, which describes urbanization as an alternative process of diffusion and coalescence (Figure 1a) [18,19];
(2) the three growth mode hypothesis, which characterizes urbanization as a wax and wane process of infilling, edge expansion and leapfrogging [20];
(3) the "landscape modification gradient" hypothesis, which states that, with increasing human modification to the landscape from natural to urban, patch

density and shape regularity increase, whereas mean patch size and landscape connectivity decrease (Figure 1c) [27]; and

(4) the diversity-complexity hypothesis that urbanization results in landscape patterns with increasing compositional diversity, geospatial complexity and ecological fragmentation (Figure 1b) [10,21–23].

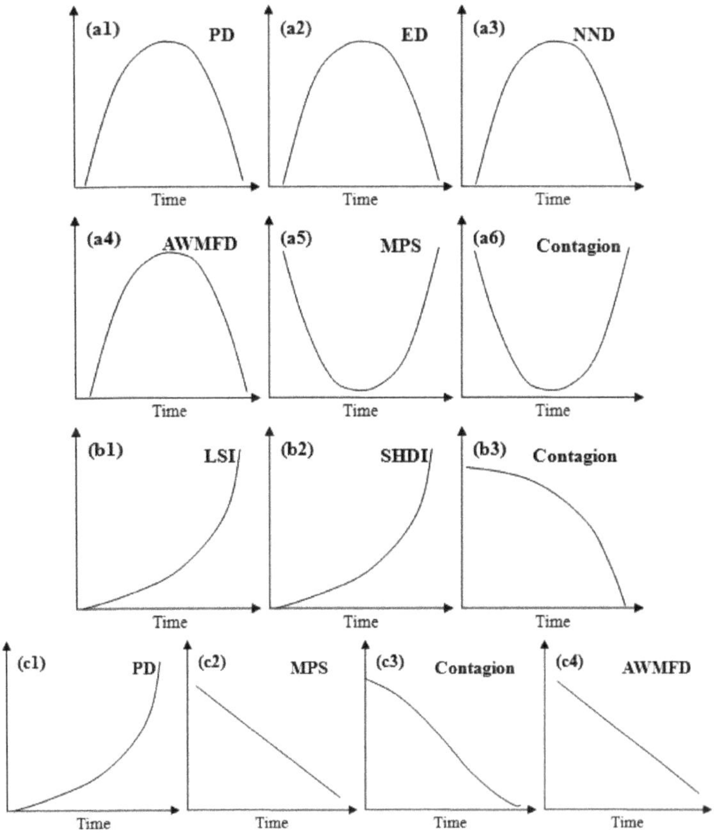

Figure 1. Changes in landscape metrics measuring urbanization patterns on the basis of three hypotheses. (**a1**–**a6**) Based on the "diffusion and coalescence" hypothesis [18,19]; (**b1**–**b3**) based on the "diversity-complexity" hypothesis [10,21–23]; (**c1**–**c4**) based on the "landscape modification gradient" hypothesis [27]. The X axis denotes the process of urbanization; the Y axis denotes the value of landscape metrics measuring urbanization patterns. Landscape metrics include: patch density (PD), edge density (ED), mean Euclidean nearest neighbor distance (NND), area-weighted mean fractal dimension (AWMFD), mean patch size (MPS), contagion, the landscape shape index (LSI) and Shannon's diversity index (SHDI) (see Table 3 for details).

2. Materials and Methods

2.1. Data Acquisition and Processing

The primary data source for our study was the Dataset of the Global Historical Sample of 30 Cities in the Atlas of Urban Expansion, produced by the Lincoln Institute of Land Policy [24] (available at the institute's website) (Table 1, Figure 2) [28]. Among the 30 sample cities, we chose 16 major cities around the world whose data on built-up areas from 1800–2000 were most complete, including Algiers, Beijing, Buenos Aires, Cairo, Guatemala City, Istanbul, London, Manila, Mexico City, Moscow, Mumbai, Paris, Santiago, Shanghai, Sydney and Warsaw (Table 1). To produce the built-up areas over the 200-year period, both remote sensing imageries and historical maps were used [29]. For the period of 1990–2000, the built-up areas were extracted from two Landsat satellite images at a resolution of 28.5 m, one as close as possible to 1990 and one as close as possible to 2000. The extracted built-up areas were further validated using high-resolution remote sensing data in Google Earth, with a producer accuracy of 91% and a user accuracy of 89% [29]. Before 1990, the built-up areas were interpreted from historical maps (Supplementary S1). These historical maps were first geo-referenced and then converted to digital formats. After that, the built-up areas in these maps were digitized to vector files. Besides the built-up areas, the administrative boundaries in 2000, the locations of central business districts (CBD) in 2000 and the urban population from 1800–2000 of 16 cities were also obtained from the Atlas of Urban Expansion (Figure 2b).

Table 1. List of the 16 study cities.

World Region *	Host Country	City	Urban Population in 2000 (millions)	Years with Data on Built-up Area **
South-Eastern Asia	Philippines	Manila	17.34	1802, 1831, 1842, 1884, 1898, 1918, 1945, 1971, 1993, 2002
South-Central Asia	India	Mumbai	16.16	1814, 1849, 1865, 1888, 1909, 1931, 1955, 1968, 1992, 2001
Eastern Asia	China	Shanghai	14.13	1810, 1853, 1875, 1902, 1914, 1944, 1973, 1989, 2001
	China	Beijing	11.87	1800, 1875, 1900, 1929, 1959, 1978, 1988, 1999
Western Asia	Turkey	Istanbul	8.83	1807, 1840, 1872, 1899, 1916, 1934, 1960, 1987, 2000

Table 1. *Cont.*

World Region *	Host Country	City	Urban Population in 2000 (millions)	Years with Data on Built-up Area **
Central America	Mexico	Mexico City	17.22	1807, 1830, 1861, 1886, 1910, 1929, 1950, 1970, 1989, 2000
	Guatemala	Guatemala City	1.77	1800, 1850, 1900, 1936, 1950, 1976, 1993, 2000
South America	Argentina	Buenos Aires	11.92	1809, 1836, 1867, 1887, 1918, 1943, 1964, 1987, 2000
	Chile	Santiago	5.34	1800, 1850, 1875, 1900, 1930, 1950, 1970, 1989, 2000
Northern Europe	United Kingdom	London	10.03	1800, 1830, 1860, 1880, 1914, 1929, 1955, 1978, 1989, 2000
Western Europe	France	Paris	9.52	1800, 1832, 1855, 1880, 1900, 1928, 1955, 1974, 1987, 2000
Eastern Europe	Russia	Moscow	9.14	1808, 1836, 1893, 1914, 1939, 1957, 1978, 1991, 2002
	Poland	Warsaw	2.00	1794, 1831, 1867, 1888, 1915, 1936, 1958, 1979, 1992, 2002
Northern Africa	Egypt	Cairo	13.08	1800, 1846, 1874, 1897, 1917, 1927, 1947, 1960, 1984, 2000
	Algeria	Algiers	3.63	1800, 1828, 1858, 1888, 1903, 1929, 1955, 1972, 1987, 2000
Oceania	Australia	Sydney	4.23	1808, 1833, 1860, 1883, 1895, 1917, 1945, 1975, 1993, 2002

* Please refer to the World Urbanization Prospects from the United Nations for the details of world region classification (http://esa.un.org/unpd/wup/); ** please refer to Supplementary S1 for the details of the data source.

After the data acquisition, all of the vector files of historical built-up areas were converted to raster data at a resolution of 28.5 m to be consistent with the data from 1990–2000. Then, we performed a time series correction to improve the continuity and comparability of the built-up areas in different years (Figure 3). According to Angel *et al.* [8], built-up areas could be assumed to continuously increase from 1800–2000, and a pixel of the built-up area detected in an earlier year would remain as built-up area in a later year. Based on this assumption, we implemented the time series correction as summarized by the following formula:

$$DN_{(n,i)} = \begin{cases} 0 & DN_{(n+1,i)} = 0 \\ DN_{(n,i)} & otherwise \end{cases} \quad (1)$$

where $DN_{(n,i)}$ and $DN_{(n+1,i)}$ are the class values (built-up area or not) at the i-th pixel in the n-th and n + 1th years, respectively. A class value of 1 represents built-up area, while 0 represents non-built-up area.

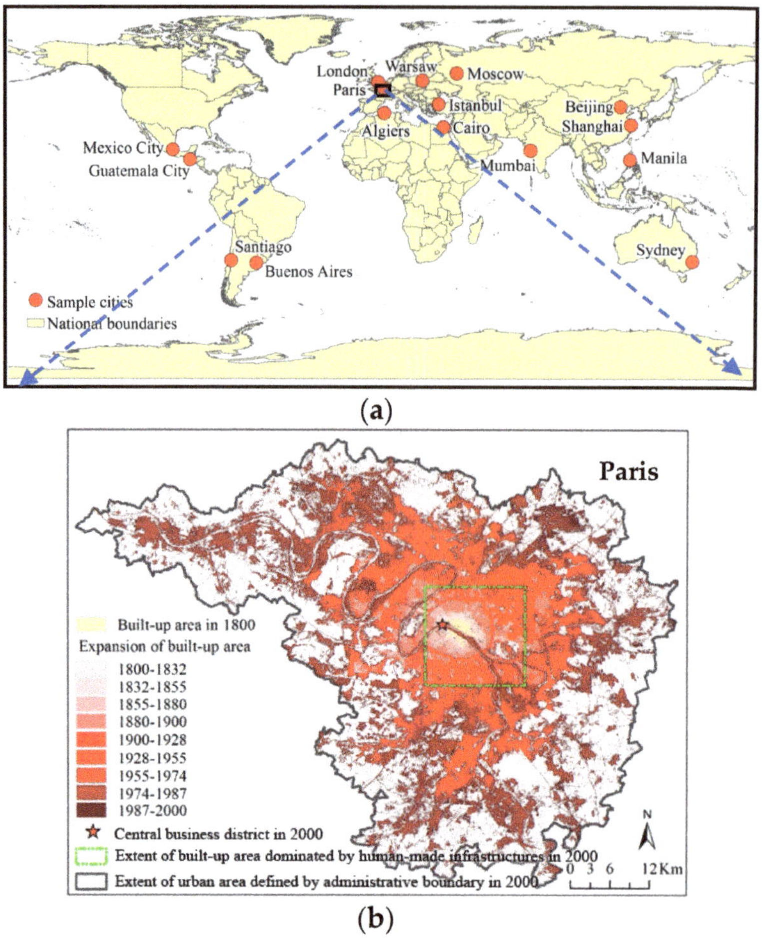

Figure 2. The locations of the 16 study cities (**a**) and, as an example, the urbanization map of Paris, illustrating urban expansion from 1800–2000 (**b**) (see Table 1 for details).

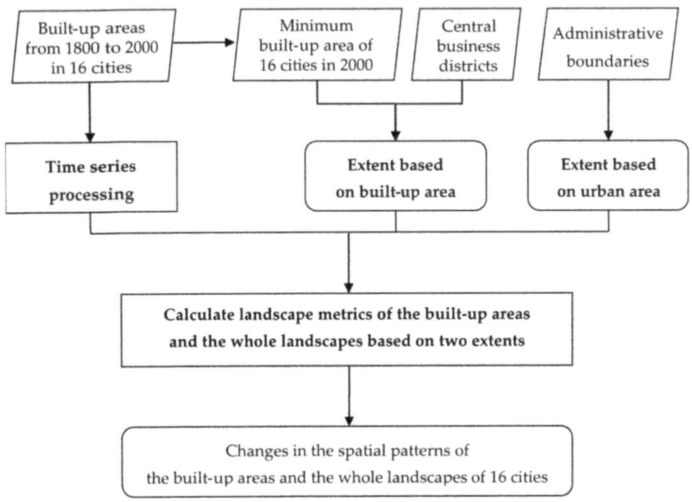

Figure 3. A flowchart of the key steps in data acquisition, processing and analysis.

2.2. *Verification of Data Consistency*

As described above, we unified the data format and the spatial resolution to improve the consistency of the two types of data and then performed a time series correction. To test the consistency of the data with correction, we further employed an indirect assessment method used by Zhang and Seto [30], Liu *et al.* [31] and He *et al.* [32], which uses information on urban population to assess the built-up area. In the assessment process, the changes in urban population from 1800–2000 and their correlations with built-up area were analyzed in all 16 cities (Supplementary S2). The results showed that the changes in urban population and built-up area represented similarly exponential trends from 1800–2000, and the turning points of urban population and built-up area were found in the same year in most cities, where increases of both urban population and built-up area mainly occurred after 1900 (Supplementary S2). The correlation analysis showed that the built-up area had a strong correlation (R of 0.85) with the urban population at a significance level of 0.01 in each city (Supplementary S2). The assessment revealed that the built-up area with correction, which was highly consistent with urban population, could represent the trend of urbanization in a reliable way.

2.3. *Methods for Quantifying Urbanization Patterns*

We analyzed the spatiotemporal patterns of urbanization at two extents: the smaller extent of built-up area dominated by human-made infrastructures and the larger extent of urban area defined by administrative boundaries [33] (Figure 3). We chose the two spatial extents to explore the possible effects of changing spatial scales reported in previous studies [22,34]. The built-up area extent of a city was drawn using a square around the city center (Figure 2b). Specifically, the city center was determined by the CBD, while the area of the square was fixed as 15 km^2, which was approximately equal to the minimum built-up area of 16 cities in 2000, to ensure the extent was almost filled with built-up areas in 16 cities in that time (Figure 2b).

To facilitate the comparison and the test of existing hypotheses, we chose a set of landscape metrics used in several previous studies [10,18–21,34–36] (Table 3). To quantify the spatiotemporal patterns of the built-up areas, we selected seven landscape metrics, including: the percentage of landscape (PLAND), patch density (PD), edge density (ED), mean Euclidean nearest neighbor distance (NND), mean patch size (MPS), area-weighted mean fractal dimension (AWMFD) and the landscape shape index (LSI). The landscape expansion index (LEI), developed by Liu *et al.* (2010), was also calculated to detect the different urban growth modes (infilling, edge expansion and leapfrog). To quantify the spatiotemporal patterns of the whole landscapes during urbanization, we used the three landscape metrics (*i.e.*, PD, MPS and AWMFD) to reflect the patch attributes and three metrics, including LSI, Shannon's diversity index (SHDI) and contagion, to represent the landscape-level pattern attributes. Based on the two spatial extents in each city, all of the landscape metrics were computed using the FRSGSTATS software (v4.1) [37], while the LEI of four cities (*i.e.*, Beijing, Mexico City, Paris and Warsaw) was computed using ArcGIS software (v10) [38].

Table 2. List of landscape metrics used in the study, all of which, except landscape expansion index (LEI), were based on McGarigal *et al.* [37] and Wu *et al.* [10].

Landscape Metric	Abbreviation	Description
Area-Weighted Mean Fractal Dimension *	AWMFD	The patch fractal dimension weighted by relative patch area, which measures the average shape complexity of individual patches for the whole landscape or a specific patch type.
Contagion *	Contagion	An information theory-based index that measures the extent to which patches are spatially aggregated in a landscape.
Edge Density *	ED	The total length of all edge segments per hectare for the class or landscape of consideration (unit: m/ha).
Landscape Expansion Index **	LEI	An indicator used for interpretation of landscape expansion types (*i.e.*, infilling edge expansion and leapfrog).
Landscape Shape Index *	LSI	A modified perimeter-area ratio of the form that measures the shape complexity of the whole landscape or a specific patch type.

Table 3. List of landscape metrics used in the study, all of which, except landscape expansion index (LEI), were based on McGarigal *et al.* [37] and Wu *et al.* [10].

Landscape Metric	Abbreviation	Description
Mean Patch Size *	MPS	The average area of all patches in the landscape (unit: ha).
Mean Euclidean Nearest Neighbor Distance *	NND	The distance to the nearest neighboring patch of the same type, based on the shortest edge-to-edge distance (unit: m).
Patch Density *	PD	The number of patches per square kilometer (*i.e.*, 100 ha).
Percentage of Landscape *	PLAND	Relative area of a specific patch type in a landscape (unit: %).
Shannon's Diversity Index *	SHDI	A measure of the diversity of patch types in a landscape that is determined by both the number of different patch types and the proportional distribution of area among patch types.

* The mathematical formulations can be found in McGarigal *et al.* [37]. ** The mathematical formulations can be found in Liu *et al.* [35]. The urban growth mode of infilling refers to the gap between old urban patches being filled up with the new urban patch (*i.e.*, LEI is between 50 and 100). The urban growth mode of edge expansion means the new urban patch is expanded from an edge of existing urban patch (*i.e.*, LEI is between 0 and 50). The urban growth mode of leapfrog means the new urban patch is isolated from the old ones (*i.e.*, LEI is equal to 0).

3. Results

3.1. Changes in the Spatial Pattern of the Built-Up Area

The percentages of built-up areas (PLAND in the software of FRAGSTATS [37]) increased exponentially in all 16 cities from 1800–2000 at both extents (Figure 4). At the urban area extent, PLAND was less than 0.5% in each city in 1800, while this value was still less than 1% in most of the cities in 1900 (Figure 4a). In 2000, PLAND was close to or greater than 5% in all of the cities (Figure 4a). The increase of PLAND from 1900–2000 was more than five-times the increase from 1800–1900 in all of the cities (Figure 4a). In addition, PLAND was less than 25% at the built-up area extent in 1800, while it was close to 100% in most cases in 2000 (Figure 4b). This increase of PLAND mainly occurred after 1900 (Figure 4b).

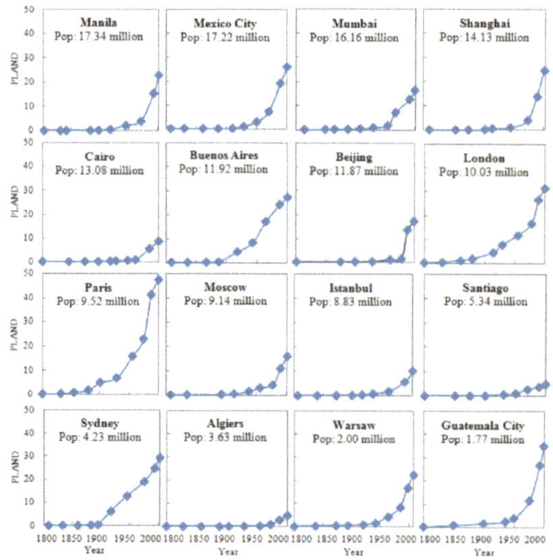

(a)

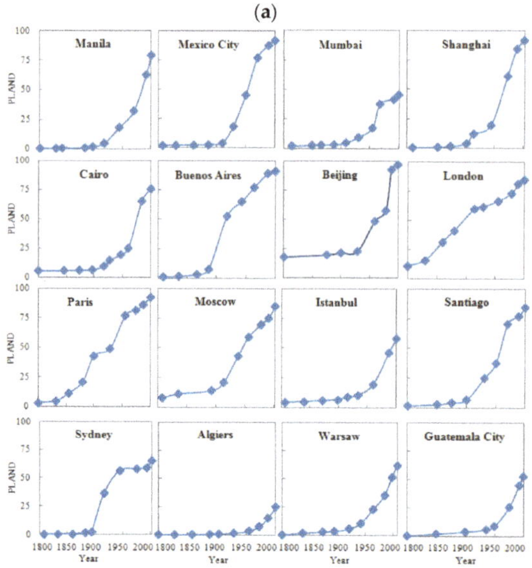

(b)

Figure 4. Changes in the percentage of the built-up area (PLAND) in the 16 study cities from 1800–2000 at the two extents: (**a**) the urban area; and (**b**) the built-up area. The study cities are ordered by the urban population in 2000.

PD, ED, LSI and AWMFD of built-up areas generally increased during urbanization from 1800–2000 (Figure 5). The values of PD, ED and LSI were all close to zero in 1800, but increased to more than 2 num/km^2, 10 m/ha and 60, respectively, in 2000, with the largest increases occurring between the 1970s and the 1990s for most of the cities (Figures S1a, S2a and S6a in Supplementary S3). The values of AWMFD increased from about 1.1 in 1800 to nearly 1.4 in 2000 in most of the cities (Figure S4a in Supplementary S3). At the built-up area extent, PD, ED, LSI and AWMFD also showed increasing trends from 1800–2000 in general (Figures S1b, S2b, S4b and S6b in Supplementary S3). In contrast, PD, ED and LSI of the built-up areas in Mumbai decreased from 1931–1955 at the two extents (Figures S1, S2 and S6 in Supplementary S3).

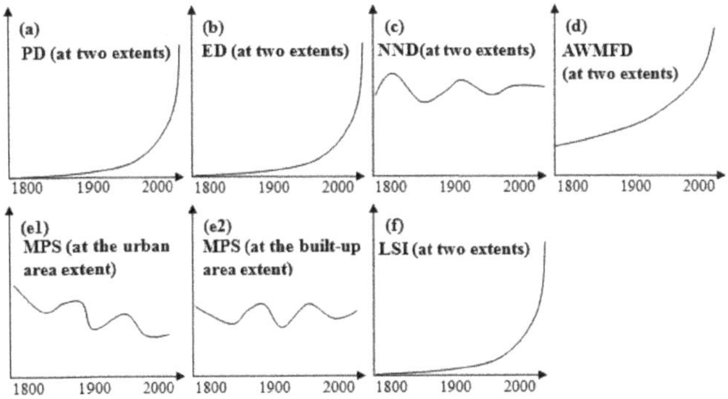

Figure 5. General trends of landscape metrics measuring the spatial pattern of the built-up area in the 16 study cities from 1800–2000 at the two extents (see Figures S1–S6 in Supplementary S3 for details). The X axis denotes the process of urbanization, and the Y axis denotes the value of landscape metrics measuring urbanization patterns. Landscape metrics include: patch density (PD), edge density (ED), mean Euclidean nearest neighbor distance (NND), area-weighted mean fractal dimension (AWMFD), mean patch size (MPS) and landscape shape index (LSI) (see Table 3 for details).

Two metrics, MPS and NND, changed as wave-shaped patterns from 1800–2000 (Figure 5). Specifically, the MPS revealed fluctuating changes with a general decreasing trend from 1800–2000 at the urban area extent; the values decreased from near 100 ha to less than 10 ha in most of the cities (Figure 5e1 and Figure S5a in Supplementary S3). At the built-up area extent, the values of MPS changed slightly between 1800 and 2000 (Figure 5e2 and Figure S5b in Supplementary S3). Furthermore, NND also changed a little, with values of about 100 m, between 1800 and 2000, in most of the cities at the two extents (Figure 5c and Figure S3 in Supplementary S3).

All three urban growth modes, infilling, edge expansion and leapfrog development, occurred simultaneously during urbanization, with changing relative dominance over time (Figure 6). Specifically, the proportions of the numbers and the areas of new urban patches for each growth mode both varied considerably among different time periods (Figure 6). Generally, leapfrog and infilling had a tendency to interchange their relative dominance, while edge expansion kept its importance in both the patch number and the patch area (Figure 6).

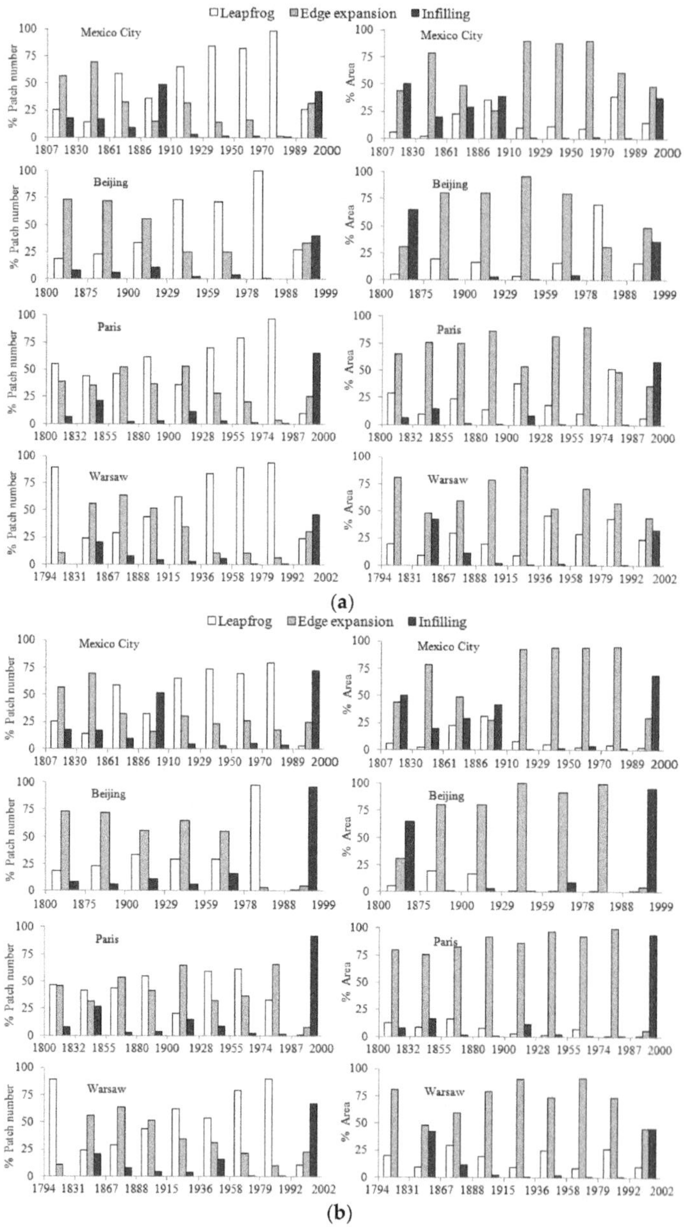

Figure 6. Changes in the relative dominance of three growth modes of urbanization (infilling, edge expansion and leapfrog) from 1800–2000 at the urban area extent (**a**) and the built-up area extent (**b**).

3.2. Changes in the Spatial Pattern of the Whole Urban Landscape

Among the four landscape metrics showing monotonic trends, PD, AWMFD and LSI increased, while MPS decreased during urbanization from 1800–2000 (Figure 7). Specifically, the increases of PD, AWMFD and LSI mainly took place after 1900. The values of PD increased from less than 1 num/km^2 in 1900 to more than 4 num/km^2 at the urban area extent and to above 10 num/km^2 at the built-up area extent in 2000 in most cases (Figure S7 in Supplementary S3). The values of AWMFD increased from less than 1.1 in 1900 to about 1.3 in 2000 at the two extents in general (Figure S8 in Supplementary S3). LSI commonly increased from about zero in 1900 to greater than 15 in 2000 at the two extents (Figure S10 in Supplementary S3). In contrast to the increases in PD, AWMFD and LSI, MPS decreased from above 1000 ha in 1800 to below 10 ha in 2000 at the urban area extent and declined from more than 100 ha in 1800 to less than 10 ha in 2000 at the built-up area extent in most of the cities (Figure S9 in Supplementary S3). In addition to the general trends, the decreases of PD, AWMFD and LSI and the increases of MPS were found at the two extents for Mumbai during the period of 1931–1955 (Figures S7–S10 in Supplementary S3).

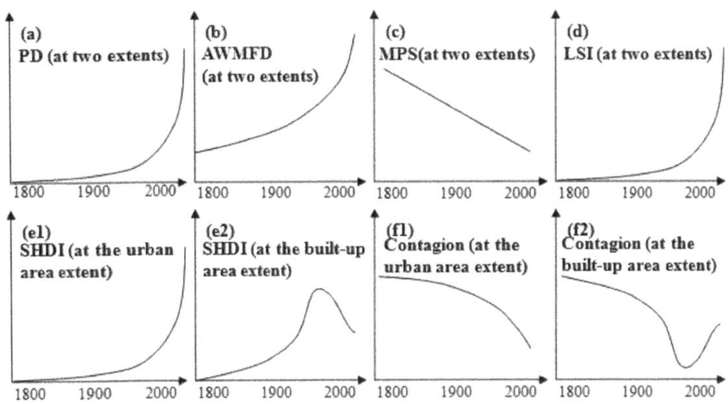

Figure 7. General trends of landscape metrics measuring the spatial pattern of the whole urban landscape in the 16 study cities from 1800–2000 at the two extents (see Figures S7–S12 in Supplementary S3 for details). The X axis denotes the process of urbanization, and the Y axis denotes the value of landscape metrics measuring urbanization patterns. Landscape metrics include: patch density (PD), area-weighted mean fractal dimension (AWMFD), mean patch size (MPS), landscape shape index (LSI), Shannon's diversity index (SHDI) and contagion (see Table 3 for details).

Changes in SHDI and contagion were generally different between the two extents (Figure 7e–f and Figures S11 and S12 in Supplementary S3). At the urban area extent, changes in the two metrics showed monotonic trends in most cases (Figures

S11a and S12a in Supplementary S3). To be specific, SHDI increased from *circa* zero in 1800 to above 0.2 in 2000 (Figure S11a in Supplementary S3), and contagion decreased from about 100 in 1800 to below 80 in 2000 (Figure S12a in Supplementary S3). At the built-up area extent, non-monotonic trends were generally found in changes of SHDI and contagion, and the thresholds of SHDI (about 0.7) and contagion (about 25) were respectively found in the same year in most of the cities (Figures S11b and S12b in Supplementary S3). Before approaching the corresponding thresholds, SHDI continuously increased, and contagion decreased, while they changed directions after reaching their thresholds, resulting in somewhat mirror images between SHDI and contagion (Figures S11b and S12b in Supplementary S3). In addition to the hump-shaped patterns and the U-shaped patterns, SHDI and contagion showed monotonic trends from 1800–2000 at the built-up area extent in Algiers (Figures S11b and S12b in Supplementary S3).

4. Discussion

4.1. Generalities and Idiosyncrasies in Urbanization Patterns

All 16 cities had experienced an exponential increase of built-up areas, resulting in dramatic changes in spatial patterns from 1800–2000. Specifically, the values of PLAND, PD, ED, LSI and AWMFD of the built-up areas all increased from 1800–2000 (Figure 5); changes in MPS and NND of the built-up areas, however, both revealed wave-shaped patterns (Figure 5). In the same period, the PD, AWMFD and LSI of the whole landscapes increased, as well, while the values of MPS decreased at the landscape level (Figure 7a–d). However, changes in SHDI and contagion differed between the two extents (Figure 7e–f). SHDI increased continuously, and contagion decreased monotonically at the urban area extent, while SHDI and contagion showed, respectively, a hump-shaped pattern and a U-shaped pattern at the built-up area extent (Figure 7e–f).

Cities in the same world region often revealed similar urbanization patterns, which usually varied slightly among different regions (Figure S13 in Supplementary S3). For instance, obvious increases of PLAND were found after the 1970s in the cities in Eastern Asia (*i.e.*, Shanghai and Beijing), which were found after the 1950s in the cities in Central America (*i.e.*, Mexico City and Guatemala City) and after the 1880s in the cities in Northern/Western Europe (*i.e.*, London and Paris) (Figure S13 in Supplementary S3). In addition, the increases of PLAND in the cities in Northern Africa (*i.e.*, Cairo and Algiers) were much less than the values in most cities in other regions (Figure S13 in Supplementary S3). The variation of urbanization patterns may result from the different urbanization levels among world regions.

The different patterns of changes mentioned above indicate that the choice of landscape metrics and the scale of analysis matter in charactering spatiotemporal patterns of urbanization [9,10,21,34]. NND of the built-up areas changed little between 1800 and 2000, which may be attributable to the distinguishing feature of urban growth that new urban patches were commonly built near existing urban patches [39] (Figure 5c); even though PD continued to increase, NND was not likely to change. Moreover, all of the changes in landscape metrics resulted from the integration of three urban growth modes: infilling, edge expansion and leapfrog development, with changing relative dominance over time (Figure 6). During the urbanization from 1800–2000, all of the three urban growth modes occurred simultaneously, while infilling and leapfrog generally showed opposite trends: when infilling increased its share in the number or the area of new urban patches, the corresponding value of leapfrog decreased, and *vice versa* (Figure 6). In addition to the alternation of the dominance of infilling and leapfrog, edge expansion remained relatively important during urbanization (Figure 6).

Besides the general patterns of urbanization, some idiosyncrasies also existed in several cities. For instance, PD, ED, LSI and AWMFD all decreased in Mumbai in the period of 1931–1955 at both extents (Figures S1, S2, S4 and S6 in Supplementary S3), and SHDI and contagion changed monotonically in Algiers from 1800–2000 at the built-up area extent (Figures S11b and S12b in Supplementary S3). The monotonic changes in SHDI and contagion in Algiers were apparently due to the much slower urbanization rate (Figure 4), with less than 25% of the built-up area as urban land in 2000, as compared to greater than 50% for other cities (Figure 4b). In addition, the unusual pattern of urbanization in Mumbai in the period of 1931–1955 may be attributable to the land policy change after the independence of India in 1947 [40].

4.2. Testing Hypotheses of Urbanization Patterns

4.2.1. The Diffusion and Coalescence Hypothesis

Diffusion here refers to the urban growth mode of leapfrogging, while coalescence represents the commingling of infilling and edge expansion [18,19,35]. According to Dietzel *et al.* [18,19], this hypothesis can be translated into testable temporal patterns of landscape metrics: the urban land area increases monotonically; PD, ED, AWMFD and NND of urban patches increase first, then peak at different times and, finally, decrease, exhibiting unimodal shapes (Figure 1a); MPS of urban patches and contagion of the whole landscape are highest at the beginning of the diffusion process and the end of the coalescence process and reach their lowest values in between, thus exhibiting somewhat mirror images of PD, ED, AWMFD and NND (Figure 1a). Our results indicate that the "diffusion and coalescence" hypothesis may manifest in many cities during a relatively short duration, but is too simplistic over a few centuries. The wave-shaped trends of MPS and NND at the two extents (Figure 5c,e) and the unimodal-shaped trends of contagion at the built-up area extent

(Figure 7f) supported this hypothesis. However, in terms of either the number or the area of new urban patches, all three urban growth modes (*i.e.*, infilling, edge expansion and leapfrog) occurred simultaneously during urbanization (Figure 6); no single process of diffusion or coalescence occurred alone during urbanization.

4.2.2. The Three Growth Modes Hypothesis

As already mentioned above, our results are more consistent with the three growth modes hypothesis than the diffusion-coalescence hypothesis. In a study of the central Yangtze River Delta region, China, Li *et al.* [20] concluded that "it is more plausible to view urbanization as a spiraling process that involves three growth modes of leapfrogging, edge-expanding, and infilling", and that "leapfrog and infilling tend to alternate in their relative dominance while edge-expansion is likely to maintain its importance throughout much of the urbanization process". Our results of four world cities fully corroborate this hypothesis (Figure 6).

4.2.3. The Landscape Modification Gradient Hypothesis

Forman and Godron [27] postulated that PD and the regularity of patch shape would increase, while MPS and landscape connectivity would decrease, along a landscape modification gradient. From a time-for-space perspective, the natural-to-urban gradient in space can be approximated by the temporal urbanization pattern of cities (Figure 1c). Taking such a perspective, our results of PD, MPS and landscape connectivity generally support the landscape modification gradient hypothesis (Figure 7a,c,f), assuming that landscape connectivity is positively related to contagion and negatively to PD [10]. However, our study shows that the regularity of patch shape, conversely related to AWMFD, decreased as urbanization progressed for the 16 cities (Figure 7b).

4.2.4. The Diversity-Complexity Hypothesis

According to Wu *et al.* [10,21–23], as urbanization unfolds, LSI and SHDI increase, whereas contagion decreases, *i.e.*, urbanization increases the compositional diversity and structural complexity (Figure 1b). In this study, we found that at the urban area extent, LSI and SHDI increased, whereas contagion decreased, monotonically, confirming the diversity-complexity hypothesis (Figure 7d,e1,f1). At the smaller extent of the urban core area, however, the monotonic changes may hit the ceiling or change directions when much of the landscape is urbanized (Figure 7e2,f2).

4.3. Suggestions for Choosing Landscape Metrics in Quantifying Urbanization Patterns

In this study, we selected landscape metrics to quantify urbanization patterns based on two principles. First, the metrics were used in the four prominent hypotheses of urbanization patterns [10,18–23,27]. Second, the metrics measured urbanization patterns well according to our previous study in Phoenix and Las Vegas [10]. Specifically, seven class-level metrics (PLAND, PD, ED, NND, MPS, AWMFD and LSI) and six landscape-level metrics (PD, MPS, AWMFD, LSI, SHDI and contagion) were used.

It is well known that many landscape metrics are closely related, resulting in redundancy in quantifying landscape patterns [37]. In our research, we also found that several metrics showed similar or inverse trends, which were seemingly related. Among class-level metrics, PD, ED, AWMFD and LSI showed similar trends during urbanization (Figure 5). Among the landscape-level metrics, PD, AWMFD and LSI represented similar trends, while SHDI and contagion showed inverse trends (Figure 7). These related metrics may cause redundancy in quantifying urbanization patterns.

To quantify landscape patterns without redundancy, eliminating related metrics is needed. To achieve this, Riitters *et al.* [41] identified six common dimensions of pattern and structure (*i.e.*, composite measures of average patch compaction, overall image texture, average patch shape, patch perimeter area scaling, the number of attribute classes and large patch density area scaling) and found that the information of 55 metrics could be reduced to six univariate metrics (*i.e.*, average perimeter-area ratio, contagion, standardized patch shape, patch perimeter area scaling, the number of attribute classes and large patch density area scaling) to represent these dimensions in a simpler way. Frohn and Hao [42] divided 16 metrics into four individual groups, including class metrics, shape metrics, patch metrics and edge metrics, and found that LSI, square pixel (SqP), ED, PD and NND were appropriate to measure spatial aggregation. According to the previous studies, we suggested that at least one of the metrics representing different aspects of landscape pattern should be selected, and two or three metrics describing patterns in the same aspect were recommended to be calculated to confirm each other. Thus, six metrics (*i.e.*, PLAND, PD, LSI, MPS, NND and contagion), which described urbanization patterns in various aspects, should be selected. In addition, ED, AWMFD and SHDI were recommended to confirm these in quantifying urbanization patterns.

5. Conclusions

Cities differ in many ways, from climate to architecture, ecology, economy and culture. While idiosyncratic differences abound, our study of 16 world cities reveals several general spatiotemporal patterns of urbanization common to seemingly disparate cities. First, urbanization generally leads to an increasingly diverse and complex cityscape. As Wu *et al.* [10,21–23] put it, as urbanization continued to unfold, "landscapes became increasingly more diverse in land use, more fragmented in structure, and more complex in shape". Second, the spatial dynamics of urbanization often operates through three growth modes, infilling, edge-expanding and leapfrogging, which commingle, but with shifting dominance [20]. Third, behind these general patterns are complex details that distinguish individual cities, which are closely related to specific environmental and socioeconomic settings. Our study demonstrates these patterns, for the first time, on long-term and global scales. These global trends help us better understand the patterns and processes of urbanization and suggest some fundamental problems with global urbanization so far. For example, urbanization around the world has led to increasing landscape fragmentation that has negative impacts on ecosystem function and services, urban sprawl that encroaches into agricultural land and natural habitat and homogenizing urban morphology, which diminishes the cultural identities of cities. Future urban studies need to consider both generalities and idiosyncrasies and relate them directly to measures of urban sustainability [5].

Supplementary Materials: The following are available online at www.mdpi.com/2071-1050/8/1/41/s1, Supplementary S1: historical map references. Supplementary S2: verification of data consistency. Supplementary S3: supporting figures on urbanization patterns.

Acknowledgments: We thank four anonymous reviewers for their valuable comments on an earlier version of the paper. This research was supported in part by the National Basic Research Programs of China (Grant Nos. 2014CB954303 and 2014CB954302), the National Natural Science Foundation of China (Grant No. 41222003, No. 41321001 and No. 41501195), the 111 project of "Hazard and Risk Science Base at Beijing Normal University" (Grant No. B08008), the Youth Scholars Program of Beijing Normal University (Grant No. 2014NT02) and the State Key Laboratory of Earth Surface Processes and Resource Ecology (Grant No. 2015-RC-01).

Author Contributions: Jianguo Wu and Chunyang He developed the original idea and contributed to the research design, writing and revision. Zhifeng Liu was responsible for data collection and processing, research design, writing and revision.

Conflicts of Interest: The authors declare no conflict of interest.

References

1. Forman, R.T.T. The urban region: Natural systems in our place, our nourishment, our home range, our future. *Landsc. Ecol.* **2008**, *23*, 251–253.
2. Grimm, N.B.; Faeth, S.H.; Golubiewski, N.E.; Redman, C.L.; Wu, J.; Bai, X.; Briggs, J.M. Global change and the ecology of cities. *Science* **2008**, *319*, 756–760.
3. Schneider, A.; Mertes, C.M.; Tatem, A.J.; Tan, B.; Sulla-Menashe, D.; Graves, S.J.; Patel, N.N.; Horton, J.A.; Gaughan, A.E.; Rollo, J.T.; *et al.* A new urban landscape in east-southeast asia, 2000–2010. *Environ. Res. Lett.* **2015**, *10*, 034002.
4. Wu, J. Urban sustainability: An inevitable goal of landscape research. *Landsc. Ecol.* **2010**, *25*, 1–4.
5. Wu, J. Urban ecology and sustainability: The state-of-the-science and future directions. *Landsc. Urban Plan.* **2014**, *125*, 209–221.

6. Zhou, Y.Y.; Smith, S.J.; Zhao, K.G.; Imhoff, M.; Thomson, A.; Bond-Lamberty, B.; Asrar, G.; Zhang, X.S.; He, C.Y.; Elvidge, C.D. A global map of urban extent from nightlights. *Environ.Res. Lett.* **2015**, *10*, 054011.
7. UN. *World Urbanization Prospects: The 2011 Revision*; United Nations, Department of Economic and Social Affairs, Population Division: New York, NY, USA, 2012.
8. Angel, S.; Parent, J.; Civco, D.L.; Blei, A.; Potere, D. The dimensions of global urban expansion: Estimates and projections of all countries, 2000–2050. *Prog. Plan.* **2011**, *75*, 53–107.
9. Luck, M.; Wu, J. A gradient analysis of urban landscape pattern: A case study from the phoenix metropolitan region, arizona, USA. *Landsc. Ecol.* **2002**, *17*, 327–339.
10. Wu, J.; Jenerette, G.D.; Buyantuyev, A.; Redman, C.L. Quantifying spatiotemporal patterns of urbanization: The case of the two fastest growing metropolitan regions in the united states. *Ecol. Complex.* **2011**, *8*, 1–8.
11. Angel, S.; Parent, J.; Civco, D.L. The fragmentation of urban landscapes: Global evidence of a key attribute of the spatial structure of cities, 1990–2000. *Environ. Urban.* **2012**, *24*, 249–283.
12. Frolking, S.; Milliman, T.; Seto, K.C.; Friedl, M.A. A global fingerprint of macro-scale changes in urban structure from 1999 to 2009. *Environ. Res. Let.* **2013**, *8*, 024004.
13. Taubenbock, H.; Esch, T.; Felbier, A.; Wiesner, M.; Roth, A.; Dech, S. Monitoring urbanization in mega cities from space. *Remote Sens. Environ.* **2012**, *117*, 162–176.
14. Li, H.; Wei, Y.H.D.; Huang, Z.J. Urban land expansion and spatial dynamics in globalizing shanghai. *Sustainability* **2014**, *6*, 8856–8875.
15. Lu, S.; Guan, X.; He, C.; Zhang, J. Spatio-temporal patterns and policy implications of urban land expansion in expansion in metropolitan areas: A case study of wuhan urban agglomeration, central China. *Sustainability* **2014**, *6*, 4723–4748.
16. Gao, J.L.; Wei, Y.D.; Chen, W.; Yenneti, K. Urban land expansion and structural change in the yangtze river delta, China. *Sustainability* **2015**, *7*, 10281–10307.
17. You, H. Quantifying urban fragmentation under economic transition in shanghai city, China. *Sustainability* **2016**, *8*, 21.
18. Dietzel, C.; Herold, M.; Hemphill, J.J.; Clarke, K.C. Spatio-temporal dynamics in california's central valley: Empirical links to urban theory. *Int. J. Geogr. Inf. Sci.* **2005**, *19*, 175–195.
19. Dietzel, C.; Oguz, H.; Hemphill, J.J.; Clarke, K.C.; Gazulis, N. Diffusion and coalescence of the houston metropolitan area: Evidence supporting a new urban theory. *Environ. B Plan. Des.* **2005**, *32*, 231–246.
20. Li, C.; Li, J.; Wu, J. Quantifying the speed, growth modes, and landscape pattern changes of urbanization: A hierarchical patch dynamics approach. *Landsc. Ecol.* **2013**, *28*, 1875–1888.
21. Wu, J. *Landscape Ecology: Pattern, Process, Scale, and Hierarchy*; Higher Education Press: Beijing, China, 2000.

22. Wu, J. Modeling. In *CAP-LTER 2001–2002 Annual Progress Report to NSF*; 2002. Available online: http://caplter.asu.edu/docs/reports/2002AnnRept/2002CAPLTER AnnualReport.pdf (accessed on 31 December 2015).
23. Wu, J.; Jenerette, G.D.; David, J.L. Linking land-use change with ecosystem processes: A hierarchical patch dynamic model. In *Integrated Land Use and Environmental Models—A Survey of Current Applications and Research*; Guhathakurta, S., Ed.; Springer: Berlin, German, 2003.
24. Jenerette, G.D.; Potere, D. Global analysis and simulation of land-use change associated with urbanization. *Landsc. Ecol.* **2010**, *25*, 657–670.
25. Schneider, A.; Woodcock, C.E. Compact, dispersed, fragmented, extensive? A comparison of urban growth in twenty-five global cities using remotely sensed data, pattern metrics and census information. *Urban Stud.* **2008**, *45*, 659–692.
26. Seto, K.C.; Fragkias, M. Quantifying spatiotemporal patterns of urban land-use change in four cities of china with time series landscape metrics. *Landsc. Ecol.* **2005**, *20*, 871–888.
27. Forman, R.T.T.; Godron, M. *Landscape Ecology*; Wiley: New York, NY, USA, 1986.
28. Lincoln Institute of Land Policy. Atlas of Urban Expansion. Available online: http://www.lincolninst.edu/ subcenters/atlas-urban-expansion/ (accessed on 7 January 2014).
29. Angel, S.; Parent, J.; Civco, D.L.; Blei, A.M. *The Persistent Decline of Urban Densities: Global and Historical Evidence of Sprawl*; Lincoln Institute of Land Policy: Cambridge, MA, USA, 2010.
30. Zhang, Q.; Seto, K.C. Mapping urbanization dynamics at regional and global scales using multi-temporal dmsp/ols nighttime light data. *Remote Sens. Environ.* **2011**, *115*, 2320–2329.
31. Liu, Z.; He, C.; Zhang, Q.; Huang, Q.; Yang, Y. Extracting the dynamics of urban expansion in china using dmsp-ols nighttime light data from 1992 to 2008. *Landsc. Urban Plan.* **2012**, *106*, 62–72.
32. He, C.; Liu, Z.; Tian, J.; Ma, Q. Urban expansion dynamics and natural habitat loss in china: A multiscale landscape perspective. *Glob. Chang. Biol.* **2014**, *20*, 2886–2902.
33. Liu, Z.; He, C.; Zhou, Y.; Wu, J. How much of the world's land has been urbanized, really? A hierarchical framework for evading confusion. *Landsc. Ecol.* **2014**, *29*, 763–771.
34. Wu, J. Effects of changing scale on landscape pattern analysis: Scaling relations. *Landsc. Ecol.* **2004**, *19*, 125–138.
35. Liu, X.; Li, X.; Chen, Y.; Tan, Z.; Li, S.; Ai, B. A new landscape index for quantifying urban expansion using multi-temporal remotely sensed data. *Landsc. Ecol.* **2010**, *25*, 671–682.
36. Weng, Y.C. Spatiotemporal changes of landscape pattern in response to urbanization. *Landsc. Urban Plan.* **2007**, *81*, 341–353.
37. McGarigal, K.; Cushman, S.A.; Neel, M.C.; Ene, E. *Fragstats: Spatial Pattern Analysis Program for Categorical Maps, 3.1st edn*; University of Massachusetts: Amherst, MA, USA, 2002.
38. ESRI. *Arcgis desktop: Release 10*; Environmental Systems Research Institute: Redlands, CA, USA, 2011.

39. He, C.; Okada, N.; Zhang, Q.; Shi, P.; Li, J. Modelling dynamic urban expansion processes incorporating a potential model with cellular automata. *Landsc.Urban Plan.* **2008**, *86*, 79–91.
40. Wissink, B. Enclave urbanism in mumbai: An actor-network-theory analysis of urban (dis)connection. *Geoforum* **2013**, *47*, 1–11.
41. Riitters, K.H.; Oneill, R.V.; Hunsaker, C.T.; Wickham, J.D.; Yankee, D.H.; Timmins, S.P.; Jones, K.B.; Jackson, B.L. A factor-analysis of landscape pattern and structure metrics. *Landsc. Ecol.* **1995**, *10*, 23–39.
42. Frohn, R.; Hao, Y. Landscape metric performance in analyzing two decades of deforestation in the amazon basin of rondonia, brazil. *Remote Sens. Environ.* **2006**, *100*, 237–251.

Urban Land Expansion and Structural Change in the Yangtze River Delta, China

Jinlong Gao, Yehua Dennis Wei, Wen Chen and Komali Yenneti

Abstract: Urban development in China has attracted considerable scholarly attention. However, more work is still needed to examine and understand the mechanisms of urban land expansion, especially within the context of globalization/marketization, decentralization and urbanization. This paper analyzes urban land expansion and structural changes in the Yangtze River Delta (YRD). We find that cities in the YRD are experiencing urban land expansion mainly characterized by the growth of residential and industrial land. The dominant characteristics of urban land expansion in cities have also varied within different development and administrative levels. Based on our conceptual framework, we have used multi-models to investigate the driving forces of urban land expansion and structural changes in the YRD. The results reveal that six influencing factors—foreign direct investment (FDI), labor, government competition, institution, population, and job-housing relations—facilitate land use change in the economic transition process. However, their impacts differ in cities in different geographical locations, as well as with different administrative levels. Finally, this paper discusses policies to promote sustainable urban land use in the YRD.

Reprinted from *Sustainability*. Cite as: Gao, J.; Wei, Y.D.; Chen, W.; Yenneti, K. Urban Land Expansion and Structural Change in the Yangtze River Delta, China. *Sustainability* **2015**, *7*, 10281–10307.

1. Introduction

Over the past few decades, unprecedented urbanization, mainly driven by globalization, centralization, and marketization, has taken place in China. Between 1978 and 2013, the nation's urbanization rate has increased from 17.9% to 53.7%. The rapid pace of urbanization has resulted in not only dramatic urban land expansion, but also issues such as environmental resources depletion, food deficits, and land degradation [1–3]. The socioeconomic and environmental problems, characterized by rapid urban land expansion, have become major challenges for achieving sustainable urban development (SUD) in China. In a step towards addressing the challenges and achieving SUD, the government of China has introduced the "National New-type Urbanization Plan" (NNUP), the country's first official plan on urbanization, in March 2014. The concept covers a wide ranging issues, including rural–urban integration, efficient use of land, and coordination between the drivers of regional development [4]. One of the major principles of the NNUP is to achieve sustainable urban land expansion, while emphasizing economic growth. A crucial prerequisite in the process of formulating sustainable urban development strategies within the NNUP is to understand the existing spatial development of urban land and the driving forces underlying the expansion [5]. It is, thus, significantly important to undertake a timely research that can not only document the structural changes of urban land, but also reveal the underlying mechanisms.

A rich body of literature is explored to uncover the patterns of urban land expansion in different cities of China [6]. The findings of the literature cover issues such as the patterns of built-up land expansion [7], the process of spatial restructuring [8], the mechanisms of urban growth [9], and the policy outcomes [10]. Although urban land expansion in China is well-documented throughout this academic literature, most of the literature is based either on single city or single province case-study contexts. Cross-city comparative studies on characterizing urban land expansion and its driving forces remain limited to date. Even the traditional theories have provided limited understanding on the rapid evolution of China's urban land use structure. Considering the significant heterogeneity, especially the different development and administrative levels, among the cities in China, it is important to conduct a comparative research on urban land expansion in different transitional cities. In context of the dearth of literature, the primary goal of this research is to examine and analyze urban land expansion and structural changes in the Yangtze River Delta (YRD), China, by focusing on the heterogeneity of cities in terms of the different development and administrative levels. The region's rapid urban development makes it an excellent case study to explore spatiotemporal urban land expansion patterns and retrospectively cross-compare urban growth drivers.

Apart from undertaking a study on the dynamics and mechanisms of urban land expansion in contemporary China, it is also essentially important to take into account the combined effect of influencing factors such as globalization/marketization, decentralization and urbanization. These triple processes can help to systematically (re)understand the mechanisms of urban land expansion during the transition era [11]. Given that NNUP is merely a start of the path towards reaching China's sustainable urban development destination, it is an opportune time to examine and reflect on these issues. To summarize, characterizing urban land expansion is a prerequisite to not only understand the urbanization process in China, its driving forces, and the socioeconomic consequences, but also to support the development of sustainable urban planning and management strategies [12].

2. Literature Review

2.1. Sustainable Land Use in Regional Urban Development

Since the 1980s, global economic transformations, stimulated by foreign investments, have not only "leap-frogged" economic development but also triggered significant changes in the process of urbanization. These transformations, through redefining the economic base of cities and recasting the spatial patterns in relation to new industrial and productive conditions, have facilitated the emergence of "region-based urbanization" [13]. A large amount of research within the new economic geography field has also emerged to understand urban land expansion in regions. Regional urban land expansion has attracted considerable scholarly attention both internationally and in China since it is threatening sustainable urban and regional development [14]. The research has provided an understanding on how the city influence is moving to wider regions, thanks to advances in technology, circulation of commodities, de-concentration of manufacturing, and diffusion between the urban and rural. The primary focus of this research area has largely been on three issues. First, to account the enduring urban land expansion and spatial changes, in the light of global economic transformations and information flows [15,16]. Second, to document the regional dispersion of population and manufacturing activities [13,17]. Last, to review the implications of the spatial structural changes on poverty and social functioning [18,19]. While sustainable land use development has been an important component in debates surrounding urban expansion [20], the literature provide limited guidance on what constitutes sustainable land use and how to achieve it.

Some of the clearer debates around sustainable land use are found in the academic work drawn on the "New Urbanism" movement. The movement, initiated at the Congress for the New Urbanism held in 1993, gives significant impetus to creating regions which improve the quality of life, while protecting the natural environment [21]. Since then, a large amount of academic scholarship concerning sustainable land use development has developed. Leccese *et al.* [22], in their Charter for the New Urbanism, identifies infill development as a means of conservation of environmental resources, attracting economic investment, and balancing social fabric—the three interacting elements of sustainable land use development. They go further and argue that neighborhood mixed-use compactness allows for locally embedded institutional and commercial activities. According to Newman and Kenworthy [23], sustainable land use seeks to reconnect transport with other developments, and establish transit-oriented development where high-density and mixed-use areas are built around high-quality transit systems. Such a focused urban land use structure can help to loosen the grasp of automobile dependence and eventually achieve smart urban growth. In brief, the different arguments on New Urbanism emphasize high density, mixing of commercial and residential land use, convenient public transit, safe and public-friendly street networks, strategically placed open spaces, and neighborhoods designed to foster social interaction and a strong sense of place as common traits of sustainable land use development in regions [24].

Though these scholarly arguments based on different theoretical contexts have provided a new direction to the literature on "sustainable land use", there is a significant scope for exploring the following points: (i) the need for comprehensive frameworks that provide a meaningful understanding on urban expansion in multiple cities of regions; (ii) the complex dynamics of urban expansion and structural change in regions, and their linkages to patterns of economic activities and urban population; and (iii) the necessity for understanding the role of multiple forces in the expansion and spatial changes in regions.

2.2. *The Driving Forces of Urban Land Expansion*

In the arguments of urban land use and land expansion, different driving forces underlying the urban land expansion are identified. These include (i) demography or demands for space by people and employment [25,26], (ii) the changing industrial landscape, with the growth of manufacturing and production facilities and associated working-class housing, services, and commerce to cater to those facilities [27,28], (iii) improvements in transportation networks (roads, public transport, *etc.*), changes in mobility patterns and transit-oriented development [29,30], and (iv) institutional factors, such as fragmentation in municipal jurisdictions and planning, and variations in tax structures [20,31].

Along with the international literature, wide ranging studies have emerged over the years to investigate the relationship between urban land expansion and its driving forces in China [32,33]. In spite of the large number of possible drivers, the majority of these studies operationalize on economic development and population growth, and to some extent, traffic and physical conditions [34,35]. Considering the complexities within China, it is preeminent to understand urban land expansion through multiple driving forces and in multiple city contexts. Friedmann [36] emphasizes that understanding the driving forces of urban land expansion in China only from an economic perspective is obviously simplistic and deterministic since it overlooks some of the fundamental social and political conditions. With the growing institutionalism and localism in economic development, scholars have attempted to apply elite and urban regime theories to understand urban land expansion in China [37]. They argue that there are three major interest groups and two coalitions in the process of urban land use development: (i) economic coalitions between urban government and market developers or investors, and (ii) political coalitions between the central and local governments [38].

Moving away from the Western urban expansion studies, scholars have recently provided new explanations regarding the process of urban land expansion in China. For example, Wei [11,39] has proposed a framework within which China's economic transition is a triple process of decentralization, marketization, and globalization. This framework provides an innovative basis for China's spatial structuring theory. Additionally, various other researchers have also investigated the effects of other factors on urban land expansion. While some scholars have demonstrated that China's urban growth is driven by economic, fiscal, and political incentives from local governments and local leaders who control land use in their jurisdictions [35,40,41], others have examined the processes of urban land expansion from the perspective of institutions, and foreign direct investment (FDI) [7,42]. Scholars also argue that the growth and spatial expansion of China's cities takes a path-dependent trajectory influenced by the legacy of socialist political and planning systems [43,44].

Though this line of research is very fruitful, and has documented patterns of urban land expansion in China, more efforts are still needed to examine the mechanisms of urban land expansion in regions under the new context of globalization/marketization, decentralization and urbanization. Most importantly, as driving forces may differ in various geographical scales and locations due to scale-dependency and geographical non-stationarity [45–47], a systematic understanding on the relation between urban land expansion and its multiple driving forces is necessary to deal with the challenges of rapid urbanization and sustainability facing China. Moving away from the traditional debates which emphasize urban expansion as a process that involves population redistribution and restructuring [13], we argue that urban expansion is a process that involves reconfiguration of multiple forces at multiple scales (local, regional, and global).

3. Materials and Methods

3.1. The Conceptual Framework

While different studies have identified a wide ranging external and internal forces that drive the mechanisms of urban land expansion processes [5,34,48,49], we have developed an integrated conceptual framework for this research based on Wei's analytical framework [11,50] because it provides a comprehensive model with different mechanisms and driving forces. The urban land expansion, namely the outward expansion/accretion and the inner reconstruction/replacement, is related to the drivers of local citizen, urban government, and global market, and their interactions (Figure 1). First, within the context of urbanization, local citizens exert an impact on the residential and industrial land by way of population growth and "jobs-housing separation". The large assortment of consumption goods and services that cities offer leads to migration of people from small towns. With the increase of urban citizens, more land is required for residential areas [51]. Meanwhile, the development of transportation and communication technologies might lead to urban spatial mismatch by the way of "jobs-housing separation" [52–54]. All these forces, thus, result in changes in the demand structures of land.

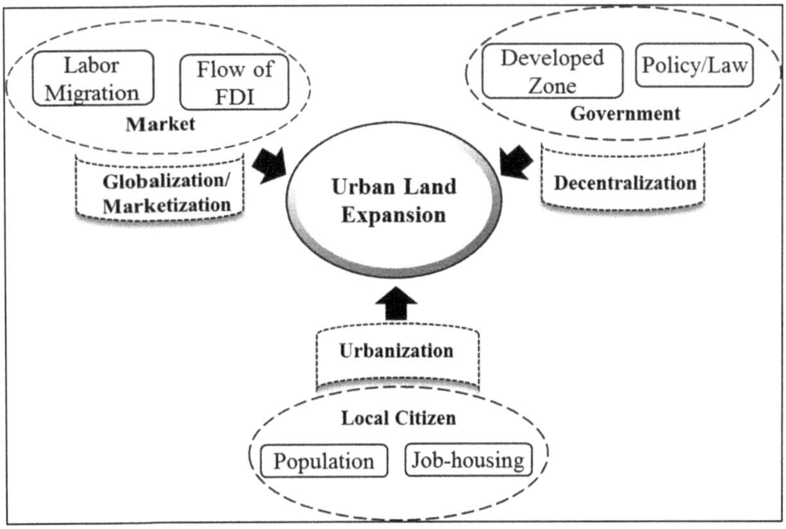

Figure 1. The conceptual framework of urban land expansion and restructuring.

Second, the drive for development zones (DZ) is also an important factor in urban growth in China [55,56]. In the process of attracting both foreign and domestic investment, urban governments tend to establish more and more DZs in the outskirts, which often lead to the agglomerated development of industrial land [57]. Along with the establishing of DZs, various scholars have confirmed that institutional changes, through land use "policies" [58] such as land use right and ownership [59], the tax-sharing reform of 1994 [60], and the housing commercialization reform [61], have great impacts on urban land expansion process in China [33].

Third, due to inter-regional trade in the context of globalization and marketization, an increasing amount of FDI is being attracted into China. Consequently, industrial production spaces have sprawled at an alarming speed, which then stimulated the demand for real estate and public utilities at the newly expanded outskirts [62]. To cater the labor-intensive industrial activities, labor has migrated from both large cities and rural areas. Coincidently, the deepening of new international division of labor (NIDL), promoted by the formation of global production networks (GPNs), has also contributed to the transformation of regional economy from an industrial composition to a value-added economy in China [63]. As a result, not only local economy structures have drastically transformed, but also the urban land use structures.

The hierarchical structure of China's urban administrative system and urban land use, which are very different from the western counterparts, exerts strong influence on local development [64]. This strong role of urban hierarchical administrative system on the local land use planning and land administration emphasizes how the significant driving forces of the land development are also influenced by the urban hierarchical system or administration levels [14]. Hence, in this research, we have also taken geographical location and administration level into consideration.

Based on this conceptual framework, this paper is one of our systematic efforts to integrate globalization, institutional change, and China's new-type urbanization to better explain urban land expansion in the Yangtze River Delta (YRD) region, China's emerging global city region.

3.2. *Study Area, Data, and Methods*

Located in the YRD and adjacent to Shanghai, the study area is one of the three largest urban agglomerations, the other two being the Pearl River Delta and the Jing-Jin-Ji regions. The study area covers about 76,322 square kilometers, which is approximately 68% of the YRD territory. The sample cities are located in central and southern Jiangsu (JS), and northern Zhejiang (ZJ), two of the most developed provinces in China. In 2009, the size of the area's economy, measured by Gross Domestic Product (GDP), has surpassed other metropolitan areas, such as the Pearl River Delta and the Beijing-Tianjin-Hebei metropolis. The land administrative system is a strategic component of land policy in most developing countries [65]. The specific question about administrative ranking and urban expansion in China is legitimate since Chinese cities are institutionalized under different administrative authorities [14]. According to the Constitution of China, administrative units are based on a three tier-system, dividing the nation into provinces, counties, and townships. However, two more levels are inserted in actual implementation: the prefectures, under provinces; and the villages, under townships. In terms of urban administration, the country is divided into provincial level cities directly administered by the Central Government, prefectures, and county-level cities. The four centrally administrated cities, Beijing, Shanghai, Tianjin, and Chongqing, are further subdivided into urban districts and counties. Prefectural level cities are subdivided into county-level cities, counties, and urban districts. Local governments at different structural hierarchies of the administrative divisions have different legislative powers on land administration. Counties are generally considered as rural areas in China. Thus, we have selected all of the cities in the YRD, excluding Shanghai and other counties (Figure 2).

Figure 2. Location & organization of study samples in the Yangtze River Delta, China.

As the administrative divisions have merged between 2000 and 2005, we have considered the sample cities consistent with the latest 2010 administrative units. The study area includes 15 prefectural level cities and 36 county level cities with a permanent population of 68.42 million in 2011 (Table 1). In 2011, the 51 sample cities together have produced more than 60,000 billion yuan of GDP (9523 billion USD).

With an annual growth rate of 19%, the regions' GDP per capita increased from 16,795 yuan (2029 USD) in 2000 to 87,700 yuan (13,920 USD).

Table 1. Cities of prefectural and county levels in the YRD.

Regions	Prefectural level cities	County level cities
Central Jiangsu (*Suzhong*)	Nantong, Taizhou, Yangzhou	Gaoyou, Haimen, Jiangdu, Jiangyan, Jingjiang, Qidong, Rugao, Taixing, Xinghua, Yizheng
Southern Jiangsu (*Sunan*)	Changzhou, Nanjing, Suzhou, Wuxi, Zhenjiang	Changshu, Danyang, Jiangyin, Jintan, Jurong, Kunshan, Liyang, Taicang, Wujiang, Yangzhong, Yixing, Zhangjiagang
Northern Zhejiang (*Zhebei*)	Hangzhou, Huzhou, Jiaxing, Ningbo, Shaoxing, Taizhou, Zhoushan	Cixi, Fenghua, Fuyang, Haining, Jiande, Lin'an, Linhai, Pinghu, Shangyu, Shengzhou, Tongxiang, Wenling, Yuyao, Zhuji

To analyze the urban land expansion process in the study area, various social and economic data from 2000 and 2010 statistical year books are used. We have also used the population data from the 2000 and 2010 Census of the People's Republic of China by county. The basic land use data is obtained from China Urban Construction Statistical Yearbooks (CUCSY) from 2000 to 2010. The CUCSY consists of nine land use types (residential, commercial & business facilities, administration & public services, industrial & manufacturing, logistics & warehouse, transportation, municipal utilities, green space and special-purpose land). Out of these, we have considered the structural changes and the growth of residential and industrial land due to their large shares in the overall urban land use.

According to our conceptual framework (Figure 1), variables representing urban land expansion, global market and local demographic drivers, and the proxy of government power are selected (Table 2). The areas of residential land, industrial land, and the whole built-up land for all prefectural and county level cities are used as dependent variables.

Based on the primary concerns of previous research [14,32], the log of foreign direct investment (FDI) is selected to understand if urban land use development is driven by the capital inflows or not. The log of labor reflects the migration of workers. We have used the level weighted (national level = 1.5, provincial level = 1) number of development zones to characterize the motivation of urban competition. The measurement of the strength of regional policies is calculated by the weighted sum strength of the different levels of policy. The number of permanent citizens is used to illustrate the level of urbanization. The jobs-housing separation index (the ratio of a city's share in the total of 51 sample cities to its share in residents) reflects the quality of urbanization. The variables for Southern Jiangsu (JS), Northern Zhejiang (ZJ), and prefecture are dummy variables and measure whether the city is located in the region itself, or is a prefecture. For *Sunan*, 0 means the city is not located in the Southern JS and 1 means the city is located in the Southern JS region; similarly for *Zhebei*. For Prefecture, 0 represents a county level city, and 1 represents that the city is a prefectural level city.

Table 2. The definitions of independent variables.

Categories	Definitions	Abbreviation
Globalization/Marketization	Ln (foreign direct investment) Ln (labor)	FDI Labor
Decentralization	Ln (level weighted number of DZs) Ln (weighted sum strength of policy)	DZs Policy
Urbanization	Ln (permanent citizens) Ln (jobs-housing separation index)	Citizen J-H
Spatial & Administration	Southern Jiangsu (dummy) Northern Zhejiang (dummy) Prefectural level (dummy)	Sunan Zhebei Prefecture

We have undertaken three modeling methods to spatially and temporally analyze the dynamics of urban land expansion in the YRD. First, *Shannon*'s entropy (H) and equilibrium degree (E), denoted as Equations (1) and (2), are applied to measure the degree of dispersion or concentration of urban land use.

$$H = -\sum_{i}^{n}(P_i)\ln(P_i) = -\sum_{i}^{n}\left(S_i/\sum_{i}^{n} S_i \right)\ln\left(S_i/\sum_{i}^{n} S_i \right) \tag{1}$$

$$E = -\sum_{i=1}^{n}(P_i)\ln(P_i)/\ln(n) \tag{2}$$

where, P_i is the probability or proportion of occurrence of a phenomenon in the ith spatial unit out of n units and, S_i is the area of the ith land use type in one city with a total of n types of land use. The equilibrium degree is the ratio of the actual value of entropy to its maximum value, and varies from 0 to 1. Furthermore, the difference of an index is defined as $D = 1 - E$, which can be used to reflect the dominance of urban land use. Cities with more land use types have higher entropy values.

Second, to understand the patterns of urban land expansion, regional land use change rates for both single land use type and comprehensive land use are computed [32]. In addition, Getis-Ord General G and Getis-Ord G_i^* are employed to analyze the global and local patterns of urban land expansion process. The spatial analysis methods are employed to investigate the general dynamics of urban land expansion in the YRD. To explore the impacts of different factors in different regions and administration levels, we set three dummy variables as Southern Jiangsu, Northern Zhejiang, and Prefecture. The global pooled regression model is framed as the following Equation (3):

$$y_{it}=\sum_{k=1}^{K}\beta_{ki}\times x_{kit}+\sum_{l=1}^{3}\sum_{m=1}^{M}\alpha_{li}\times d_{li}\times x_{mit}+\sum_{l=1}^{3}\lambda_{li}\times d_{li}+u_{it} \tag{3}$$

where, y_{it} is the dependent variable value of region i in the year t; x_{kit} is the value of variable k of region i in the year t; d_{li} is the regional or administration dummy variable of region i; $d_{li} \times x_{mit}$ is the cross variable; β, α and λ are the coefficients to be estimated; and u is the error.

To further identify the unique coefficient of each geographical unit, to avoid the heterogeneity of spatial effects, and to capture the spatially varying relationship between probability of land use conversion and explanatory variables in different years, we have applied geographically and temporally weighted regression (GTWR) [66]. The GTWR is calculated using the following Equation (4):

$$y_i=\beta_i(u_i,v_i,t_i)+\sum_{k}\beta_k(u_i,v_i,t_i)X_{ik}+\varepsilon_i \tag{4}$$

where β_i is the constant parameter specific to each space-time location i with a spatio-temporal coordinates of (u_i, v_i, t_i). β_k is the parameter of independent variable X_k at location i.

The mechanisms of urban land expansion in the YRD are assumed to be sensitive to the cities' administrative ranks. Thus, the spatial regime model (SRM), a technique that can explicitly recognize the heterogeneity of land expansion mechanisms in different levels of cities, is employed [14]. Following the work of Cravo and Resende [67], we have allowed the coefficients to vary across two different regimes as shown in Equation (5):

$$\begin{bmatrix} gr^*_{i,tA} \\ gr^*_{i,tB} \end{bmatrix}=\begin{bmatrix} Y^*_{i,t-1A} & X^*_{i,t-1A} & 0 & 0 \\ 0 & 0 & Y^*_{i,t-1B} & X^*_{i,t-1B} \end{bmatrix}\begin{bmatrix} b_A \\ \varphi_A \\ b_B \\ \varphi_B \end{bmatrix}+\begin{bmatrix} v_{i,tA} \\ v_{i,tB} \end{bmatrix} \tag{5}$$

where the subscripts A and B indicate different regimes of prefectural level cities and county level cities, respectively; $gr^*_{i,tA}$ and $gr^*_{i,tB}$ are N × 1 column vectors with observations for urban land expansion for spatial regimes A and B; $Y^*_{i,t-1A}$ and $Y^*_{i,t-1B}$ are N × 2 matrices including the constant term and initial urban land expansion of each regime; $X^*_{i,t-1A}$ and $X^*_{i,t-1B}$ are the N × K matrices of observations on other explanatory variables for each regime; $v_{i,tA}$ and $v_{i,tB}$ are the N × 1 vectors of error terms.

4. Results

4.1. *Temporal Evolution of Urban Land Use Structure in the YRD*

In this section, we mainly analyze the temporal changes in residential, industrial, and the whole built-up land between 2000 and 2010. At an overall level, it can be observed that the residential land has increased from 448 km^2 in 2000 to 868 km^2 and 1180 km^2 in 2005 and 2010, respectively (Table 3). However, the overall percentage remains unchanged at 28.1%, thereby indicating the relatively consistent growth of residential land within the total urban sprawl. The proportion of industrial and manufacturing land to the overall land use has increased from 28.6% in 2000 to 29.3% in 2010. This signifies the rapid expansion and high growth process of industries in the YRD. Another significant observation is that the land under municipal utilities has increased from 9.5% in 2000 to 10.8% and 12% in 2005 and 2010, respectively, indicating the development of both physical and social infrastructure.

Administratively, the industrial area increase of 363 km^2 during 2000 to 2005, and the residential area increase of 221 km^2 during 2005 to 2010 in prefectural level cities are the dominant observations. However, a decline in the land use under industrial land from 30.5% in 2005 to 29.7% in 2010 in prefecture level cities can also be observed. In county level cities, the residential land has decreased from 31.1% in 2000 to 30.5% in 2010 and the industrial land has increased from 28.1% to 29.0%. It is also observed that the increase of municipal utilities in prefectural level cities is much faster than that in the county-level cities (Table 3). All these features indicate that cities with higher administrative hierarchies are in transition from producing to service cities, while the county level ones are still in the process of rapid industrialization.

Table 3. Areas of main urban land use in 51 sample cities in 2000, 2005, and 2010 (km^2).

Urban Land Use Type	51 Cities	Geographical Location			Administration Hierarchy	
		Southern Jiangsu	Central Jiangsu	Northern Zhejiang	Prefecture	County Level
Year 2000						
Residential	448 (28.1)	185 (26.3)	85 (31.8)	179 (28.7)	301 (26.9)	148 (31.1)
Administration &Public Services	185 (11.6)	88 (12.5)	29 (10.8)	68 (10.9)	128 (11.5)	57 (12.0)
Industrial & Manufacturing	455 (28.6)	209 (29.7)	85 (31.8)	161 (25.9)	322 (28.8)	133 (28.1)
Street and Transportation	89 (5.6)	37 (5.3)	15 (5.6)	36 (5.8)	65 (5.8)	23 (4.9)
Municipal Utilities	151 (9.5)	63 (9.0)	22 (8.3)	66 (10.5)	107 (9.6)	44 (9.3)
Green Space	164 (10.3)	85 (12.1)	16 (6.0)	63 (10.0)	115 (10.2)	49 (10.4)
Year 2005						
Residential	868 (28.1)	427 (27.5)	129 (32.5)	312 (27.3)	601 (26.8)	266 (31.1)
Administration & Public Services	351 (11.4)	174 (11.2)	44 (11.1)	133 (12.6)	259 (11.6)	92 (10.9)
Industrial & Manufacturing	927 (30.0)	489 (31.6)	119 (30.1)	318 (27.8)	685 (30.5)	242 (28.6)
Street and Transportation	137 (4.4)	68 (4.4)	18 (4.7)	50 (4.4)	102 (4.6)	35 (4.1)
Municipal Utilities	334 (10.8)	161 (10.4)	39 (10.0)	133 (11.7)	243 (10.8)	91 (10.8)
Green Space	321.3 (10.4)	161 (10.4)	28 (7.1)	133 (11.7)	236 (10.5)	85 (10.1)
Year 2010						
Residential	1180 (28.1)	551 (27.5)	194 (32.5)	436 (27.3)	822 (27.2)	359 (30.5)
Administration & Public Services	477 (11.4)	229 (11.4)	71 (11.9)	178 (11.2)	351 (11.6)	127 (10.8)
Industrial & Manufacturing	1229 (29.3)	601 (30.0)	180 (30.2)	447 (28.0)	897 (29.7)	332 (29.0)
Street and Transportation	172 (4.1)	83 (4.2)	16 (2.6)	73 (4.6)	131 (4.3)	41 (3.5)
Municipal Utilities	506 (12.0)	226 (11.3)	69 (11.5)	211 (13.3)	375 (12.4)	131 (11.1)
Green Space	439 (10.5)	224 (11.2)	47 (7.9)	167 (10.5)	300 (9.9)	139 (11.8)

Note: Numbers in parentheses are the proportions (%). Source: [68].

The development level analysis infer that the cities in Central Jiangsu (JS) always have a higher percentage (more than 30%) of residential land than that of cities in the Southern JS and Northern Zhejiang (ZJ) provinces. The domination of residential land expansion indicates the relatively slow industrial development in Central JS. The increase of industrial land percentage from 25.9% in 2000 to 28% 2010 in Northern ZJ (Table 3) shows the rapid industrial expansion of the cities in this sub-region. While in the most developed Southern JS, industrial land along with the residential land has sharply increased during the first five years and then decreased in the following years. This demonstrates that the Southern JS province is leading the YRD region in terms of both industrialization and urbanization. Geographically, three significant growth poles are observed in the study area during 2000 to 2005. They are Suzhou and Nanjing in the Southern JS and Hangzhou in the Northern ZJ province. In the period of 2005 to 2010, the hot spot of Hangzhou shifted to Ningbo (Figure 3).

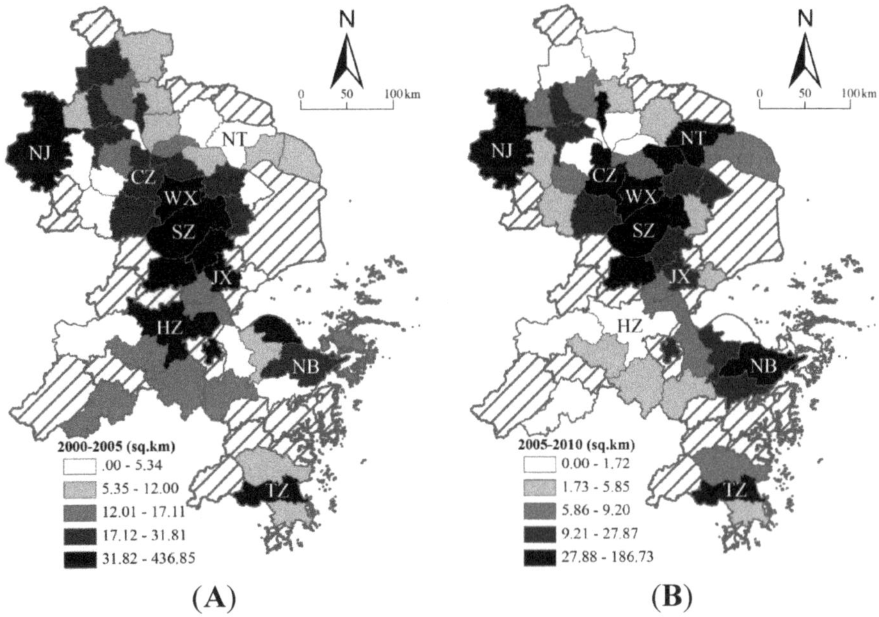

Figure 3. Built-up area growth in 51 sample cities, (**A**) 2000 to 2005 and (**B**) 2005 to 2010. (NJ = Nanjing, HZ = Hangzhou, SZ = Suzhou, WX = Wuxi, CZ = Changzhou, NB = Ningbo, JX = Jiaxing, NT = Nantong, TZ = Taizhou.)

In order to unfold the overall changes of urban land use structure in the YRD, we have further computed the *Shannon*'s entropy (H), equilibrium degree (E) and dominance (D). Through this, we find the equilibrium and complexity degree of urban land use structure in cities with different administrative hierarchies and geographical locations for the period 2000 to 2010. The results (Table 4) show that, for all the 51 cities, the values of H and E have decreased from 1.92 and 0.87 in 2000 to 1.83 and 0.83 in 2005, respectively. However, for the same period, the D value has increased from 0.13 to 0.17. A similar trend can be observed in both the prefecture and county level cities. This indicates that though the dominant trends vary in different cities, urban land change has always been dominated by one or two types. Similarly, between 2000 and 2005, the values of H decreased by 0.07, 0.01 and 0.14 in Southern JS, Central JS, and Northern ZJ, respectively. However, the dominant trend of urban land growth has stopped or even changed in Southern JS and Northern ZJ after 2005 with the *Shannon*'s entropy remaining unchanged and increasing by 0.05, respectively. It manifests that cities in the developed Southern JS are gradually optimizing their urban land use structure to be more coordinated and mature. The gaps between cities with different development levels have begun to shrink.

Table 4. The entropy of urban construction land in the sample cities, 2000 to 2010.

Region	*Shannon's* Entropy			Equilibrium Degree			Dominance Index		
	2000	2005	2010	2000	2005	2010	2000	2005	2010
Southern JS	1.91	1.84	1.84	0.87	0.84	0.84	0.13	0.16	0.16
Central JS	1.82	1.81	1.75	0.83	0.82	0.80	0.17	0.18	0.20
Northern ZJ	1.95	1.81	1.86	0.89	0.83	0.85	0.11	0.17	0.15
Total 51 Cities	1.92	1.83	1.84	0.87	0.83	0.84	0.13	0.17	0.16
Prefecture	1.95	1.86	1.85	0.89	0.85	0.84	0.12	0.15	0.16
County level	1.85	1.73	1.80	0.84	0.79	0.82	0.16	0.21	0.18

Coincidentally, the values of H for the prefecture and county level cities decreased from 1.95 and 1.85 in 2000 to 1.86 and 1.73 in 2005, respectively. The increase of the value of county level cities to 1.80 in 2010 indicates the halt of the dominant trend of urban land expansion. Furthermore, the fact that the prefecture level cities have bigger entropies than their county level counterparts confirms the hypothesis that land use structures in bigger cities are more complicated and less dominated [69]. However, the entropy of prefecture level cities decreased from 1.95 in 2000 to 1.86 and 1.85 in 2005 and 2010, respectively. On the contrary, the entropy of county level cities declined from 1.85 in 2000 to 1.73 in 2005 and then increased to 1.80 in 2010 (Table 4), leading to a narrowing gap.

4.2. Spatial Pattern of Urban Land Expansion in the YRD

In this section, the results from the analysis of spatial patterns of urban land use change, and in particular the changes in residential and industrial land in the 51 sample cities, through the global Getis-Ord G, are discussed. As illustrated in Table 5, the observed values of G for all three change rates are greater than their mathematical expectations. Furthermore, while the Z scores for change rates of the total urban land and residential land during 2000 to 2005 are significant, the change rate of industrial land is not. This result implies that in the first five years change rate of urban land is clustered, and urban land use change, especially for the residential land, has mainly occurred in the several active cores. Due to the widespread expansion of industrial land in the YRD, the clustering pattern disappeared in the later periods. After 2005, the Z score for the change rate of the total urban land became insignificant, thereby indicating that a particular hot spot of urban land use change gradually disappeared. In contrast, the hot spots of residential and industrial land changes during this period still exist.

Table 5. The Global Getis-Ord G of urban land expansion in 51 sample cities.

Change Rate	2000–2005			2005–2010		
	G ($\times10^{-2}$)	E ($\times10^{-2}$)	Z Score	G ($\times10^{-2}$)	E ($\times10^{-2}$)	Z Score
Residential land	2.15	2.00	1.20	2.25	2.00	1.25
Industrial land	2.09	2.00	0.02	2.29	2.00	1.13
Build-up areas	2.19	2.00	2.01	2.10	2.00	0.96

Note: G = observed value of G; E = mathematical expectation of G; Z= *standardized form of* G.

In order to uncover the spatial distribution of the hot spots and their change, the local Getis-Ord G_i^* values are calculated. The following are the results from the spatial analysis of the Getis-Ord G_i^*(Figures 4 and 5):

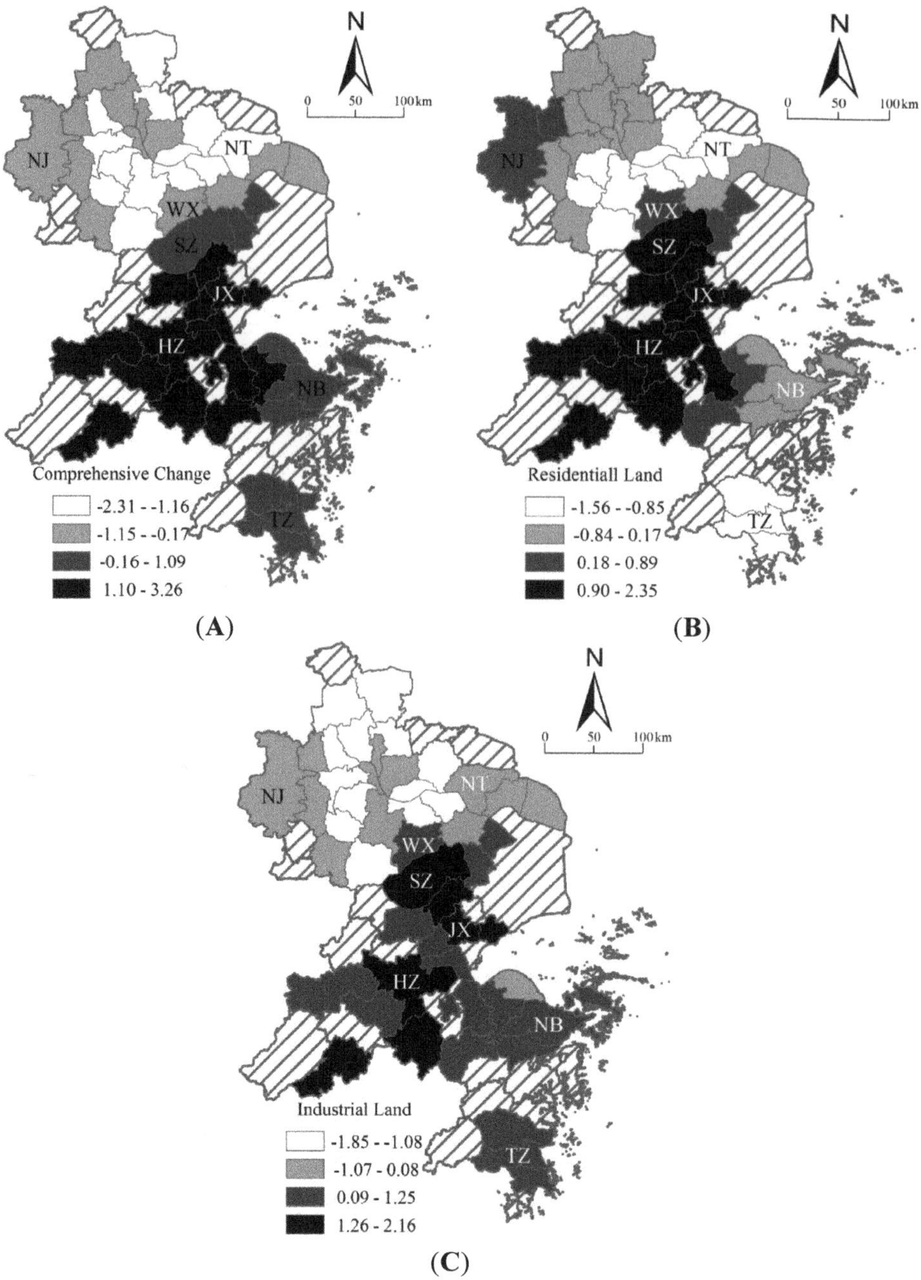

Figure 4. Hot spots of urban land use change in 51 sample cities, 2000 to 2005. Note: NJ = Nanjing, HZ = Hangzhou, SZ = Suzhou, WX = Wuxi, NB = Ningbo, NT = Nantong, JX = Jiaxing, TZ = Taizhou. (**A**) Comprehensive land change. (**B**) Residential land change. (**C**) Industrial land change.

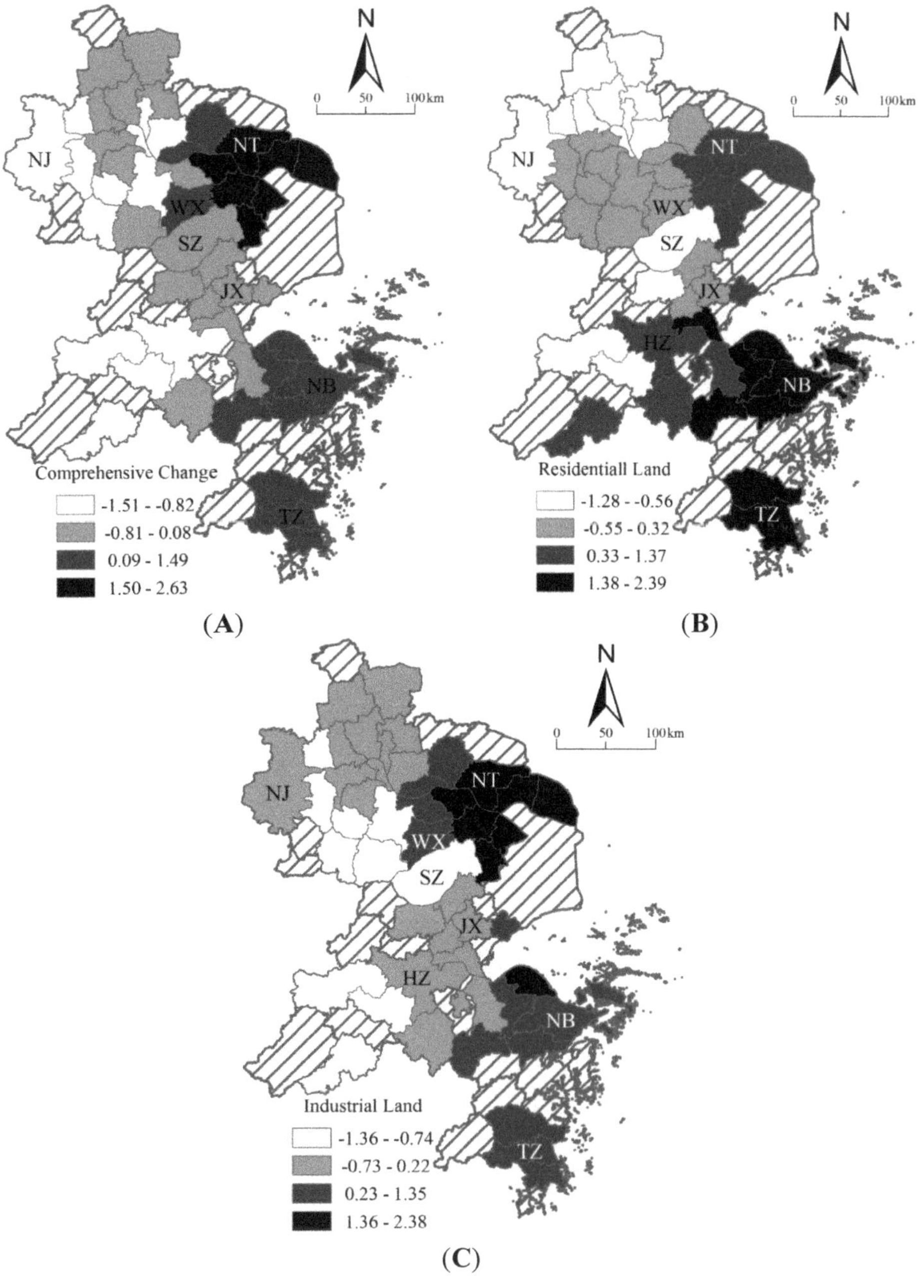

Figure 5. Hot spots of urban land use change in 51 sample cities, 2005 to 2010. Note: NJ = Nanjing, HZ = Hangzhou, SZ = Suzhou, WX = Wuxi, NB = Ningbo, NT = Nantong, JX = Jiaxing, TZ = Taizhou. (**A**) Comprehensive land change. (**B**) Residential land change. (**C**) Industrial land change.

(a) Since 2000, urban land use structure in the YRD has been maturing gradually. A shift in significant hotspots can be observed. For example, while the Northern ZJ (around Hangzhou) is a hotspot in 2000 to 2005, the region of Suzhou-Nantong is a hotspot in 2005 to 2010. There are more residential hot spots than industrial ones in the 51 sample cities, which indicate the relatively concentrated growth of residential land and the widespread sprawl of industrial land;

(b) During 2000 to 2005, the hot spots of residential land are mainly located at the junction of Suzhou-Jiaxing-Hangzhou (*Sujiahang*) expressway, followed by Nanjing. Outside these high value areas ($\geq$0.9) are the median value (0.18 to 0.89) and the low value ($\leq$0.17) areas, which correspond to a core-periphery structure. The two hotspots of industrial land are Suzhou and Hangzhou. We can also observe a more significant core-periphery structure in the region of Northern ZJ;

(c) During 2005 to 2010, hot spots of residential land located at Hangzhou have become weaker, along with Nanjing. Instead, hot spots at Ningbo and Taizhou have emerged. Hot spots of industrial land located at Hangzhou, Suzhou and their surrounding cities have almost disappeared. Noticeably, a zonal hot spot has grown from Qidong-Haimen, through Changshu-Taicang, to Kunshan surrounding the core of Shanghai. Additionally, the inland cities have a decreasing number of active spots than the coastal cities during this period.

4.3. Driving Forces of Urban Land Expansion

To some extent, the process of globalization, marketization, decentralization and urbanization in the YRD could adequately explain the urban land expansion during the transition era, with three adjusted R-squared values of 0.87, 0.78, and 0.74, respectively (Table 6). First, from the perspective of globalization and marketization, we find that the coefficients of FDI are significant at 90%, 95% and 90% levels for the overall built-up area, the industrial land, and the residential land respectively. The factor of labor has significant coefficients for the whole built-up expansion and industrial land change at 99% and 95% levels. This signifies that the deepening new international division of labor (NIDL) promoted by the formation of global production networks (GPNs) has resulted in the transformation of regional economy from an industrial composition into a value-added hierarchy. As a result, not only have the local economy structures transformed drastically, but also the urban land use structures of cities in the YRD. For the residential land change, the values are not as significant as expected and are negative, which signifies that people might be working in one city without actually living there, and thus contributing to the "jobs-housing imbalance" [70].

Table 6. Result of panel data regressions.

Variable	Built-up Areas Expansion		Residential Land Change		Industrial Land Change	
	Coefficient	*t*-Statistic	Coefficient	*t*-Statistic	Coefficient	*t*-Statistic
Global factors						
Ln (FDI)	0.06	1.63 *	0.11	2.34 **	0.09	1.61 *
Ln (Labor)	0.09	3.67 ***	−0.09	−0.83	0.07	1.77 **
Ln (DZs)	0.18	3.32 ***	−0.06	−0.83	0.30	3.55 ***
Ln (Policy)	0.70	4.11 ***	0.88	3.85 ***	0.52	1.88 *
Ln (Citizens)	0.71	12.99 ***	0.89	12.37 ***	0.46	5.24 ***
Ln (Job-housing)	0.02	0.38	0.16	2.86 ***	0.15	2.14 **
Region impact						
Ln (FDI) × Sunan	0.21	4.19 ***	0.11	1.67 *	0.17	2.13 **
Ln (Labor) × Sunan	0.02	0.80	0.02	0.81	0.02	0.75
Ln (FDI) × Zhebei	0.10	2.31 **	−0.02	−0.4	0.01	0.14
Ln (Labor) × Zhebei	0.06	4.91 ***	0.02	1.24	0.05	2.36 **
Administration impact						
Ln (FDI) × Prefecture	−0.21	−5.55 ***	−0.28	−5.58 ***	−0.05	−0.73
Ln (Labor) × Prefecture	0.22	10.75 ***	0.28	10.72 ***	0.19	5.94 ***
Adjusted R^2	0.868538		0.779129		0.737394	
Durbin-Watson stat	1.961137		2.35877		1.858222	

Note: *** Significant at 0.01 level; ** Significant at 0.05 level; * Significant at 0.1 level.

Second, from the decentralization perspective, in line with the findings of previous studies [14,33,62], "development zones" (DZs) have significant and positive impact on both the whole built-up area expansion and industrial land change. The insignificant impact of DZs on residential land change implies higher density of industries and a worse habitat environment. Along with the competition for local development, government policy and institutional changes have more influences on urban land structural changes, particularly the change of residential land (Table 6). These results imply that structural changes in urban space have facilitated the adjustment of urban land as a proactive strategy of city competition.

Third, since the 1980s, urbanization has greatly flourished urban real estate markets in the YRD. The increase of total housing construction area from 138.6 million square meters in 1984 to 347.6 million square meters in 1995—with the urban settlements getting doubled from about 21 million to about 44 million—has consequently accelerated the increase in urban residential land. The regression results (Table 6) also indicate that the number of "population", as an influencing factor of urbanization, has significant impact on the growth of built-up area ($c = 0.71$), residential land ($c = 0.89$), and industrial land ($c = 0.46$).

The results of this research indicate that for sustainable growth of residential and industrial land, "jobs–housing" relationship is a fundamental influencing factor [71]. With massive investment in urban transport infrastructure since the early 1990s, residential land has extended into suburbs or counties, which further has pushed the industrial land towards the outlying urban locations [33,72,73]. Another important finding is that the impact of the global market force, reflected by the FDI, on the urban land expansion in Southern JS is the greatest, implying a clustered and uneven development pattern in the YRD. The influence of labor is significant ($p < 0.01$) in Northern ZJ but insignificant in Southern JS, especially for industrial land expansion. In other words, the labor-intensive industries still play a major role and have a greater effect on urban land use structure in Northern ZJ than in Southern JS.

Finally, in terms of the administration level, both FDI and labor have significant impact on urban land expansion, especially in prefectural level cities. Interestingly, while FDI has negative impact, labor exerts positive impact. The underlying reason for this could be attributed to two interrelated aspects (a) the strict control on urban land by the government in prefectural level cities and their sub-districts, and (b) a little lower stringency in the county level cities that has been driving foreign enterprises' manufacturing units to be spatially located in counties and their headquarters in the prefectural districts. Due to this, the amount of FDI is high in prefectural cities with little change in urban land. With the urban land increase, the residential land growth in prefectural level cities has a more significant coefficient of labor than the whole YRD region. It implies that people tend to live in the prefectural districts while working in county level cities or even counties.

4.4. Significance of Geographical Location and Administrative Hierarchy in Urban Land Expansion

This section analyzes how different dynamics of urban land expansion are sensitive to the geographical location and hierarchical structure of China's urban administrative system. By applying GTWR and spatial regime models, we further produce, in a more rigorous multi-variant environment, a series of coefficients for cities located in different regions (Table 7) and two sets of comparable coefficients dedicated to each level of cities (Table 8).

The results imply that the urban land expansion mechanisms of cities in the YRD are sensitive to the geographical location. In different regions, the selected factors have dissimilar impacts on urban land expansion (Table 7). For instance, FDI apparently has more positive influence on urban land expansion in Southern JS and Northern ZJ than in Central JS. This result confirms that the developed cities with better infrastructure systems have yielded more success in drawing FDI and that the power of globalization is allocated by the level of development in the YRD. By contrast, there are more positive coefficients of labor in Central JS than the other two regions. This result demonstrates that increasing interest of labor to work in developed cities is leading to a frequent labor migration from the less developed regions of Central JS. A similar situation has also appeared for variables of DZ and government policy. This implies that urban growth and land expansion in the developing cities rely on both the local and upper-level government to a greater extent. Population growth is evidently a positive impetus after urban land expansion in China [43]. Notably, the effect of permanent citizens, which is characterized by the population living in the urban area for six months or longer, is more intensive for urban land expansion in developed Southern JS and Northern ZJ than Central JS. The job–housing relationship is more positive in Northern ZJ, where the private economies are more active.

Table 7. GTWR parameter estimate summaries.

Variables	Built-up Areas		Residential Land		Industrial Land	
	Mean	Positive	Mean	Positive	Mean	Positive
Southern Jiangsu (Observations = 153)						
Ln (FDI)	0.15	81.05%	0.18	72.55%	0.16	76.47%
Ln (Labor)	0.57	79.74%	0.83	83.66%	0.81	85.62%
Ln (DZs)	0.34	69.93%	−0.08	43.79%	0.26	67.97%
Ln (Policy)	−0.10	49.02%	−0.72	7.19%	−0.54	30.07%
Ln (Citizens)	0.28	75.82%	0.20	72.55%	0.05	57.52%
Ln (Job-housing)	−0.31	30.72%	−0.72	13.07%	−1.04	6.54%
Central Jiangsu (Observations = 117)						
Ln (FDI)	−0.02	47.01%	−0.03	47.86%	−0.05	38.46%
Ln (Labor)	0.55	83.76%	1.44	94.02%	1.28	98.29%
Ln (DZs)	0.99	96.58%	0.45	92.31%	0.82	88.03%
Ln (Policy)	0.40	88.03%	−0.13	23.93%	−0.12	76.92%
Ln (Citizens)	−0.08	47.86%	−0.68	26.50%	−0.64	11.97%
Ln (Job-housing)	−0.63	16.24%	−1.16	5.13%	−1.28	2.56%
Northern Zhejiang (Observations = 189)						
Ln (FDI)	0.12	78.84%	0.06	64.02%	0.21	77.25%
Ln (Labor)	0.03	51.85%	−0.03	53.44%	−0.01	52.38%
Ln (DZs)	0.57	89.42%	0.40	70.37%	0.40	85.71%
Ln (Policy)	0.21	54.50%	−0.77	35.45%	−0.45	21.69%
Ln (Citizens)	0.52	84.13%	0.82	83.07%	0.65	91.01%
Ln (Job-housing)	0.12	72.49%	0.19	70.90%	0.22	64.02%
Bandwidth	0.4661		0.5047		0.6389	
Adjusted R^2	0.9896		0.9559		0.9289	

Note: "Mean" denotes the average value of all the coefficients. "Positive" denotes the proportion of positive ones.

Table 8 reports the spatial regime model outputs of estimation of parameters for two different administrative hierarchies. Coefficients of all variables for two regimes have different values, which implies that the urban land expansion mechanisms of cities are sensitive to administrative hierarchies, in addition to the geographical location. For prefecture cities, we find that the build-up area expansion is more sensitive to variables such as the number of permanent citizens and the job–housing relation. The model of residential land change indicates that the DZs have greatly promoted the growth of residential land areas. The industrial land change is sensitive to all variables, in which the government impact of DZs and policy are only marginally positive.

Table 8. Results of spatial regime model.

Variables	Build-up Areas		Residential Land		Industrial Land	
	Prefecture	County	Prefecture	County	Prefecture	County
Ln (FDI)	0.05 **	0.14 ***	0.01	0.15 ***	0.28 ***	0.11 ***
Ln (Labor)	0.12 *	0.59 ***	0.18 *	0.64 ***	0.34 ***	0.54 ***
Ln (DZs)	0.24 ***	0.08 *	0.50 ***	−0.22 **	0.19 *	0.35 ***
Ln (Policy)	0.17 **	−0.11 *	0.17 *	−0.58 ***	0.14 *	−0.39 ***
Ln (Citizens)	0.83 ***	0.27 ***	0.27 **	0.24 **	0.75 ***	0.13 *
Ln (Job-housing)	0.10 ***	−0.09*	0.01	−0.01	0.23 ***	0.02
Adjusted R^2	0.844425		0.637504		0.755844	

Note: *** Significant at 0.01 level; ** Significant at 0.05 level; * Significant at 0.1 level.

For county level cities, we have observed that there is an apparent dissimilarity between the prefectures. The build-up area expansion, as well as the growth of residential land and industrial land, are most influenced by the global market. Results shown in Table 8 imply that the FDI and Labor have significant positive influence on urban land expansion in the county level cities, demonstrating active economy in the county level cities of the YRD [40]. The coefficient of DZs is significantly negative for residential land growth but positive for industrial land, which means that the establishment of DZs in county level cities is largely to increase GDP and is still far from the integration of industrial and urban functions [74]. The urban land expansion in county level cities is less sensitive to the variables of urbanization, in which only the number of citizens is marginally positive.

5. Discussion

Regarding urban land expansion and its driving mechanisms, several important discussion points have emerged from the findings of this paper. First, the overall level urban expansion in the Yangtze River Delta (YRD) (in the period 2000 to 2010) is characterized by a relatively consistent growth in residential land, and a triple growth rate in industrial land, demonstrating the industrialization driven urban development path of the overall region. While these patterns could be similar to the spatial structural changes in cities across the world [12], unlike the cities in the Western world, a large number of sub-regional spatial differences exist in the YRD. A deeper spatial analysis reveals that while cities in the higher level areas, such as the less developed Central Jiangsu (JS), are experiencing rapid increase in residential land and relatively slow industrial development, the cities in the lower level regions, such as the Northern Zhejiang (ZJ), are witnessing rapid industrialization and a sharp increase in industrial land. While in the most developed Southern JS, urban expansion is characterized by a balanced growth of residential and industrial land. In brief, while the prefectural level cities are transforming from production to service-based, the county level ones still focus on industrialization. The transformations in the YRD, as observed in the findings of this research, could be largely attributed to two aspects: (i) as the prefectural level cities are highly urbanized, the local governments have been focusing on de-concentration of manufacturing, and providing more land for neighborhood development and service based centres; and (ii), the policy of "Increase Power of the County" (*qiangxian kuoquan*) and the development competition between county level cities and prefectures has recently led to the allocation of more industrial land in the county level cities [75].

Second, the cities in the developed areas (e.g., Southern JS and Northern ZJ) are gradually optimizing their urban land use structure to be more coordinated and compact. On the other hand, the continuously dominant urban land expansion in the less developed Central JS is not only leading to simple and low-density land use patterns but also, to some extent, is widening the gap between the region and the developed areas in terms of attracting economic investment. As emphasized by the previous literature [22], while the patterns of compact development, as seen in the developed areas of the YRD, can promote sustainable land use, the sprawl and low-density patterns of the less developed areas can reinforce unsustainable urban land use in the YRD. An in-depth analysis of the evolution of urban land expansion further discloses an emerging core-periphery spatial pattern in the YRD. The hot spots of urban land use change have been concentrated at the core cities, which are more developed or have higher administrative hierarchies, in the initial stage. Then, the hot spots have gradually spread to counties in the surrounding periphery. With the increasing urbanization and industrialization in the past decade, and the narrowing gaps between both different administration and development levels, urban land expansion in the YRD shows a convergence trajectory and regional integration.

Third, by integrating the global market functions to the existing framework [11,50], and incorporating the six influencing factors, our conceptual framework can not only contribute to an improved understanding of the transitional nature of structural changes in China's urban land expansion, but also use Western theories to explain China's urban development. Specifically, as an emerging global city region in China and the world's sixth-largest economic center, the YRD region has been facing fierce global and regional competition for investment from the Transnational Corporations (TNCs) [75]. A variety of development zones (DZs) and new towns are being established to attract economic investment and absorb skilled migrants [57]. By the end of 2010, 132 provincial and national DZs were established in the 51 sample cities, which account for about 1/10th of the country. Due to this increasing investments, cities in this region have been witnessing large migration of labor from rural areas [76], which consequently has accelerated urban land expansion. In addition to the active market in the YRD, governmental policies have also been playing important roles in the processes of urbanization and industrialization [35]. The central development strategies, such as the integration plan of the YRD, the modernization plan of the Southern JS, and the development plans of the Coastal JS and ZJ, have obviously accelerated urban expansion in the YRD, although the policies need to be more conducive to promoting sustainable regional development. Due to the influence of these different forces, urban population has increased in the YRD, which in turn has accelerated the process of urbanization and greatly flourished the urban real estate markets. The total urban settlements in the 51 sample cities have increased from about 31 million in 2000 to 49 million in 2010, with the housing construction area almost doubling from 662 million square meters to about 1230 million square meters. Similarly, with massive investment in regional transport infrastructure since the early 1990s [1], the jobs–housing relationship in the YRD has changed and led to the restructuring of the current urban land use.

More generally, this finer-scale research reveals that in comparison to Western countries, urban land expansion mechanisms in China are sensitive to both development changes and administrative hierarchies. The reforms are implemented through the central government-directed top-down process and are largely beyond the control of local governments, although policy formation involves central–local interactions [43]. The process of industrialization, however, takes place within a more bottom-up approach in the YRD, particularly in the developed Southern JS [59]. Consequently, the driving forces of the global market and urban institutions are more significant in active county level cities in Southern JS and higher hierarchical prefectures than the other counterparts. The urban growth and land expansion in less active developing cities, on the other hand, rely on both the local and upper-level government to a great extent. Due to the regional inequality in both the urban economy and the standard of living of the residents [77], urban land expansion is not so sensitive to the urbanization process in the less developed and urbanized county level cities.

These findings imply that the governments need to better balance the allocation of capital and resources. A list of guidelines should be developed to promote sustainable urban land use instead of allowing the urban land expansion driven by natural forces. Admittedly, sustainable urban development (SUD) strategies, such as arable land and ecology protection, economical and intensive land use, coordination of economy, food security, have already been at the centre of China's policy [76]. However, the current land-use policies target different problems, which unexpectedly result in limited success or even in some cases, adverse consequences [4]. For the sustainable development of urban land, two policy recommendations are proposed. First, the local governments need to better balance the allocation of capital and resources at the finer scale. To do so, terms and conditions for siting industrial projects should be more rigorous in the active county level cities of Southern JS and Northern ZJ. For instance, ZJ provincial government has established a market to trade land development rights across localities in the province. This has helped to evade the constraints of the construction land quota in the fast developing regions and protect the ecological land and farmland in the agricultural and/or ecological counties of ZJ [78]. More of such kinds of initiatives could not only lead to controlling the rapid industrial land expansion and improving land use density, but also coordinating the land use structure in the less developed Central JS as well as narrowing the development gaps within the region. Second, as core cities (both the developed and higher hierarchy) have been playing leading roles in regional urban land use, further reforms might drive sustainable development of these cities and their surrounding areas. For example, the redevelopment or renewal of the initially allocated state-owned land by employing land leasing fees in urban centers, the rigorous control of the real estimate market in higher hierarchical prefectural cities, and the mixed-use of urban land instead of specific single land use zoning (e.g., residential and commercial/office, industrial and storage *etc.*) cannot only help to achieve SUD but also address the challenges associated with rapid urban land expansion.

6. Conclusions

This paper has studied the urban expansion and structural change in the YRD, and examined the driving forces based on the developed conceptual framework. Drawing on the accounts of central findings of this research, we find that there is a strong case for claims of regional inequality in this urban land expansion picture. We can make this conclusion with reference to the notions of sustainable land use, whose conditions are not met [24]. We can also argue that the findings of this research provide a rich narrative and novel perspectives to the debates on urban land expansion in regions, by capturing regional temporal and spatial patterns of changes in urban land use and its driving forces in the YRD region.

The research has several policy and theoretical implications. First, the outcomes of this research could provide immense foresight and empirical bases for the "National New-type Urbanization Plan (NNUP)", which aims to contribute to regional land management and promote sustainable urban development in China. This study also attempts to provide an empirical foundation so as to improve policymakers' understanding of the nature of interactions of these contested phenomena as basis for evolving appropriate policies to mitigate the challenges involved. Second, our conceptual framework by integrating the market function to the framework on the transitional nature of China's development cannot only better describe China's urban restructuring process, but also assist in using Western theories to explain China's urban development. Addressing and exploring the implications of multiple driving forces on urban land expansion is needed for informed decisions and to develop urban regional policies that ensure greater sustainability. Finally, this research could also be improved and extended by considering other natural indicators of urban development, such as topography as well as the influence of urban land use inequality on regional industrial development.

Acknowledgments: We would like to acknowledge the funding of the National Natural Science Foundation of China (41329001; 41130750; 41201111) and Ford Foundation. We would also like to express our great gratitude to Bo Huang at CUHK for his generous help with the Matlab package.

Author Contributions: Jinlong Gao was responsible for the data collection and analysis. Jinlong Gao and Yehua Dennis Wei contributed to the conceptual development and the research design. Jinlong Gao, Yehua Dennis Wei and Komali Yenneti wrote the paper. Yehua Dennis Wei and Wen Chen provided guidance and suggestions. All authors have read and approved the final manuscript.

Conflicts of Interest: The authors declare no conflict of interest.

References

1. Zhang, H.; Uwasu, M.; Hara, K.; Yabar, H. Sustainable urban development and land use change—A case study of the Yangtze River Delta in China. *Sustainability* **2011**, *3*, 1074–1089.
2. Tan, R.; Wang, R.; Sedlin, T. Land-development offset policies in the quest for sustainability: What can China learn from Germany? *Sustainability* **2014**, *6*, 3400–3430.
3. Xie, H.; Zou, J.; Jiang, H.; Zhang, N.; Choi, Y. Spatiotemporal Pattern and Driving Forces of Arable Land-Use Intensity in China: Toward Sustainable Land Management Using Emergy Analysis. *Sustainability* **2014**, *6*, 3504–3520.
4. Bai, X.; Shi, P.; Liu, Y. Realizing China's urban dream. *Nature* **2014**, *509*, 158–160.
5. Shafizadeh-Moghadam, H.; Helbich, M. Spatiotemporal variability of urban growth factors: A global and local perspective on the megacity of Mumbai. *Int. J. Appl. Earth Obs. Geoinf.* **2015**, *35*, 187–198.
6. Mcgee, T.G.; Lin, G.C.S.; Marton, A.M.; Wang, M.Y.L.; Wu, J. *China's Urban Space: Development under Market Socialism*; Routledge: London, UK, 2007.
7. Wu, K.-Y.; Zhang, H. Land use dynamics, built-up land expansion patterns, and driving forces analysis of the fast-growing Hangzhou metropolitan area, eastern China (1978–2008). *Appl. Geogr.* **2012**, *34*, 137–145.
8. Xu, J.; Liao, B.; Shen, Q.; Zhang, F.; Mei, A. Urban Spatial Restructuring in Transitional Economy—Changing Land Use Pattern in Shanghai China. *Chin. Geogr. Sci.* **2007**, *17*, 19–27.
9. Anderson, G.; Ge, Y. Do economic reforms accelerate urban growth? The case of China. *Urban Stud.* **2004**, *41*, 2197–2210.

10. Schneider, A.; Seto, K.C.; Webster, D.R. Urban growth in Chengdu, Western China: Application of remote sensing to assess planning and policy outcomes. *Environ. Plan. B* **2005**, *32*, 323–345.
11. Wei, Y.D. Decentralization, marketization, and globalization: The triple processes underlying regional development in China. *Asian Geogr.* **2001**, *20*, 7–23.
12. Wu, W.; Zhao, S.; Zhu, C.; Jiang, J. A comparative study of urban expansion in Beijing, Tianjin and Shijiazhuang over the past three decades. *Landsc. Urban Plan.* **2015**, *134*, 93–106.
13. Aguilar, A.G.; Ward, P.M. Globalization, regional development, and mega-city expansion in Latin America: Analyzing Mexico City's periurban hinterland. *Cities* **2003**, *20*, 3–21.
14. Li, H.; Wei, Y.D.; Liao, F.H.; Huang, Z. Administrative hierarchy and urban land expansion in transitional China. *Appl. Geogr.* **2015**, *56*, 177–186.
15. Sassen, S. *Losing Control? Sovereignty in An Age of Globalization*; Stanford University Press: Redwood City, CA, USA, 1996.
16. Mittelman, J. Rethinking the international division of labour in the context of globalization. *Third World Q.* **1995**, *16*, 273–295.
17. Aguilar, A.G. Mexico City growth and regional dispersal: the expansion of largest cities and new spatial forms. *Habitat Int.* **1999**, *23*, 391–412.
18. Mollenkopf, J.; Castells, M. (Eds.) *Dual City: The Restructuring of New York*; Russell Sage: New York, NY, USA, 1991.
19. Sernau, S. Economies of Exclusion: Economic Change and the Global Underclass. In *At the Crossroads of Development. Transnational Challenges to Developed and Developing Societies*; Behar, J.E., Cuzan, A.G., Brill, E.J., Eds.; BRILL: Leiden, The Netherlands, 1997.
20. Camagni, R.; Gibelli, C.M.; Rigamonti, P. Urban mobility and urban form: the social and environmental costs of different patterns of urban expansion. *Ecol. Econ.* **2002**, *40*, 199–216.
21. CNU. Charter of the New Urbanism, 2002. Available online: http://cnu.org/sites/default/files/charter_english.pdf (accessed on 29 July 2015).
22. Ligmann-Zielinska, A.; Church, R.; Jankowski, P. Sustainable urban land use allocation with spatial optimization. In Proceedings of the 8th International Conference on Geocomputation, Ann Arbor, MI, USA, 31 July–3 August 2005.
23. Newman, P.W.; Kenworthy, J.R. The land use-transport connection: An overview. *Land Use Policy* **1996**, *13*, 1–22.
24. Song, Y.; Knaap, G.J. New urbanism and housing values: A disaggregate assessment. *J. Urban Econ.* **2003**, *54*, 218–238.
25. Li, L.; Sato, Y.; Zhub, H. Simulating spatial urban expansion based on a physical process. *Landsc. Urban Plan.* **2003**, *64*, 67–76.
26. Stewart, J. Hot 'Lanta's Urban Expansion and Cultural Landscape Change. *Geogr. Rev.* **1999**, *89*, 132–140.
27. Hise, G. Nature's workshop' industry and urban expansion in Southern California, 1900–1950. *J. Hist. Geogr.* **2001**, *27*, 74–92.

28. Bradshaw, T.; Muller, B. Impacts of rapid urban growth on Farmland conversion: Application of new regional land use policy models and geographical information systems. *Rural Sociol.* **1998**, *63*, 1–25.
29. Cervero, R. Road expansion, urban growth, and induced travel. *J. Am. Plan. Assoc.* **2003**, *69*, 145–163.
30. Northrup, M.; Duket, L. *Smart Growth: Designating Priority Funding Areas;* Maryland Office of Planning: Annapolis, MD, USA, 1997.
31. Grout, A.C.; Jaeger, K.W.; Plantinga, J.A. Land-use regulations and property values in Portland, Oregon: A regression discontinuity design approach. *Reg. Sci. Urban Econ.* **2011**, *41*, 98–107.
32. Gao, J.; Wei, Y.D.; Chen, W.; Chen, J. Economic transition and urban land expansion in Provincial China. *Habitat Int.* **2014**, *44*, 461–473.
33. Liu, Y.; Yue, W.; Fan, P.; Song, Y. Suburban residential development in the era of market-oriented land reform: The case of Hangzhou, China. *Land Use Policy* **2015**, *42*, 233–243.
34. Liu, Y.; Yue, W.; Fan, P. Spatial determinants of urban land conversion in large Chinese cities: A case of Hangzhou. *Environ. Plan. B* **2011**, *38*, 706–725.
35. Zhang, Q.; Wallace, J.; Deng, X.; Seto, K.C. Central *versus* local states: Which matters more in affecting China's urban growth? *Land Use Policy* **2014**, *38*, 487–496.
36. Friedmann, J. Four theses in the study of China's urbanization. *Int. J. Urban Reg. Res.* **2006**, *30*, 440–451.
37. Zhang, T. Urban development and a socialist pro-growth coalition in Shanghai. *Urban Aff. Rev.* **2002**, *37*, 475–499.
38. Zhang, S. Land-centered urban politics in transitional China—Can they be explained by Growth Machine Theory? *Cities* **2014**, *41*, 179–186.
39. Wei, Y.D. Urban land use transformation and determinants of urban land use size in China. *GeoJournal* **1993**, *30*, 435–440.
40. Zhang, J.; Wu, F. China's changing economic governance: Administrative annexation and the reorganization of local governments in the Yangtze River Delta. *Reg. Stud.* **2006**, *40*, 3–21.
41. Ding, C.; Lichtenberg, E. Land and urban economic growth in China. *J. Reg. Sci.* **2011**, *51*, 299–317.
42. Ding, C. Urban spatial development in the land policy reform era: Evidence from Beijing. *Urban Stud.* **2004**, *41*, 1889–1907.
43. Wei, Y.D. Restructuring for growth in urban China: Transitional institutions, urban development, and spatial transformation. *Habitat Int.* **2012**, *36*, 396–405.
44. Lin, G.C. Peri-urbanism in globalizing China: A study of new urbanism in Dongguan. *Eurasian Geogr. Econ.* **2006**, *47*, 28–53.
45. Gao, J.; Li, S. Detecting spatially non-stationary and scale-dependent relationships between urban landscape fragmentation and related factors using geographically weighted regression. *Appl. Geogr.* **2011**, *31*, 292–302.

46. Liao, F.H.; Wei, Y.D. Modeling determinants of urban growth in Dongguan, China: A spatial logistic approach. *Stoch. Environ. Res. Risk Assess.* **2014**, *28*, 801–816.
47. Luo, J.; Wei, Y.D. Modeling spatial variations of urban growth patterns in Chinese cities: The case of Nanjing. *Landsc. Urban Plan.* **2009**, *91*, 51–64.
48. Wei, Y.D.; Li, W. Reforms, globalization, and urban growth in China: The case of Hangzhou. *Eurasian Geogr. Econ.* **2002**, *43*, 459–475.
49. Liu, J.; Zhan, J.; Deng, X. Spatio-temporal patterns and driving forces of urban land expansion in China during the economic reform era. *AMBIO J. Hum. Environ.* **2005**, *34*, 450–455.
50. Wei, Y.D. Regional inequality in China. *Prog. Hum. Geogr.* **1999**, *23*, 49–59.
51. Deng, X.; Huang, J.; Rozelle, S.; Uchida, E. Growth, population and industrialization, and urban land expansion of China. *J. Urban Econ.* **2008**, *63*, 96–115.
52. Wang, E.; Song, J.; Xu, T. From "spatial bond" to "spatial mismatch": An assessment of changing jobs–housing relationship in Beijing. *Habitat Int.* **2011**, *35*, 398–409.
53. Loo, B.P.; Chow, A.S. Jobs–housing balance in an era of population decentralization: An analytical framework and a case study. *J. Transp. Geogr.* **2011**, *19*, 552–562.
54. Zhou, J.; Zhang, C.; Chen, X.; Huang, W.; Yu, P. Has the legacy of Danwei persisted in transformations? the jobs-housing balance and commuting efficiency in Xi'an. *J. Transp. Geogr.* **2014**, *40*, 64–76.
55. Wei, Y.D.; Leung, C.K.; Luo, J. Globalizing Shanghai: foreign investment and urban restructuring. *Habitat Int.* **2006**, *30*, 231–244.
56. Wei, Y.D.; Leung, C.K. Development Zones, Foreign Investment, and Global City Formation in Shanghai. *Growth Chang.* **2005**, *36*, 16–40.
57. Wei, Y.D. Zone fever, project fever: Economic transition, development policy, and urban expansion in China. *Geogr. Rev.* **2015**, *105*, 156–177.
58. Shahtahmassebi, A.; Pan, Y.; Lin, L.; Shortridge, A.; Wang, K.; Wu, J.X.; Wu, D.; Zhang, J. Implications of land use policy on impervious surface cover change in Cixi County, Zhejiang Province, China. *Cities* **2014**, *39*, 21–36.
59. Zhu, J. Local growth coalition: The context and implications of China's gradualist urban land reforms. *Int. J. Urban Reg. Res.* **1999**, *23*, 534–548.
60. Wong, C.P. Central-local Relations Revisited the 1994 tax-sharing reform and public expenditure management in China. *China Perspect.* **2000**, *31*, 52–63.
61. Li, J.; Chiang, Y.-H.; Choy, L. Central-local conflict and property cycle: A Chinese style. *Habitat Int.* **2011**, *35*, 126–132.
62. Zheng, X.; Geng, B.; Wu, X.; Lv, L.; Hu, Y. Performance Evaluation of Industrial Land Policy in China. *Sustainability* **2014**, *6*, 4823–4838.
63. Taylor, P.J. *World City Network: A Global Urban Analysis*; Routledge: London, UK, 2004.
64. Chen, A.; Partridge, M.D. When are cities engines of growth in China? Spread and backwash effects across the urban hierarchy. *Reg. Stud.* **2013**, *47*, 1313–1331.
65. Steudler, D.; Rajabifard, A.; Williamson, I.P. Evaluation of land administration systems. *Land Use Policy* **2004**, *21*, 371–380.

66. Liao, F.H.; Wei, Y.D. Dynamics, Space, and Regional Inequality in Provincial China: A Case Study of Guangdong Province. *Appl. Geogr.* **2012**, *35*, 71–83.
67. Cravo, T.A.; Resende, G.M. Economic growth in Brazil: A spatial filtering approach. *Ann. Reg. Sci.* **2013**, *50*, 555–575.
68. MHURD. *China Urban Construction Stastical Yearbook, 2000, 2005, 2010*; China Planning Press: Beijing, China, 2001, 2006, 2011. (In Chinese)
69. Xu, X.Q.; Chu, D.K.Y. *Modern Urban Geography*; China Architecture & Building Press: Beijing, China, 1988. (In Chinese)
70. Ding, C.; Zhao, X. Assessment of Urban Spatial-Growth Patterns in China During Rapid Urbanization. *Chin. Econ.* **2011**, *44*, 46–71.
71. Giuliano, G.; Small, K.A. Is the journey to work explained by urban structure? *Urban Stud.* **1993**, *30*, 1485–1500.
72. Yuan, F.; Wei, Y.D.; Chen, W. Economic transition, industrial location and corporate networks: Remaking the Sunan Model in Wuxi City, China. *Habitat Int.* **2014**, *42*, 58–68.
73. Zheng, S.; Fu, Y.; Liu, H. Housing-choice hindrances and urban spatial structure: Evidence from matched location and location-preference data in Chinese cities. *J. Urban Econ.* **2006**, *60*, 535–557.
74. Van Oort, F.; Burger, M.; Raspe, O. On the economic foundation of the urban network paradigm: Spatial integration, functional integration and economic complementarities within the Dutch Randstad. *Urban Stud.* **2010**, *47*, 725–748.
75. Wei, Y.D.; Liao, F.H. The embeddedness of transnational corporations in Chinese cities: Strategic coupling in global production networks? *Habitat Int.* **2013**, *40*, 82–90.
76. Liu, Y.; Fang, F.; Li, Y. Key issues of land use in China and implications for policy making. *Land Use Policy* **2014**, *40*, 6–12.
77. Wei, Y.D.; Yu, D.; Chen, X. Scale, agglomeration, and regional inequality in provincial China. *Tijdschrift voor Economische en Sociale Geografie* **2011**, *102*, 406–425.
78. Zhang, W.; Wang, W.; Li, X.; Ye, F. Economic development and farmland protection: An assessment of rewarded land conversion quotas trading in Zhejiang, China. *Land Use Policy* **2014**, *38*, 467–476.

Scenario Simulation and the Prediction of Land Use and Land Cover Change in Beijing, China

Huiran Han, Chengfeng Yang and Jinping Song

Abstract: Land use and land cover (LULC) models are essential for analyzing LULC change and predicting land use requirements and are valuable for guiding reasonable land use planning and management. However, each LULC model has its own advantages and constraints. In this paper, we explore the characteristics of LULC change and simulate future land use demand by combining a CLUE-S model with a Markov model to deal with some shortcomings of existing LULC models. Using Beijing as a case study, we describe the related driving factors from land-adaptive variables, regional spatial variables and socio-economic variables and then simulate future land use scenarios from 2010 to 2020, which include a development scenario (natural development and rapid development) and protection scenarios (ecological and cultivated land protection). The results indicate good consistency between predicted results and actual land use situations according to a Kappa statistic. The conversion of cultivated land to urban built-up land will form the primary features of LULC change in the future. The prediction for land use demand shows the differences under different scenarios. At higher elevations, the geographical environment limits the expansion of urban built-up land, but the conversion of cultivated land to built-up land in mountainous areas will be more prevalent by 2020; Beijing, however, still faces the most pressure in terms of ecological and cultivated land protection.

Reprinted from *Sustainability*. Cite as: Han, H.; Yang, C.; Song, J. Scenario Simulation and the Prediction of Land Use and Land Cover Change in Beijing, China. *Sustainability* **2015**, *7*, 4260–4279.

1. Introduction

Since the land use and land cover (LULC) change project was launched by the International Geosphere and Biosphere Program (IGBP) and the International Human Dimensions Program (IHDP) on Global Change, land use research programs on a global scale have become central to international climate and environmental change research [1–3]. In recent years, the land use change process on a regional and local scale has also drawn the interest of scholars, who argue that the more local scales may be capable of testing and verifying the spatial patterns of LULC change on a global scale and that they also reveal the connection between land use change and human activity [4]. Some scholars focused on the impact of urbanization on LULC change [5–8], who considered that population growth and economic development drove urban land expansion and resulted in a great quantity of water bodies and agricultural lands being converted to built-up areas, which significantly affected the local, regional and global environment, including habitat quality [5], green spaces [7], environmental degradation [6,8], water quality [9,10], and so on. For instance, taking the Greater Dhaka of Bangladesh as the study area, Dewan *et al.* found the increasing contribution of GDP to the national economy and industrial growth was the major factor driving rapid LULC change, which led to environmental degradation and landscape fragmentation [6,11,12]. We know the LULC change process is dynamic and resulted from the interaction between natural and socio-economic elements at different scales. Because of the scale effect and the scale sensitivity of land system patterns [13], the related factors affecting LULC change differ, so it is important to understand their interaction and reasonably predict the future demand of land, which is key in land use planning and management.

Land use models are powerful tools that can be used to analyze the causes of LULC change and to evaluate land use policy [14]. Based on model analysis and the simulation of land use spatial patterns, the driving factors of LULC change can be revealed, clarifying the rate of land use and making possible multiple LULC scenarios in order to predict future land use demand. At present, models of LULC change have been developed to explore where, when and why it occurs based on the goals of a particular study. Existing studies on the LULC offer both global scale [15], national and regional scale [12,16–18], as well as some other research on different basins, such as Erhai Basin [19], the Mississippi River Basin [20], and different climate zones [18,21]. In these studies, the models involved mainly refer to the agricultural land dynamics (ALADYN) model [18], ant colony optimization [19], artificial neural networks and cellular automata [20], and so on. According to different research objectives, the relevant LULC models can be divided into three categories. Firstly are empirical-statistical models, such as regression models, *etc.* These models are developed based on mathematical equations to carry out statistical analysis on the factors affecting LULC change; however, they lack the consideration of social factors [22]. Secondly are spatially explicit models or rule-based models, such as the cellular automata model [23,24]. These spatial models are primarily used to determine the pattern and process of LULC change and to project the locations of future changes; however, it is still difficult to simulate the effect of human activities on LULC change [25]. Thirdly, agent-based models have been developed to simulate LULC change by individual agents [26]. However, due to the large number of interacting agents that need to be taken into account, most current multi-agent LULC models are only able to simulate simplified landscapes [26].

Previous models have attempted to incorporate biophysical and socio-economic data into land use simulations, but it still remains a major research challenge, because of the discrepancy among these different datasets [27,28]. As noted by Verburg *et al.*, no single model is capable of considering all of the processes of LULC change at different scales [29]. Some current LULC models have improved the analysis of the influence of a single factor on LULC change, but are unable to consider the effects of multiple factors and different processes that reflect spatial dimensions [30]. In view of the stability and resilience of LULC systems [31], LULC models should not only analyze land use at a single scale, but should also address the multiscale characteristics of land use systems and pay more attention to the interaction of the driving factors. The method for the selection of an appropriate integrated LULC model has therefore become important in LULC simulation studies.

The Conversion of Land Use and its Effects at Small regional extent (CLUE-S) model is a scale dynamic model with multiscale characteristics based on system theory. The scale is better suited than others to understand the relationship between LULC change and its driving factors and to explore possible process changes in various LULC scenarios at different spatial scales [11]. Compared with the above-mentioned models, the CLUE-S model incorporates the natural factors and social-economic factors, spatial and non-spatial distribution by combining a top-down with a bottom-up method, so the model is more comprehensive, open and extensive [32]. The CLUE-S model is preferred for addressing the spatial allocation of LULC change [14,31], but has some limitations in that it requires another mathematical model to calculate future land use demand.

The Markov model can most importantly depict the direction of LULC shifts, providing a framework for analyzing future land use demand. The model has been widely used to predict LULC change [33]. However, the traditional Markov model's ability to provide spatial analysis is weak, and it therefore has some difficulties in allocating predicted land requirements to geographical space. Moreover, existing research on LULC prediction, for the most part, takes the transition probabilities for land use types as a constant value, with little adjustment for socio-economic development and is thus also limited [34]. Based on the above discussion, we propose an approach by combining the Markov model and the CLUE-S model to deal with some shortcomings of the existing LULC models, taking the LULC prediction module and spatial analysis module as a whole to achieve the spatio-temporal simulation and prediction of LULC change. In this model, the Markov model assigns a transition probability to each single pixel at the time steps [35], and the CLUE-S model undertakes the simulation of the spatio-temporal dynamics of each single pixel. The CLUE-S and Markov model integrate human decision-making to achieve a more realistic simulation of LULC changes, which can provide a scientific method for LULC planning and management.

China has experienced a transition from its planned economy to a market economy, the rapid development of the economy and society, the acceleration of the urbanization process and the implementation of a regional development and ecological protection strategy, all of which significantly influence the spatial pattern of its LULC change [3]. Since the establishment of the market economic system, the real estate market has continued to develop, inspired by housing demand, leading to a rational readjusting of urban industrial structure, which, in turn, results in a large amount of agricultural land that has been transformed into urban land. As a result, the landscape of Chinese cities has changed dramatically. As the capital of China, the characteristics of LULC change in Beijing are typical and representative as a result of the faster economic and population growth. The objectives of the study are to: (1) propose an approach by developing a CLUE-S model that will transfer and allocate land demand from the Markov model to improve LULC change projection; (2) describe the temporal and spatial characteristics and identify the primary driving factors of LULC change patterns; and (3) predict and simulate the evolution tendency of land use spatial patterns under different scenarios.

2. Materials and Methods

2.1. Study Area

Beijing, the capital of China and one of the largest cities in the world, covers 16 districts and two counties with a land area of 16,410 km^2. The administrative region of Beijing is traditionally divided into three parts: the central city, the suburbs and the outer suburbs. The central city includes two districts, Xicheng and Dongcheng. The suburbs are made up of four districts, Haidian, Chaoyang, Fengtai and Shijingshan, and the outer suburbs are made up of eight districts and two counties, Tongzhou, Shunyi, Fangshan, Daxing, Changping, Mentougou, Huairou and Pinggu districts and Miyun and Yanqing counties (Figure 1). Beijing has undergone rapid urbanization and economic growth since the economic reforms of 1978, and in 2013, the population was approximately 21.15 million. Within the administrative region, the urban population is 18.25 million, which accounts for 86.31% of permanent residents. The new population will need new space and so creates a higher demand for residential land, thus encouraging rapid expansion of the region's urbanized area. Beijing's built-up areas grew from 397 km^2 to 1268 km^2 from 1980 to 2013, and the expansion of urban built-up land areas led to the decrease of cultivated land from 4258 km^2 in 1980 to 2317 km^2 in 2008. According to the overall land utilization plan in Beijing (2006–2020), the amount of reserved cultivated land cannot be less than 2147 km^2 by 2020, which means that the amount of cultivated land converted to built-up land is very limited. However, rapid population growth and expansion of built-up areas represent a daunting challenge to cultivated land protection, all of which lead to some traffic congestion, pollution and other issues.

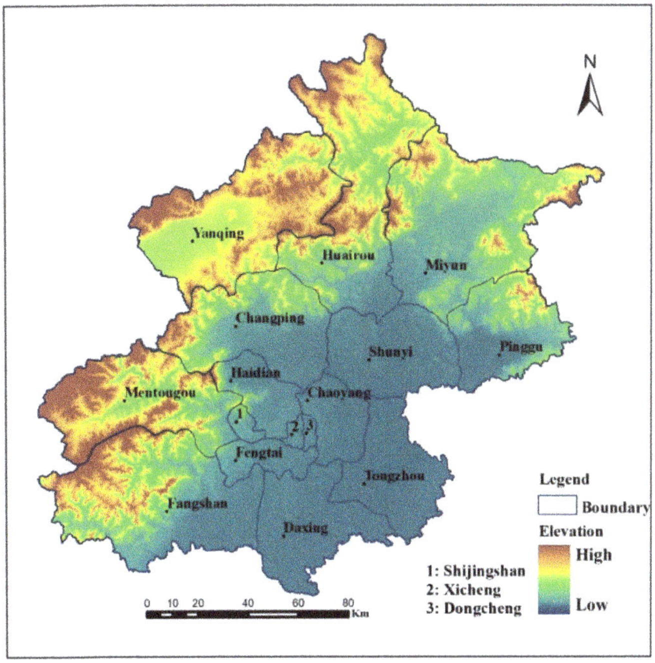

Figure 1. A map of the study area.

2.2. Data

Multi-annual socio-economic statistical data and spatial data were collected for evaluating the LULC change process. Among these data, the socio-economic statistical data came from the Beijing statistical yearbook, the Beijing area statistical yearbook and population census data and included permanent resident population, population density, GDP, per-capita living area, urbanization level and investment in fixed assets. Spatial data included remotely-sensed data that were obtained from Landsat TM images in 1985, 2000 and 2010. The TM image has a spatial resolution of 30 m. In addition to the remotely-sensed data, topographic maps were collected for establishing the digital elevation model (DEM) to obtain other biophysical data and also had a spatial resolution of 30 m. Moreover, landform and soil vector data were extracted from a digital geomorphology database and soil database in China at a 1:1,000,000 scale. All of the spatial data had unified projection and spatial registration, and the remote sensing image data were pretreated using Erdas image-processing software. We cut the remote sensing image according to the administrative boundary of the study area and then extracted the remote sensing information using the supervised classification method and a man-machine interactive interpretation.

Based on the characteristics of LULC change in the study area, we followed the classification system of the national resources and environmental background dynamic remote sensing survey database. Six LULC types were identified, including cultivated land, woodland, grassland, construction land (mainly referring to built-up land, that is land for the construction of buildings, fixtures and their auxiliary facilities, including commercial and industrial land, residential land, warehousing land, traffic-dominated land and other infrastructure, such as roads), water and unused land. After that, field investigation and Google Earth were used to check and correct the accuracy of the interpreted images, which allowed obtaining the land use maps for 1985, 2000 and 2010.

The classification accuracy of LULC types by Kappa coefficients for 1985, 2000 and 2010 had values that were greater than 0.80, which confirmed that the interpretation results met the analysis requirements. We estimated the slope direction of Beijing based on the DEM, and then, we calculated the spatial accessibility of LULC types to the administrative center, to roads and to main water bodies, as well as the corresponding distance layers. In addition, the spatial distribution of population and socio-economic data could be obtained by spatial interpolation using the method of kriging and natural neighbor. According to the requirements of the CLUE-S model combined with the actual situation of Beijing, the LULC maps and spatial layers were transformed to a grid format with a resolution of 300 m × 300 m in the same projection coordinates. Finally, the grid maps, including three periods of land use and some driving factors, were prepared for the model. Considering the periodic variation of LULC change and the rate of socio-economic development, the time span of the prediction could not be too short, so we selected 1985, 2000 and 2010 as the time nodes to simulate Beijing's LULC pattern in 2020.

2.3. Methods

2.3.1. Land Use Dynamic Degree

The land use dynamic degree refers to the rate of change in land use types for a specific time horizon, including quantitative changes in land resources and spatial changes in land use patterns [36]. It is an index that describes the regional difference of the rate of LULC change and reflects the comprehensive influence of social and economic activities on LULC change. The land use dynamic degree (S) in the period t is calculated as follows:

$$S = \left\{ \sum_{ij}^{n} \left(\Delta S_{i-j} / S_i \right) \right\} \times (1/t) \times 100\% \quad (1)$$

where S_i is the area of land type i in the beginning of the period, $\Delta S_{i\text{-}j}$ is the total area of land type i converted into other types and t is the study period.

2.3.2. Markov Model

LULC change is not unidirectional in nature; a given land type might theoretically change from one category of land use to any other. The Markov model can represent all of the possible directions of LULC change among all of the land use categories [23]. The Markov model is closely related to dynamic distributed lag models and consists of two primary land components: the transition matrix and the transition probability matrix, which represent the number and the probability of land shifting from one land use group to other groups in some observed period [34]. The transition probability matrix can be expressed as follows:

$$P_{ij} = \begin{vmatrix} P_{11} & \cdots & P_{1n} \\ \vdots & \vdots & \vdots \\ P_{m1} & \cdots & P_{mn} \end{vmatrix} \left(0 \leq P_{ij} \leq 1, \sum_{j=1}^{n} P_{ij} = 1\right) \tag{2}$$

where P_{ij} denotes the probability of land use i shifting to land use j; m and n are the number of land use types.

The Markov model is a random process of a state with no aftereffect characteristics, that is to say, a state of the system at some point $t + 1$ is only relevant to the state that is currently known at time t, but is unrelated to the moment before t. Land use demand is closely related to regional socio-economic development, which is mainly driven by human factors. The Markov model effectively combines human factors and land use. The land use distributions at the beginning (S_t) and at the end (S_{t+1}) of a time period, as well as a transition probability matrix (P_{ij}) representing the LULC change that occurred during that period are used to construct the Markov model [23], which is expressed as follows:

$$S_{t+1} = P_{ij} \times S_t \tag{3}$$

where S_{t+1} and S_t represent the states of land use at given point $t + 1$ and t, respectively.

Based on local socio-economic information, land use planning and related policies in Beijing from 1985 to 2010, three future socio-economic scenarios have been defined to predict land use demand for 2020 using the Markov model. The first scenario is characterized by natural development in which we assume that the factors that currently influence land use keep pace with the trend of LULC change from 1985 to 2010 and will not have changed greatly from 2010 to 2020. The second scenario is characterized by rapid development, in which we assume that areas of LULC change quickly by considering the growth rate of GDP and population, urbanization level, per capita living space and the floating population to obtain the demand for land use in 2020. The third scenario is characterized by ecological and cultivated land protection. In this scenario, woodlands and water bodies are designated as nature reserves that play an important role in ecological security and cannot change to any other land use categories, and the basic farmland is then taken as a restricted area in which the cultivated land cannot be converted into other land use types.

2.3.3. CLUE-S Model

The CLUE-S model was developed by Dutch scholars for the spatial simulation of LULC change on a small and medium-sized scale [31]. It was primarily used to explore the relationship between LULC change and related natural and socio-economic factors using a quantitative method, which was then used to simulate the LULC change process and to analyze the basic rule of the spatial-temporal evolution of land use [37].

The CLUE-S model has two parts: a spatial analysis module and a non-spatial analysis module. The spatial analysis module allocates land demands from the Markov model to spatially explicit land use patterns according to different transition probabilities and transition rules in land use categories and then will conduct a spatial simulation of land requirements in different scenarios [31]. The non-spatial analysis module focuses on factors that influence the spatial pattern of LULC change, such as socio-economic variables, regional spatial variables and land-adaptive variables. Socio-economic variables include population, economic factors, marketing conditions and related policies that are used to discuss the relationship between LULC change and economic development. Regional spatial variables mainly refer to the accessibility of land use types to the administrative center, roads and main water bodies, which are used to restrain the land use demand. The location suitability is a weighted average of suitability based on empirical analysis and capturing the historic and current location preferences in response to location characteristics [29]. Land-adaptive variables in this study include slope, soil, temperature and altitude, which are regarded as the potential LULC change factors. The contribution of different variables to the driving effect for one specific land use type presents the difference, so in this study, a logistic regression model is selected to estimate the influence of related factors on different land use types. The equation is expressed as follows [31,37]:

$$Log\left(\frac{P_i}{1-P_i}\right) = \beta_0 + \beta_1 X_{1,i} + \beta_2 X_{2,i} \cdots\cdots + \beta_n X_{n,i} \quad (4)$$

where P_i is the probability of a grid cell for the occurrence of land use type i. X is the related driving factors; β represents the regression coefficients between land use types and driving factors.

3. Results and Analysis

3.1. LULC Change Characteristics

We obtain the LULC maps for 1985, 2000 and 2010 with the help of Erdas image processing software and the ArcGIS platform (see Figure 2). The rate of LULC change during the different periods is illustrated by the index of the land use dynamic degree. From 1985 to 2000, the value of the land use dynamic degree is 0.99, which indicates that the rate of LULC change is relatively slow. From 2000 to 2010, the value of the land use dynamic degree is 1.64, which demonstrates that the rate of LULC change has accelerated compared to the previous period and that the effects of socio-economic activities on the land use pattern have become more intense.

On a spatial scale, the expansion of construction land is strikingly clear from the center to the edges, which confirms a spatial pattern of urban growth. The direction of construction land expansion is primarily along the Ring road from Chaoyang and Haidian in the suburbs to Tongzhou and Changping in the outer suburbs. In addition, the area with reductions in cultivated land is primarily in the eastern and southern regions of Beijing, which coincides with the extension direction of radioactivity along the highways and national roads.

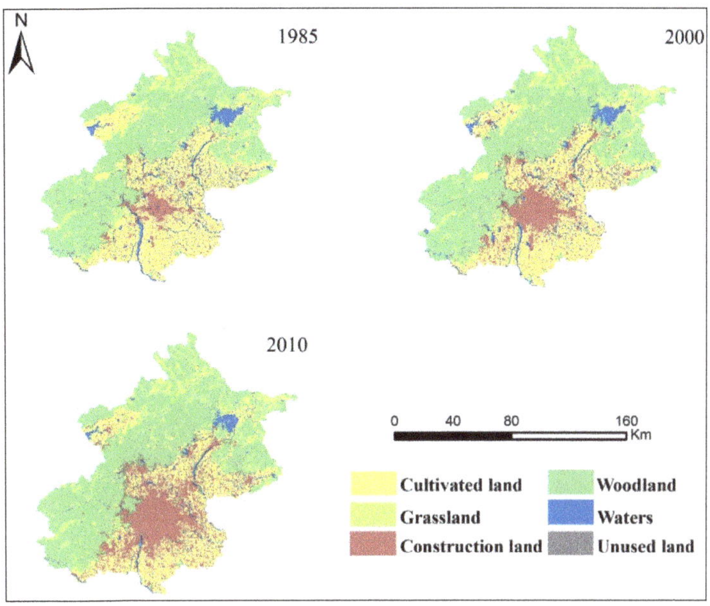

Figure 2. Land use pattern by year in Beijing (1985–2010).

3.2. Regression Analysis of the LULC Change Process

A logistic regression model was constructed to explore the relationship between LULC change and the related driving factors that play an important role in revealing the internal mechanism of LULC change. A logistic regression model is run for each of the land use types in 1985, 2000 and 2010, respectively. We test the statistics of the driving factors using the ROC curve proposed by Pontius [38], which evaluates the predicted probabilities by comparing them with the observed value [31]. Pontius Jr found that any ROC above 50% was better than random [38]. The obtained value was greater than 50%, indicating a probability distribution consistent with the actual distribution of land use types. Generally, an ROC above 70% shows that the driving factors had a greater explanatory power for a certain land use type.

Due to the statistical caliber inconsistencies for 1985 and a lack of restrictions on basic farmland, the regression results are poor. The ROC test of three land types is less than 0.6, so we list only the results for 2000 and 2010 (Table 1). Some driving factors, including altitude, slope, GDP, urbanization, population density, distance to the urban center, distance to the local road, distance to the railway, distance to the highway and distance to the river, are selected to evaluate their influence on land use types. As shown in Table 1, the ROC test statistics for various land use types is above 0.7, which shows clearly that the spatial distribution for all land use categories can be explained by the related driving factors, but different driving factors result in some differences in various land use types.

Traffic and socio-economic factors may play an important role in the conversion of land use types, especially their accessibility to the administrative center and roads, as well as their urbanization level. In terms of topographic factors, the effect of the slope on land use types is clearer, with a higher slope conducive to forest land and a lower slope good for other land types.

Table 1. Logistic regression results of the spatial distribution of land use types in Beijing from 2000 to 2010.

Driving factors	Years	Cultivated land		Woodland		Grassland		Waters		Construction land		Unused land	
		β	Exp(β)	β	Exp(β)	β	Exp(β)	β	Exp(β)	β	Exp(β)	β	Exp(β)
Elevation	2000	−0.0011	0.9989	0.0004	1.0004					0.0011	1.0011		
	2010	−0.0008	0.9992	−0.0003	0.9997					0.0015	1.0015		
Slope	2000	−0.2270	0.7734	0.0331	1.0337			−0.0264	0.9739	−0.0220	0.9783	−0.0581	0.9435
	2010	−0.1121	0.8980	0.0171	1.0173					−0.0217	0.9785		
Distance to the river	2000	−0.0614	0.9404	0.0917	1.0960	−0.0476	0.9535	−0.1784	0.8366	0.0620	1.0640		
	2010	−0.0606	0.9412	0.0954	1.1001	−0.0430	0.9579	−0.2470	0.7812	0.0260	1.0263		
Distance to the urban center	2000	0.0234	1.0669	0.0185	1.0187	0.0519	1.0533			−0.0838	0.9196	0.0609	1.0628
	2010	0.0088	1.0212	0.0155	1.0156	0.0633	1.0654			−0.0877	0.9160	−0.0859	0.9177
Distance to the local road	2000			0.0351	1.0358					−0.1481	0.8623		
	2010			0.0505	1.0518	0.0296	1.0301	0.0464	1.0475	−0.1658	0.8472		
Distance to the railway	2000	0.0504	1.0517	−0.0630	0.9389	−0.0410	0.9598	0.0367	1.0640			−0.0423	0.9585
	2010	0.0331	1.0336	−0.0491	0.9521	−0.0215	0.9787	0.0513	1.0603				
Distance to the highway	2000	0.0432	1.0577	0.0606	1.0625	0.0204	1.0206			−0.0546	0.9468		
	2010	0.0357	1.0349	0.0628	1.0648					−0.0657	0.9364		
GDP	2000	−0.0225	0.9777	−0.0185	0.9815	0.0322	1.0327						
	2010					0.0015	1.0015	0.0015	1.0015	0.0013	1.0013		
Urbanization	2000	−0.0274	0.9726	−0.0985	0.9062	−0.0807	0.9225	−0.0173	0.9828	0.0362	1.0369		
	2010	−0.0055	0.9943	−0.0146	0.9855	−0.0347	0.9659						
Population density	2000			0.0006	1.0006	0.0002	1.0002			−0.0002	0.9998		
	2010	−0.0404	0.9584	−0.0011	0.9989	−0.0005	0.9995	−0.0014	0.9986	0.0013	1.0013		
Constant	2000	−0.3677	0.6923	0.9723	2.6440	−1.7567	0.1726	−3.3856	0.0339	−1.3041	0.2714	−6.5598	0.0014
	2010	−0.0352	0.9654	−0.8484	0.4281	−3.1322	0.0436	−4.3192	0.0133	0.2457	1.2785	−4.8677	0.0077
ROC value	2000	0.7730		0.8460		0.8100		0.7470		0.8250		0.7990	
	2010	0.7080		0.8300		0.7970		0.7980		0.8390		0.7320	

Note: β represents the regression coefficients between land use types and driving factors, Exp(β) represents the probability of land use type change. All driving forces are significant at $p < 0.05$. When Exp(β) > 1, this indicates the probability increases of land use types upon an increase in the value of the driving factors, and when Exp(β) < 1, this indicates the probability of land use type reductions.

3.3. *Analysis of Transition Probability Matrix*

The transition probability matrix shows the transfer direction of land use types (see Tables 2 and 3). From 1985 to 2000, construction land, woodland and grassland are the most stable classes, with 0.99, 0.99 and 0.94 probabilities, respectively. The most dynamic classes are unused land, water and cultivated land, with transition probabilities of 0.62, 0.63 and 0.83; in these classes, cultivated land was primarily transformed into construction land, and cultivated land that is occupied for urban expansion is evident. From 2000 to 2010, the transition of various land use types is consistent with the previous period; the most stable classes are still construction land, woodland and grassland, with 0.94, 0.97 and 0.92 probabilities, respectively. The most dynamic class is unused land, which was primarily transformed into woodland with transition probabilities of 0.61, indicating that the afforestation policy played a role in Beijing. Moreover, the transition of construction land to other types of land is also clear.

Table 2. Transition probability matrix of land use types in Beijing from 1985 to 2000.

	Cultivated land	Woodland	Grassland	Waters	Construction land	Unused land
Cultivated land	0.83	0.02	0.00	0.02	0.13	0.00
Woodland	0.00	0.99	0.00	0.01	0.00	0.00
Grassland	0.01	0.03	0.94	0.02	0.00	0.00
Waters	0.21	0.06	0.01	0.63	0.09	0.00
Construction land	0.00	0.00	0.00	0.01	0.99	0.00
Unused land	0.25	0.02	0.01	0.02	0.08	0.62

Table 3. Transition probability matrix of land use types in Beijing from 2000 to 2010.

	Cultivated land	Woodland	Grassland	Waters	Construction land	Unused land
Cultivated land	0.78	0.03	0.00	0.01	0.17	0.01
Woodland	0.01	0.97	0.00	0.01	0.01	0.00
Grassland	0.02	0.04	0.92	0.00	0.02	0.00
Waters	0.17	0.05	0.04	0.58	0.15	0.01
Construction land	0.04	0.01	0.00	0.00	0.94	0.01
Unused land	0.04	0.61	0.14	0.00	0.05	0.16

3.4. *Predicting LULC Change Based on the Markov Model*

Based on LULC change from 1985 to 2010, the Markov model is used to predict the land requirements for the different land use types in 2020 according to the three future socio-economic scenarios discussed above. The prediction results are presented in Table 4. The table shows that the areas of cultivated land, grassland and water have a decreasing trend, as do woodland and construction land. A large difference is found in the land requirements for various land use types under different scenarios.

Table 4. The prediction results of land use under different scenarios, 2010–2020 (units: ha).

Years	Scenario design	Cultivated land	Woodland	Grassland	Waters	Construction land	Unused land
2010	Actual land use	406,692	741,789	119,772	44,082	311,013	10,305
2020	Natural development	333,279	756,891	115,083	21,330	400,014	7056
	Rapid development	328,535	757,223	114,975	17,333	408,672	6916
	Ecological-cultivated land protection	360,912	751,635	116,718	34,890	364,575	4923

3.5. Model Validation

Combining the results of the logistic regression model, the transition matrix of the land use types and land-type conversion elasticity, the CLUE-S model is used to simulate the distribution of land use patterns in order to derive the simulated map of land use pattern for Beijing in 2010, then the simulated land use map is compared with the actual satellite-derived land use map for 2010 based on the Kappa statistic (see Figure 3).

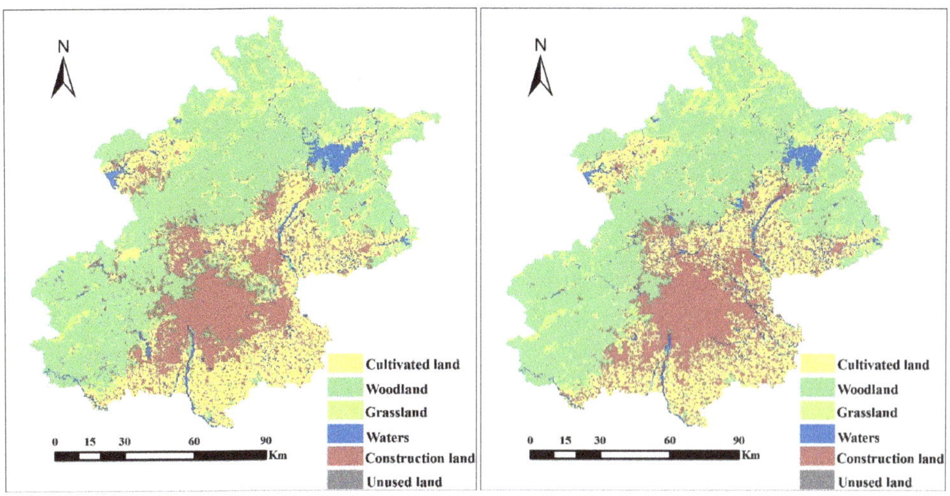

(a) (b)

Figure 3. Comparison of simulated (**a**) and observed (**b**) land use situations in Beijing, 2010.

The Kappa statistic, which can reflect the simulation accuracy of the model, was used as the validation method to evaluate the ability of the model to simulate the spatial pattern of land use [39]. Its expression is as follows:

$$Kappa = (P_o - P_c)/(P_p - P_c) \quad (5)$$

where P_o is the percent correct for the output, P_c is the expected percent correct due to chance and P_p is the percent correct when the classification is perfect.

The value of the Kappa statistic is 0.87, which indicates that the consistency is good between the predicted results and the actual land use situation, and it shows that the model is reliable for Beijing and can be used to predict LULC change under different scenarios. Figure 3 shows that although more similarity is found between the simulated map and the empirical map of land use in 2010, it still exhibits a certain bias. Cultivated land, woodland, grassland, waters and unused land in the simulated land use map are relatively similar to the corresponding classes in the actual land use map for 2010, while the construction land class is poorly simulated. Analysis of the simulated land use maps reveals that the model generally underpredicted the location of the construction land class. There are three explanations for this difference. On the one hand, the CLUE-S and Markov model employs the contiguity rule, which is used to simulate the growth of a land use type near the existing similar land use class [40]. This suggests that the model's simulation accuracy increases with the proportion of a given land class relative to others [41]. Therefore, when a few of the nearby pixels belong to a land use type, such as construction land, then the transition potential is down-weighted, which may possibly result in the poor simulation of construction land [41]. On the other hand, the constraint layer of the simulation of land use is primarily based on the appropriate adjustments to basic farmland in 2002, which cause a small discrepancy on the bound level. Besides, the socio-economic data obtained in this simulation are mainly at the county level and do not cover the *jiedao*-level data, which influences the accuracy of the simulation to some extent.

3.6. Analysis of Simulation Results

Combined with the land demand under different scenarios, land transfer rules, related driving factors and constraints, we conduct the simulation and prediction of the spatial distribution of Beijing's land use in 2020 (see Figure 4).

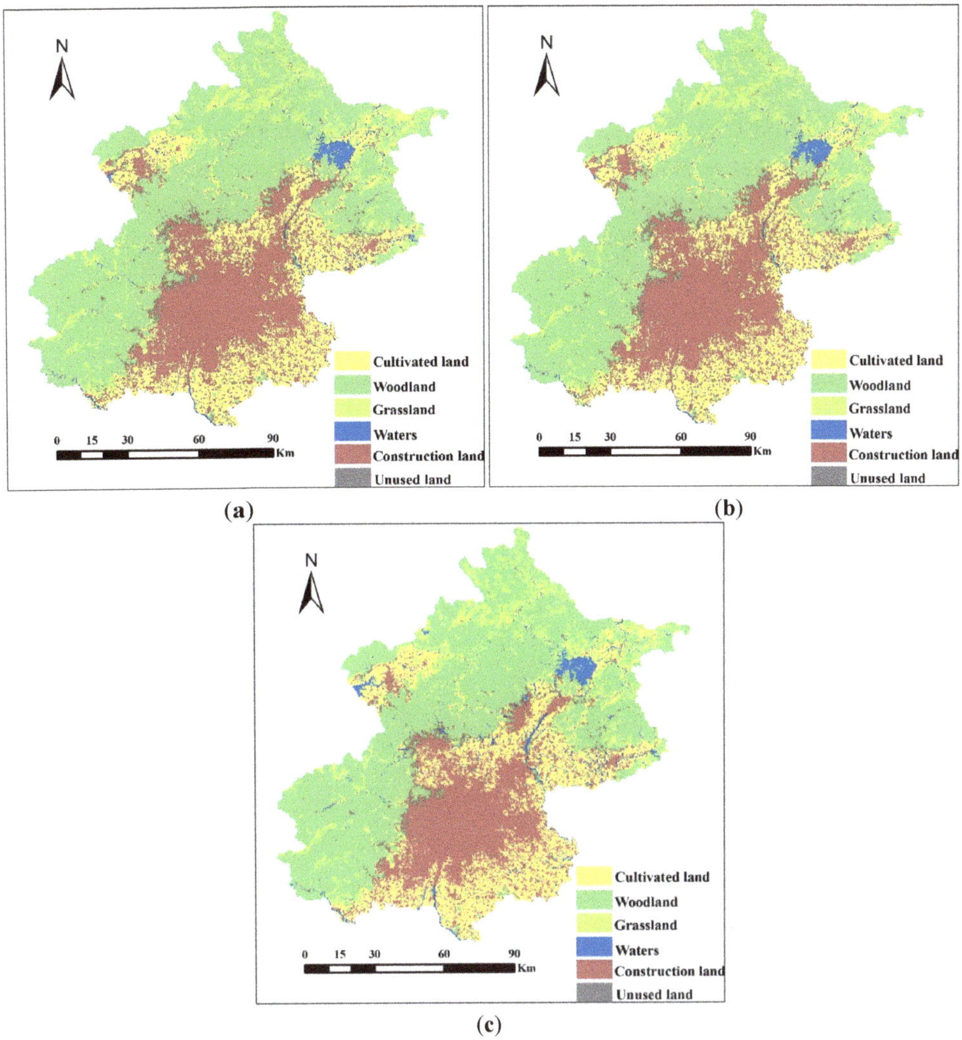

Figure 4. Simulation of land use demand under different scenarios in Beijing, 2020. (**a**) Natural development scenario (**b**) Rapid development scenario (**c**) Ecological and cultivated land protection scenario.

Under the natural development scenario (Figure 4a) and rapid development scenario (Figure 4b), the trend of construction land expansion is clearly along the direction of main traffic arteries, such as the airport express rail. Comparing the simulated scenarios for 2020 with the actual map for 2010, this conversion to urban land occurs primarily in the northeast, northwest, southwest and east of Beijing, in districts, such as Tongzhou, Shunyi, Changping, Fangshan and Huairou, and in Yanqing County. Affected by the development policy of the new town, the functions in the inner city gradually transfer to the new town, which attracts population concentrated in Changping and Shunyi, and leads to the expansion of residential land, with the area of cultivated land and waters greatly reduced. The results of the simulation in the rapid development scenario are similar to those in the natural development scenario, but the changes in land use types are more fundamental. The dramatic change in land use types, including water area reductions of 60.7% by 2020, will lead to water-resource shortages, most seriously in Beijing. In addition, construction land increases 31.4% and primarily comes from cultivated land, waters and unused land; the change trend of woodland and grassland is more subtle.

Under the ecological and cultivated land-protection scenarios (Figure 4c), the study assumes a land use that is under a strict cultivated-land protection policy from 2010 to 2020. Similar to the above scenarios, the expansion trend for construction land is relatively moderate and occurs mainly in the east, northwest and southwest, in districts, such as Tongzhou, Shunyi, Changping and Fangshan. Woodland area increases slightly, and the increase rate of construction land is 17.2%, which decreases by 11.4% compared with the natural development scenario. Waters and cultivated land areas decrease 20.9% and 11.3%, respectively, which indicates that the regulation has a significant effect on restricting the expansion of construction land. However, the conversion trend of cultivated land and woodland to construction land is also more clearly in the mountainous area of Mentougou and Shijingshan districts and in Yanqing County, which indicates that Beijing still faces the pressure of ecological and cultivated land protection.

4. Discussion

To better simulate land use patterns, a group of LULC models were specifically developed for a particular case study [14,42], which provided some opportunities to select an approach that would best address the research questions and characteristics related to the study area. It was important to acknowledge that no single model or approach can sufficiently describe the different processes at all spatial and temporal scales [42]. Because of residents' travel being determined by the land use spatial pattern, some researchers linked land use with traffic demand and put forward the land use and transportation interaction model (LUTI model) to describe the relationship between urban land use and transportation, which improved the simulation accuracy of the model. However, these models just focused on the external results and had not explained the internal reason for individual behavior choices, so it was difficult to simulate the interaction between different behaviors [43]. Since then, other scholars proposed an urban spatial equilibrium model by combining spatial economics and the LUTI model, such as the TRANUS model (an integrated land use/transportation model was developed by Barra *et al.*) [44], which was mainly used to assess the social-economic impact of planning policy in a large spatial scale [45]. Due to the lower spatial precision of the macroscopic model, it was not conducive to capturing the individual behavior activities, so someone proposed spatial non-equilibrium models with macroscopic and microscopic characteristics, such as the IRPUD model (The model was initially designed and implemented at the Institute of Spatial Planning of the University of Dortmund) [46], the DELTA model (The model has been developed by Simmonds, which was used to extend relatively conventional transport models into land-use/transport interaction) [47], the UrbanSim model [48] and TIGRIS XL model (an land-use and transport interaction model for the Netherlands is developed for and owned by the Ministry of Infrastructure and Environment and the Netherlands Environmental Assessment Agency) [49]. The research objects of these models were still urban land and traffic system, but they adopted a bottom-up method. Taking UrbanSim as the example, this system is composed of four modules, such as employment location choice model, residential location choice model, land development choice model and real estate price model [48]. Compared with the previous models, spatial non-equilibrium models had difficulty in accurately simulating the complex economic behavior and agglomeration effect driven by market price [43,45]. According to the above discussion, we know that the interaction model is an applied urban model more fit for application, which is not limited to simulating LULC change, but can also be used for location selection and policy evaluation.

Recent validation studies have indicated that most spatial models still contain a high level of uncertainty [50]. In this study, we develop an approach by combining the Markov model and CLUE-S model to deal with some shortcomings in existing LULC models and characterize the land change processes and predict LULC change under different scenarios. The Markov model can depict the direction of LULC shifts and predict the future land requirements for land use categories by taking into account the influence of related factors on land use requirements. The CLUE-S model can allocate the predicted land requirements to geographical space using the Markov model. The results of the combined Markov and CLUE-S models indicate that the model is capable of representing LULC change in Beijing, which suggests that the approach is a useful tool for the analysis of related driving factors and the estimation of their influence on LULC change. The methodology can be effective and realistic for predicting possible LULC change under different scenarios and for providing a scientific basis for land use decision making and planning.

The complexity of the LULC system requires that the selection of driving factors for the CLUE-S model be based on the theoretical relationships between driving factors and land use [31]. We select the related driving factors that affect LULC change from land-adaptive variables, regional spatial variables and socio-economic variables, but there are still some other factors that are difficult to quantify. The selection of variables and indicators, to a certain extent, may cause some differences in the simulation results or model parameters, which will produce certain effects on the driving factors and the prediction of LULC change. For example, changes in household sizes may have an impact on housing demand and the land use spatial pattern. In the planned economy era, most of the families in China were three generations, even four generations under one roof. Now, the family structures have changed gradually to a nuclear family, and every newlywed couple generally requires a separate entrance from the newborn small family or nuclear family; DINK (a DINK household in which there are two incomes and no children) families and empty nesting are growing along with the development of society. The miniaturization of household sizes must put forward more requirements on housing. Increased housing demand is largely caused by shrinking household sizes, population growth, property taxes and other factors [51]. Housing demand stimuli will certainly change the land use pattern. When urban centers still have space, increased households may cause its development to be more compact. On the contrary, increased housing demands may lead to sprawl if not accommodated by high-density or not afforded by high housing price, particularly in the city center [52]. In this situation, the rapid spatial expansion of the city will lead to more rural and arable land converting to built-up areas and simultaneously putting immense pressure on the land use pattern in future. Therefore, the identification of the methods for the selection of the more scientific variables that reflect the influence of human activities on LULC change is critical.

The method of combining the Markov model and the CLUE-S model is found to have the potential to reflect the complexity of LULC change, but if we select different forecasting models for land demand, it may produce different results. In many cases, it may even be most appropriate to use different model approaches to study the same region and to then compare the outcomes of such forecast models, which may lead to a better and more complete understanding of the LULC change process [53]. The method has therefore become an important direction for current research and could involve using a different model in combination with the CLUE-S model, such as the grey interconnect degree, the system dynamics model or multi-agent system models [26], which offer a reasonable alternative to predict and simulate land use patterns. In addition, as a megalopolis in the process of rapid urbanization, urban development and planning and major infrastructure construction, the related policies of functional dispersal from the central city could have significant impacts on the formation and evolution of land use patterns in Beijing. Now, many LULC models, for instance generic urban models [54], the TIGRIS XL model [49] and the SILENT (the Sustainable Infrastructure, Land-Use, Environment and Transport) model [55], have transformed some policies, such as infrastructure or land use zoning, into spatial parameters in simulating and predicting land use demand, so this can help decision makers to anticipate the impacts of the proposed policy. However, in this study, we do not consider the effects of various policy factors in the simulation process. A further study could employ regional spatial factors, land-adaptive factors, socio-economic and policy-related factors together in the simulation of further LULC change in Beijing, which could guide more informed decision making.

5. Conclusions

This paper explored the characteristics of LULC change and simulated future land use demand by combining a CLUE-S model with a Markov model, which overcame their respective disadvantages in demand prediction and spatial allocation and gave some insight into a better understanding of possible changes in land use. The study chose Beijing as its case study, recognizing the related driving factors from land-adaptive variables, regional spatial variables and socio-economic variables, then performed the simulation on land use demand and revealed LULC change trends in 2020.

The results suggest that following a rapid urbanization process, cultivated land converts to urban built-up land, which will become the main feature of LULC change in the future. The pattern will be more serious in the mountainous areas, such as the Mentougou, Shijingshan and Huairou districts and Yanqing County. From three scenarios, we find that the expansion trend of urban land occurs mainly northeast, northwest, southwest and east of Beijing, in districts, such as Tongzhou, Shunyi, Changping and Fangshan. The simulation of land use also shows, however, that the major difference between development scenarios (natural development and rapid development) and protection scenarios (ecological and cultivated land protection) occurs because the mountainous areas of Beijing are important ecological barriers and water conservation areas and the geographical environment with a higher elevation limits the expansion of urban land. The relationship between construction land and the ecological environment should therefore be comprehensively studied and estimated [37]. By protecting ecological and cultivated land and strictly controlling the expansion of built-up land, important adjustments can be made to the regional land use structure, the regional ecology can be accelerated and the sustainable economic development of Beijing can be prioritized.

Acknowledgments: This research was supported by the Natural Science Foundation of China (41271132) and the Non-profit Industry Financial Program of Ministry of Land and Resources of China (No: 201211001).

Author Contributions: Huiran Han played an important role in the conception of the study, performing the data analyses and drafting and revising the manuscript. Chengfeng Yang contributed to the data gathering and played an important role in interpreting the results. Jinping Song contributed to the conceptual framework of this paper. All authors read and approved the final manuscript.

Conflicts of Interest: The authors declare no conflict of interest.

References

1. Lambin, E.F.; Baulies, X.; Bockstael, N.; Fischer, G.; Krug, T.; Leemans, R.; Moran, E.F.; Rindfuss, R.R.; Skole, D.; Turner ll, B.L.; *et al. Land-Use and Land-Cover Change: Implementation Strategy*; IGBP Report No.48/IHDP Report No. 10; IGBP Secretariat: Stockholm, Sweden, 1999; pp. 125–126.
2. Turner, B.L., II; Skole, D.L.; Sanderson, S.; Fischer, G.; Fresco, L.O.; Leemans, R. *Land-Use and Land-Cover Change Science/Research Plan*; IGBP Report No.35/HDP Report No.7; IGBP Secretariat: Stockholm, Sweden, 1995; pp. 132–133.
3. Liu, J.Y.; Kuang, W.H.; Zhang, Z.X.; Xu, X.L.; Qin, Y.W.; Ning, J.; Zhou, W.C.; Zhang, S.W.; Li, R.D.; Yan, C.Z.; *et al.* Spatiotemporal characteristics, patterns, and causes of land-use changes in China since the late 1980s. *Chin. Geogr. Sci.* **2014**, *24*, 195–210.
4. Meyer, W.B.; Turner, B.L. Human population growth and global land-use/cover change. *Annu. Rev. Ecol. Syst.* **1992**, *23*, 39–61.
5. Dewan, A.M.; Yamaguchi, Y. Land use and land cover change in Greater Dhaka, Bangladesh: Using remote sensing to promote sustainable urbanization. *Appl. Geogr.* **2009**, *29*, 390–401.
6. Dewan, A.M.; Kabir, M.H.; Nahar, K.; Rahman, M.Z. Urbanization and environmental degradation in Dhaka metropolitan area of Bangladesh. *Int. J. Environ. Sustain. Dev.* **2012**, *2*, 118–147.
7. Byomkesh, T.; Nakagoshi, N.; Dewan, A.M. Urbanization and green space dynamics in Greater Dhaka, Bangladesh. *Landsc. Ecol. Eng.* **2012**, *8*, 45–58.
8. Corner, R.J.; Dewan, A.M.; Chakma, S. Monitoring and prediction of land-use and land-cover (LULC) change. In *Proceedings of Dhaka Megacity—Geospatial Perspectives on Urbanization, Environment and Health: Chapter 5*; Dewan, A.M., Corner, R.J., Eds.; Springer Geography: Berlin, Germany, 2014; pp. 75–97.
9. Yin, Z.; Walcott, S.; Kaplan, B.; Cao, J.; Lin, W.; Chen, M.; Liu, D.; Ning, Y. An analysis of the relationship between spatial patterns of water quality and urban development in Shanghai China. *Comput. Environ. Urban.* **2005**, *29*, 197–221.
10. Zhao, S.; Da, L.; Tang, Z.; Fang, H.; Song, K.; Fang, J. Ecological consequences of rapid urban expansion: Shanghai, China. *Front. Ecol. Environ.* **2006**, *4*, 341–346.
11. Dewan, A.M.; Yamaguchi, Y. Effect of land cover changes on flooding: example from Greater Dhaka of Bangladesh. *Int. J. Geoinf.* **2008**, *4*, 11–20.
12. Dewan, A.M.; Yamaguchi, Y.; Ziaur Rahman, M. Dynamics of land use/cover changes and the analysis of landscape fragmentation in Dhaka metropolitan, Bangladesh. *Geojournal* **2012**, *77*, 315–330.
13. Liu, W.D. *Thinking of Economic Geography*; Sciences Press: Beijing, China, 2013; pp. 164–165.
14. Verburg, P.H.; Schot, P.P.; Dijst, M.J.; Veldkamp, A. Land use change modelling: Current practice and research priorities. *Geojournal* **2004**, *61*, 309–324.
15. Schaldach, R.; Alcamo, J.; Koch, J.; Kölking, C.; Lapola, D.M.; Schüngel, J.; Priess, J.A. An integrated approach to modeling land-use change on continental and global scales. *Environ. Model. Softw.* **2011**, *26*, 1041–1051.

16. Imbernon, J. Pattern and development of changes in the Kenyan highlands since the 1950s. *Agr. Ecosyst. Environ.* **1999**, *176*, 67–73.
17. Tian, Y.; Yin, K.; Lu, D.; Hua, L.; Zhao, Q.; Wen, M. Examining land use and land cover spatiotemporal change and driving forces in Beijing from 1978 to 2010. *Remote Sens.* **2014**, *6*, 10593–10611.
18. Kidron, G.J.; Karnieli, A.; Benenson, I.; Grinblat, Y. Simulating land-use degradation in West Africa with the ALADYN model. *J. Arid. Environ.* **2015**, *112*, 52–63.
19. Xu, Q.; Yang, K.; Wang, G.; Yang, Y. Agent-based modeling and simulations of land-use and land-cover change according to ant colony optimization: A case study of the Erhai Lake Basin, China. *Nat. Hazards* **2015**, *75*, 95–118.
20. Yi, Q.; Lam, N.S.N. Modeling land use and land cover changes in a vulnerable coastal region using artificial neural networks and cellular automata. *Environ. Monit. Assess.* **2015**.
21. Badreldin, N.; Goossens, R. Monitoring land use/land cover change using multi-temporal Landsat satellite images in an arid environment: A case study of El-Arish, Egypt. *Arab. J. Geosci.* **2014**, *7*, 1671–1681.
22. Pei, B.; Pan, T. Land use system dynamic modeling: literature review and future research direction in China. *Prog. Geogr.* **2010**, *29*, 1060–1066.
23. Arsanjani, J.J.; Helbich, M.; Kainz, W.; Boloorani, A.D. Integration of logistic regression, Markov chain and cellular automata models to simulate urban expansion. *Int. J. Appl. Earth Obs. Geoinf.* **2013**, *21*, 265–275.
24. Ahmed, B.; Ahmed, R. Modeling urban land cover growth dynamics using multi-temporal satellite images: A case study of Dhaka, Bangladesh. *ISPRS. Int. J. Geoinf.* **2012**, *1*, 3–31.
25. Vaz, E.; Walczynska, A.; Nijkamp, P. Regional challenges in tourist wetland systems: An integrated approach to the Ria Formosa in the Algarve, Portugal, 2013. *Reg. Environ. Change* **2013**, *13*, 33–42.
26. Parker, D.C.; Manson, S.M.; Janssen, M.A.; Hoffmann, M.J.; Deadman, P. Multi-agent systems for the simulation of land-use and land-cover change: A review. *Ann. Assoc. Am. Geogr.* **2003**, *93*, 314–337.
27. Veldkamp, A.; Verbug, P.H.; Kok, K.; de Koning, G.H.J.; Priess, J.; Bergsma, A.R. The need for scale sensitive approaches in spatially explicit land use change modeling. *Environ. Model. Assess.* **2001**, *6*, 111–121.
28. Al-shalabi, M.; Billa, L.; Pradhan, B.; Mansor, S.; Al-Sharif, A.A. Modelling urban growth evolution and land-use changes using GIS based cellular automata and SLEUTH models: The case of Sana'a metropolitan city, Yemen. *Environ. Earth. Sci.* **2013**, *70*, 425–437.
29. Verburg, P.H.; Eickhout, B.; Meijl, H.V. A multi-scale, multi-model approach for analyzing the future dynamics of European land use. *Ann. Regional. Sci.* **2008**, *42*, 57–77.
30. Hubacek, K.; Sun, L. A scenario analysis of China's land use and land cover change: Incorporating biophysical information into input–output modeling. *Struct. Econ. Dynam.* **2001**, *12*, 367–397.

31. Verburg, P.H.; Soepboer, W.; Veldkamp, A.; Limpiada, R.; Espaldon, V.; Mastura, S.S. Modeling the spatial dynamics of regional land use: The CLUE-S model. *Environ. Manag.* **2002**, *30*, 391–405.
32. Wu, J.S.; Feng, Z.; Gao, Y.; Huang, X.; Liu, H.; Huang, L. Recent progresses on the application and improvement of the CLUE-S model. *Prog. Geogr.* **2012**, *31*, 3–10.
33. Wu, Q.; Li, H.Q.; Wang, R.S.; Paulussen, J.; He, Y.; Wang, M.; Wang, Z. Monitoring and predicting land use change in Beijing using remote sensing and GIS. *Landsc. Urban Plan.* **2006**, *78*, 322–333.
34. Lu, R.C.; Huang, X.J.; Zuo, T.H.; Xiao, S.S.; Zhao, X.F.; Zhang, X.Y. Land Use Scenarios Simulation Based on CLUE-S and Markov Composite Model—A Case Study of Taihu Lake Rim in Jiangsu Province. *Sci. Geol. Sinica* **2009**, *29*, 577–581.
35. Nourqolipour, R.; Shariff, A.R.B.M.; Balasundram, S.K.; Ahmad, N.B.; Sood, A.M.; Buyong, T.; Amiri, F. A GIS-based model to analyze the spatial and temporal development of oil palm land use in Kuala Langat district, Malaysia. *Environ. Earth Sci.* **2015**, *73*, 1687–1700.
36. Wang, X.L.; Bao, Y.H. Study on the methods of land use dynamic change research. *Prog. Geogr.* **1999**, *18*, 81–87.
37. Hu, Y.; Zheng, Y.; Zheng, X. Simulation of land-use scenarios for Beijing using CLUE-S and Markov composite models. *Chin. Geogr. Sci.* **2013**, *23*, 92–100.
38. Pontius, R.G., Jr.; Schneider, L.C. Land-cover change model validation by an ROC method for the Ipswich watershed, Massachusetts, USA. *Agr. Ecosyst. Environ.* **2001**, *85*, 239–248.
39. Pontius, R.G., Jr; Cornell, J.D.; Hall, C.A. Modeling the spatial pattern of land-use change with GEOMOD2: Application and validation for Costa Rica. *Agr. Ecosyst. Environ.* **2001**, *85*, 191–203.
40. Pontius, R.G., Jr.; Malanson, J. Comparison of the structure and accuracy of two land change models. *Int. J. Geogr. Inf. Sci.* **2005**, *19*, 243–265.
41. Kamusoko, C.; Aniya, M.; Adi, B.; Manjoro, M. Rural sustainability under threat in Zimbabwe—Simulation of future land use/cover changes in the Bindura district based on the Markov-cellular automata model. *Appl. Geogr.* **2009**, *29*, 435–447.
42. Verburg, P.H.; Overmars, K.P. Dynamic simulation of land-use change trajectories with the CLUE-s model. *Model. Land-Use Change* **2007**, *90*, 321–337.
43. Wan, L.; Jin, Y. Review on applied urban modeling and new trends of urban spatial policy models. *Urban Plan. Forum.* **2014**, *1*, 81–91.
44. Barra, D.L.T. *Integrated Land Use and Transport Modeling. Decision Chains and Hierarchies*; Cambridge University Press: Cambridge, UK, 1989.
45. Liu, L.; Long, Y.; Batty, M. A retrospect and prospect of urban models: Reflections after interviewing Mike Batty. *City Plan. Rev.* **2014**, *38*, 63–70.
46. Wegener, M. The IRPUD model. Spiekermann & Wegener in Dortmund. Available online: http://www.spiekermann-wegener.com/mod/pdf/AP_1101_IRPUD_Model.pdf (accessed on 1 December 2011).
47. Simmonds, D.C. The design of the delta land-use modeling package. *Gen. Inf.* **1999**, *26*, 665–684.

48. Waddell, P. Urbansim: Modeling urban development for land use, transportation, and environmental planning. *J. Am. Plann. Assoc.* **2002**, *68*, 297–314.
49. Zondag, B.; De Bok, M.; Geurs, K.T.; Molenwijk, E. Accessibility modeling and evaluation: The TIGRIS XL land-use and transport interaction model for the Netherlands. *Comput. Environ. Urban* **2015**, *49*, 115–125.
50. Pontius, R.G., Jr.; Huffaker, D.; Denman, K. Useful techniques of validation for spatially explicit land-change models. *Ecol. Model.* **2004**, *179*, 445–461.
51. Brueckner, J.K.; Kim, H. Urban sprawl and the property tax. *Int. Tax Public Finan.* **2003**, *10*, 5–23.
52. Tsai, Y. Housing demand forces and land use towards urban compactness: A push-accessibility-pull analysis framework. *Urban Stud.* **2014**, *6*, 1–17.
53. Castella, J.C.; Pheng Kam, S.; Dinh Quang, D.; Verburg, P.H.; Thai Hoanh, C. Combining top-down and bottom-up modelling approaches of land use/cover change to support public policies: Application to sustainable management of natural resources in northern Vietnam. *Land. Use. Policy* **2007**, *24*, 531–545.
54. De Bok, M.; Zondag, B.; Petersen, E. Modeling land use in the Generic Urban Model. In Proceedings of European Transport Conference (ETC), Strasbourg, France, 18–20 September 2006; pp. 15–28.
55. Yigitcanlar, T.; Dur, F. Developing a sustainability assessment model: The sustainable infrastructure, land-use, environment and transport model. *Sustainability* **2010**, *1*, 321–340.

Urban Land Expansion and Sustainable Land Use Policy in Shenzhen: A Case Study of China's Rapid Urbanization

Jing Qian, Yunfei Peng, Cheng Luo, Chao Wu and Qingyun Du

Abstract: Shenzhen is a city that is highly representative of China's rapid urbanization process. As the city rapidly expands, there are enormous challenges to the sustainable use of land resources. This paper introduces the evolution of urban land expansion and the sustainable land use policy of the Shenzhen Government since 2005. The policy covers the reduction in rural-to-urban land conversion, the delineation of urban growth boundaries, arable land reclamation and the establishment of farmland protection areas, urban redevelopment, and the investigation and prosecution of illegal construction. This paper considers the aspects of urbanization and land management systems that are unique to China. The current top-down indicative and mandatory mode of control, which relies on the central government, has very limited effects. Good results were achieved in Shenzhen for the following elements: governmental self-restraint, governmental identity change, and policy innovation. Shenzhen's sustainable land use practices can provide a reference for other cities in China.

Reprinted from *Sustainability*. Cite as: Qian, J.; Peng, Y.; Luo, C.; Wu, C.; Du, Q. Urban Land Expansion and Sustainable Land Use Policy in Shenzhen: A Case Study of China's Rapid Urbanization. *Sustainability* **2016**, *8*, 16.

1. Introduction

Since the reform and opening up, China has entered a phase of high-speed urbanization. The level of urbanization increased at an alarming rate from 17.9% in 1978 to 54.8% in 2014 [1]. Fast urbanization causes rapid growth of urban land. From 1996 to 2012, the national urban land increased by 2380 square kilometres per year on average [2]. As Friedmann has said, China has been urbanizing at breakneck speed [3]. With the rapid expansion of cities, excessive consumption of resources, regional development imbalances, food security, and other issues have become more apparent. Urban sustainable land use poses a huge challenge [4,5].

From an economic perspective, urban expansion is the result of market forces. As long as the marginal benefit of urban land use is greater than that of agricultural land use, urban expansion occurs [6]. However, a single market mechanism in the allocation of urban land resources may lead to market failures, including the unfair distribution of benefits and the loss of public land resources [7,8]. While representing public interests, the government may use certain policy instruments to directly or indirectly control urban spatial expansion in terms of location, speed, timing, quality, and cost [9]. From a global perspective, with the successive emergence of theories and ideological trends, such as the New Urbanism, Smart Growth, and the Compact City, governments have begun to promote sustainable urban development and control urban land expansion with policy instruments [10]. In the late 1970s, the United States proposed the Urban Growth Management concept, which targeted the protection of open spaces, rational urban growth, natural resource conservation, the improvement of public facilities and transportation, and coastal resources conservation [11,12]. Urban land expansion was reduced through policy tools such as the delineation of an Urban Growth Boundary (UGB), transfer of land development rights, public land purchases, building permits, infrastructure restrictions, and the establishment of development impact fees [13–15]. In the United Kingdom, the "Greenbelt" policy, which established a green span between cities and rural areas so that green space could play a role as an urban space barrier to control the unlimited expansion of the downtown area, was regarded as an important tool to control London's expansion [16]. In response to excessive agglomeration of the urban scale in Seoul, the South Korean government also adopted policy tools such as the transfer of urban functions, movements that promoted new villages, and zoning to encourage industry relocation and inhibit population influx to effectively control the size of the city [17]. The Japanese government commonly uses land readjustment as a means to suppress city size [18,19]. In order to realize the revival of the central areas of cities, urban redevelopment policy was chosen as the main tool to improve the quality of cities in the United States, Britain, Germany, and other developed countries [20,21]. On the other hand, to cope with the loss of farmland resources, which resulted from urban expansion, farmland protection policy was also adopted by most of the developed countries, such as the Farmland Preservation Program (FPP) [22] and the Conservation Easement Donating (CED) policy [23]. Most studies show that the policies of each country have led to remarkable achievements in improving the efficiency of urban land use, controlling urban sprawl, easing pressure on transportation and infrastructure, and improving environmental quality [24–29].

Urbanization in China started relatively late. Since the 1990s, urban land expansion gradually began to be taken seriously in China, and many scholars began to research Western urban growth management policies and their success in controlling urban land expansion. Based on the study of urban spatial expansion in Nanjing, Zhu found that Nanjing's experience in sustainable land use provided some effective ways to control urban expansion [30]. Chen thought that although the UGB was an effective tool to reduce urban land expansion in the USA, it still required further studies according to Chinese characteristics [31]. Besides, Feng [32] and Liu [33] argued that because the urbanization process, the foundation of land ownership system, and the driving forces of urban expansion in China were different from those of developed countries, it was improper to copy the urban land management from these countries without any innovation. Thus, we found that, comparing with developed countries, the characteristics of urban land expansion and the institutional environment in China are notably special. It is difficult for such international experiences to fully adapt to China's unique national conditions. These are mainly reflected in the following respects: (i) In the market-driven West, suburbanization is the main manifestation of low-density urban spread. In China urban land expansion is mainly led and monopolized by the local government. Explosive urban growth is caused by the rapid influx of the rural population and other urban populations. (ii) Although China's urban land use system has implemented market-oriented reforms, it is still very clearly shaped by a planned economy [34]. The central government achieves tight control over the land-use behavior of local governments via layers of top-down mandatory modes of control and a land supervision system. (iii) Because rapid urbanization has led to construction on a large quantity of arable land, national food security is threatened. Protecting arable land to ensure self-sufficiency in food production is a basic Chinese state policy and, therefore, protecting arable land is both a target for controlling urban land expansion and an important method of doing so [35,36]. (iv) China has implemented different systems of land ownership rights in urban and rural areas. The state owns urban land, whereas the collective owns rural land. Meanwhile, according to China's land management system, only state-owned land can participate in urban development. Rural land must go through government land expropriation (and some compensation must be provided) to be converted into urban land [37]. However, in the land expropriation process, the interest relationships between the government, developers, and individuals are extremely delicate and complex. For example, developers hope to get land at a low price, while land owners hope to get more land compensation, and the government hopes to get more differential income in the process of land expropriation.

Shenzhen is a city that is highly representative of China's rapid urbanization process. In this paper, the main objectives are: (i) to analyze the course and characteristics of urban land expansion over the past 35 years; (ii) to expound the problems of traditional countermeasures and to summarize Shenzhen's policy framework for sustainable land use; and (iii) to introduce several measures that Shenzhen has taken to confront the land resource crisis and to put forward some reference for other cities.

2. Study Area and Data

Shenzhen is located in southern Guangdong Province, China, and borders Hong Kong. Its administrative area is 1996 km^2 (Figure 1). After being established as

China's first special economic zone in 1980, Shenzhen quickly developed from a sleepy small border town into a large modern city, quite impressive in the history of industrialization, modernization, and urbanization all over the world. From 1979 to 2014, the total population of Shenzhen has increased from 300,000 to 18 million [38] (far higher than the official number of 10.78 million), and its total Gross Domestic Product (GDP) has increased from 200 million yuan to 1.6 trillion yuan [39], as much as Hong Kong. Meanwhile, urban land has also expanded from 3 km^2, when the Special Administrative Region was first established, to 968 km^2, an expansion of more than 300-fold. However, the rapid development of Shenzhen's social economy has far exceeded the sustainable supply capacity of resources. In 2005, in a government report, the government noted that Shenzhen was facing four "unsustainable" situations, specifically land, water resources, population, and the environment [40], of which land was considered the most important.

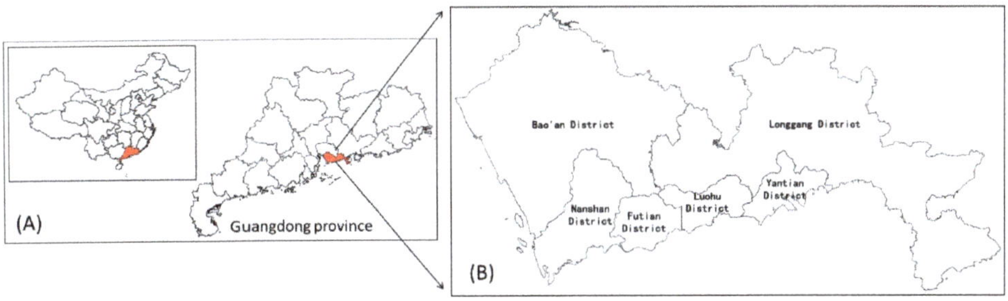

Figure 1. The location of the study area in China: (**A**) the location of Shenzhen in China; (**B**) Shenzhen administrative area map.

To analyze urban land expansion and spatial patterns in the study area, remote sensing images and land use surveys are used. Since 1995, The Urban Planning-Land & Resources Commission of Shenzhen Municipality (UPLRCSM) has carried out a land use survey each year. From the official channels we obtained land use survey results for 2005 and 2014. Restricted by technical conditions, the land use survey was not carried out in 1979 and 1986, so a Landsat Multispectral Scanner (MSS) and a Landsat Thematic Mapper (TM) images were selected. Then the construction land data were interpreted from the two remote sensing images. In fact, the results of the land use survey were also interpreted from the remote sensing images. Therefore, we think the construction land data for the four years were comparable. In addition, we have land conversion data, arable land data, and illegal construction data from the UPLRCSM. In order to compare with other cities in China, the data for GDP and fiscal revenue were collected from the Statistical yearbook in Shenzhen [41], Beijing [42], Shanghai [43], and Guangzhou [44]. The land-transferring fees data were collected from the Chinese Industry Research Network (http://www.chinairn.com).

3. Urban Land Expansion of Shenzhen

After the reform and opening up, Shenzhen developed rapidly due to the special economic zone policy advantages and the geographical advantage of being adjacent to Hong Kong. From 1979 to 2014, Shenzhen's urban land expansion can be roughly divided into three stages.

3.1. 1979–1986

In the early stage of the reform and opening up, because of a lack of construction funds, the Shenzhen local government could only rely on the land resources to attract capital to invest in setting up factories. However, because the law at that time still prevented the buying and selling of land, three development models were adopted. The government commissioned tracts of land to other developers, utilized foreign funds to set up factories, or leased land to foreign-owned enterprises for development. Although the government only received a small amount of land use fees, it successfully developed the Shekou Industrial Zone, Nanyou, Overseas Chinese Town, and other development zones by relying on the aforementioned development approaches. During this period, the construction land increased from 3 km^2 to 48 km^2, with an average annual growth of 6.9 km^2. Because transportation and location (*i.e.*, distance from the ports) has a major influence on industry, the construction land was mainly concentrated in Luohu District and the two sides of Shennan Avenue and the Guangzhou–Shenzhen Highway (Figure 2B).

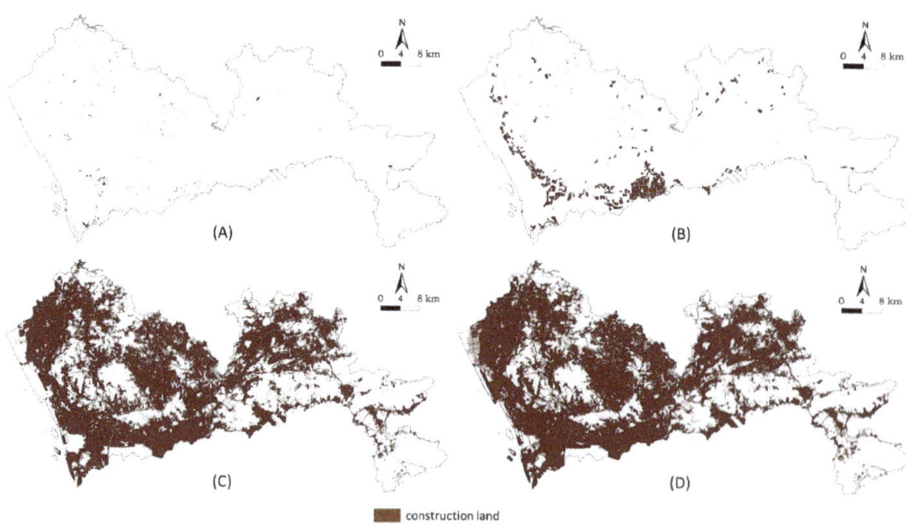

Figure 2. The evolution of Shenzhen construction land: (**A**) 1979; (**B**) 1986; (**C**) 2005; and (**D**) 2014.

3.2. 1987–2005

In 1987, Shenzhen led the country in implementing a land paid use system. Under the premise of retaining land ownership rights, the state relinquished land use rights to users by way of auctions, tenders, agreements, *etc.*, for a certain price, tenure, and purpose [45]. This model solved the construction funding problems of the Shenzhen Special Economic Zone, and it also created a good investment environment. With the rapid development of an export-oriented economy via "Enterprises of Processing Industries and Compensation Trade" and "the three types of foreign-invested enterprises or ventures", Shenzhen experienced an urban growth process based on rapid industrialization-driven urbanization. In 1984, the Shenzhen Government drafted the "Master Plan for Shenzhen Special Economic Zone (1986–2000)". According to the document, Shenzhen was designed to be a megalopolis. Under the guidance of this plan, urban function and quality gradually improved, infrastructure such as transportation and municipal administration was gradually perfected, and urban land rapidly expanded. With increases in the floating population and the demand for housing after the start of the 1990s, the real estate industry developed rapidly and gradually became one of the core industries in Shenzhen and the main driver of urban land expansion. Meanwhile, an informal construction land expansion also occurred. Because the government expropriated a large quantity of rural land, there is no land available for local farmers to plant. They thus built factories and then leased them as a major source of income. As a consequence, labor-intensive industries attracted a large number of migrant workers, whose need for housing tempted local farmers to expand their private houses and rent them out [46]. In summary, due to the excessive pursuit of investment-driven economic growth and a loose environment for development, jointly pushed by both the formal and the informal land markets, land resources in Shenzhen were quickly consumed. During this period, urban land increased by 891 km^2, and the average annual growth was 46.8 km^2. The annual growth rate of construction land was 16.9%. In 2005, the construction land area in Shenzhen reached 939 km^2, accounting for 47% of the total land area of the city, which exceeded the 20%–30% general ratio range of an international metropolis [47]. Natural conditions become the biggest determinants of spatial pattern in this rapid urban urbanization period. Until 2005, in addition to mountains, rivers, lakes, and other inappropriate construction areas, construction land has sprawled everywhere in the city (Figure 2C).

3.3. 2006–2014

The local government gradually apprehended the previous unsustainable land use patterns and started to change land use concepts and methods by using a variety of policy tools to suppress the expansion of urban construction land. During 2006–2014, when economic and population growth rates remained generally stable, the growth rate of urban construction land clearly decreased. The total urban construction land in 2014 was 968 km^2. Compared with 2005, it only increased by 29 km^2. The average annual growth was 3.2 km^2. The annual increase was less than 1/15 that of the 1987–2005 period. The increase in urban construction land averaged an annual growth rate of only 0.3%, representing negligible growth. Urban land expansion was counteracted through the series of sustainable land use policy, including UGB, Establishment of Farmland Protection Areas, Urban Redevelopment, *etc.* Urban land expansion was counteracted and the construction layout then becomes more concentrated than sprawled (Figure 2D).

4. Policy Framework for Sustainable Land Use

In China, local government behavior is one of the most important drivers of urban land expansion [48–50]. In the 1990s, China implemented a tax system reform and re-divided powers and financial power between the central government and local governments, which strengthened the financial centralization capacity of the central government and caused a huge gap in local fiscal revenue and expenditure. Local governments had to raise their own funds to maintain economic growth; thus, land grant fees became the main source of fiscal revenue [51,52]. On the other hand, the existing performance appraisal system had always used GDP and fiscal revenue-based economic development indicators as the most important appraisal indicators for measuring the performance of the local government chief executive, further exacerbating the utility of local officials as the administrative decision-makers. However, local governments attracted investment, for example, by means of low-cost land grants, establishing development zones, and increasing GDP and employment. Furthermore, local governments implemented local infrastructure construction and provided public services through low-cost land expropriation and high-price land sales to obtain profits from the land to expand fiscal revenue, eventually leading to the massive growth of urban construction land [53,54]. The central government stood for the "public interests" of all citizens, stressed regional fairness and rational resource allocation, pursued the maximization of overall benefits for the economic-social-ecological environment and national food security, and strengthened the size control over cities ranked high in size. The central government used compulsory administrative orders to intervene in and regulate urban land expansion. These administrative methods include controlling the speed and scale of rural-to-urban land conversion in various cities, issuing arable land protection indices, and even requiring each major city to delineate a UGB. All these are the most basic means of controlling urban land expansion in China. However, due to the lack of theoretical support and the inability to meet regional development needs, the results of such top-down government policy tools were not as positive as expected [55]. Given the temptation of land revenue and achievements in urban development, local governments treated central index control negatively, even finding ways to cheat the central government [56,57]. The best example is as follows: Since 1997, the central government "froze the construction occupation of arable land". In 1998, the country's arable land decreased by 120 km^2, although not all of this arable land was occupied by construction [58].

In summary, the decisions of most local governments to maximize their own self-interests reduce the overall interests of the entire society. Accordingly, overdevelopment beyond the social carrying capacity of the land and the irrational expansion of cities may occur. Relying solely on the central government's mandatory and indicative mode of control over local governments to suppress the effects of urban land expansion has a very limited effect. When the Shenzhen government was faced with a land resource crisis, it built a new policy system. These policies, aimed at different objects and behaviors, have achieved certain results by using administrative, planning, economic, legal, and other means. Compared with a traditional management and control model, Shenzhen's policy measures were more diverse and targeted (Figure 3).

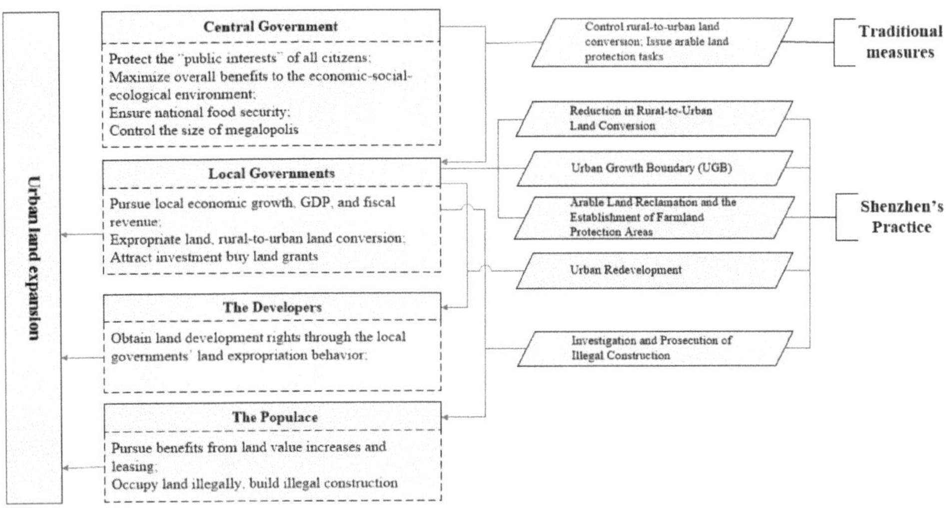

Figure 3. Shenzhen sustainable land use policy system.

5. Shenzhen Practice and Results

5.1. Reduction in Rural-to-Urban Land Conversion

Each year, based on the national economic situation, local natural endowments, and the urban development stage, the Central Government issues rural-to-urban land conversion indices to local governments to control the size of cities through index quotas [59,60]. These are especially strict for large cities. From 2009 to 2014, the central government issued a rural-to-urban land conversion quota of approximately 15 km^2 to Shenzhen. However, the Shenzhen government has not used up all of its quota. Since 2009, Shenzhen's rural-to-urban land conversion has gradually decreased. In 2013 and 2014, the size was not even 1/5 of the issued index (Figure 4). In China, in the process of converting agricultural land to construction land, obtaining land grant fees from land grants has been an important source of revenue for local governments [61,62]. From 2009 to 2013, Shenzhen has basically maintained land grant revenue ratios of less than 20%. Compared with Beijing, Shanghai, and Guangzhou—cities of approximately equal population and GDP—the ratio is at a lower level (Figure 5). The Shenzhen Municipal Government has gradually abandoned its dependence on land finances by reducing the size and speed of rural-to-urban land conversion.

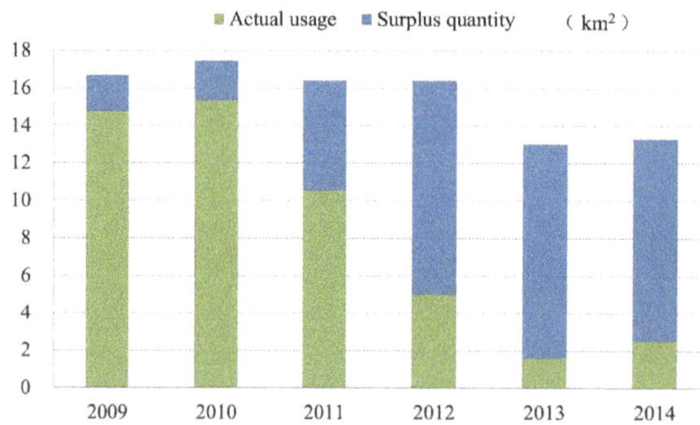

Figure 4. The 2009–2014 rural-to-urban land conversion index for Shenzhen.

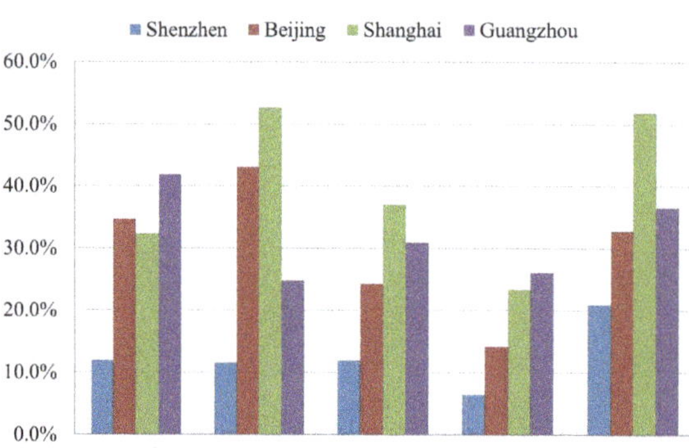

Figure 5. The ratio of land grant fees accounting for government revenues.

5.2. Urban Growth Boundary (UGB)

Rapid urban land expansion had led to a continued decline in the proportion of the city's ecological areas and the loss of self-maintenance capabilities. The ratio of land covered by crops, forests, and grasslands in 2006 was 45.72%, whereas the ratio in 1996 was 51.36% [63]. In 2005, the Shenzhen government delineated a UGB that included first-order water source protection areas, nature reserves, farmland protection areas, and forests, in addition to country parks, mountains, rivers, reservoirs and wetlands, ecological corridors, and green spaces and islands, with a total of 974 km^2 of land excluded from the scope of permitted urban development (Figure 6). The ratio of this area is nearly half the area of the entire city. The Government implemented strict building permits outside the scope of the UGB. Except for essential road transport infrastructure, public and tourism facilities, all other construction was banned.

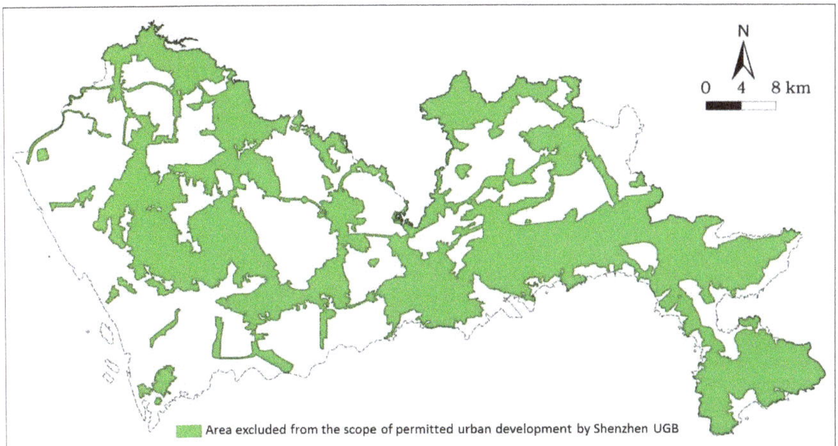

Figure 6. Shenzhen UGB.

5.3. Arable Land Reclamation and the Establishment of Farmland Protection Areas

Rapid urban land expansion has also led to a considerable amount of erosion of arable land. From 1995 to 2009, the loss of arable land was up to 51.6%, and the net loss in the city's arable land was 33.7 km^2. Since 2009, the Shenzhen government has taken a series of measures to strengthen the protection of arable land. First, it optimized as much as possible the location selection of projects so as not to occupy arable land. From 2009 to 2014, Shenzhen's annual mean occupation of arable land by construction was approximately 1 km^2, only 1/10 that of 2002. The city's loss of arable land is basically under control. Second, as the government has initiated arable land reclamation work in 2012, the area of arable land increased in 2013 (Figure 7). Third, it has delineated farmland protection areas by centralizing the distribution of the city's farmland. Prior to 2010, Shenzhen's arable land distribution was piecemeal, partly surrounded by urban land, and threatened by urban pollution. There were approximately 1500 parcels of arable land with areas of less than 400 m^2. Scattered farmland not only reduces the efficiency of agricultural production but is also facing the risk of being annexed by cities at any time [64,65]. Since 2010, the Shenzhen government has planned 27 farmland protection areas, taking a land replacement approach to bring together scattered farmland tracts, which has both enhanced the efficiency of land use and allowed farmland protection areas to play a barrier role in urban land expansion (Figure 8).

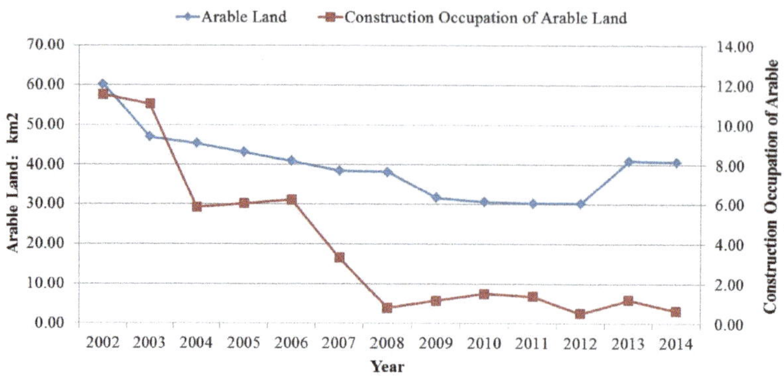

Figure 7. The 2002–2014 total amount of arable land in Shenzhen and the status of the occupation of arable land by construction land.

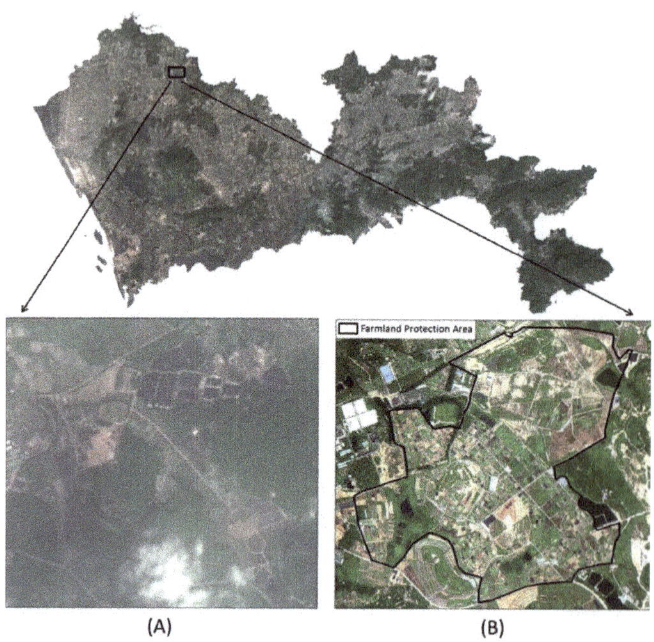

Figure 8. A before-and-after comparison of a farmland protection area establishment: (**A**) before establishment; (**B**) after establishment.

5.4. Urban Redevelopment

The urban redevelopment of old and inefficient built-up areas inside cities can provide better production space and living space and reduce the occupation of suburban farmland (Figure 9) [66,67]. After the land had been expropriated by the government, the farmers around the city could only build houses in the area delineated by the government and maintain their livelihood by leasing property. It thus formed a unique phenomenon—urban villages [68–70]. There were a total of 320 villages in the city, covering 93.49 km^2 [71]. These villages in the city crisscrossed with the newly developed city and featured outdated functions, hidden security risks, and incomplete supporting facilities. They contrasted greatly with the metropolis and did not reflect proper land values [72,73]. From the perspective of improving a city's quality, urban villages should become the area of focus for urban redevelopment. However, the various economic demands of owners within the urban villages and the huge costs of demolishing the mixed-in illegal buildings have caused the government to recoil at urban redevelopment [74]. Take the urban redevelopment of Gangxia urban village in 2009 as an example; because it is close to Shenzhen CBD, during the process of urban redevelopment, more than 20 billionaire families and 10 individual billionaires were created by the compensation. In China, currently, the local government is the only legitimate body of land expropriation [75,76]. Thus, developers cannot intervene before the land expropriation. They can only passively wait for the government to grant the land. To change this unidirectional method of land circulation, the Shenzhen government formulated policies in 2009 to encourage urban redevelopment. The biggest highlight of this policy was establishing a mechanism for coordinating the interests of the three parties—the government, developers, and owners. The developers led demolition talks, resettlement compensation, and other interest negotiation work; the government only formulated rules and served as an intermediate supervisor by enforcing related matters according to the contract after all parties had reached an agreement (Figure 10). In this process, the original owner land is converted into state-owned land (which cannot exceed a certain ratio of illegal land); the developers obtain land development rights and compensate the owners monetarily or with housing; and, meanwhile, the developers have to return to the government a certain ratio of supporting facilities, such as roads, primary schools, and kindergartens. This policy directs developers' investment enthusiasm to the old town. In 2014, 30% of the entire city's land grants came from urban redevelopment. The land grant premium reached 16.4 billion yuan, exceeding the total amount of 2008 land grant fees.

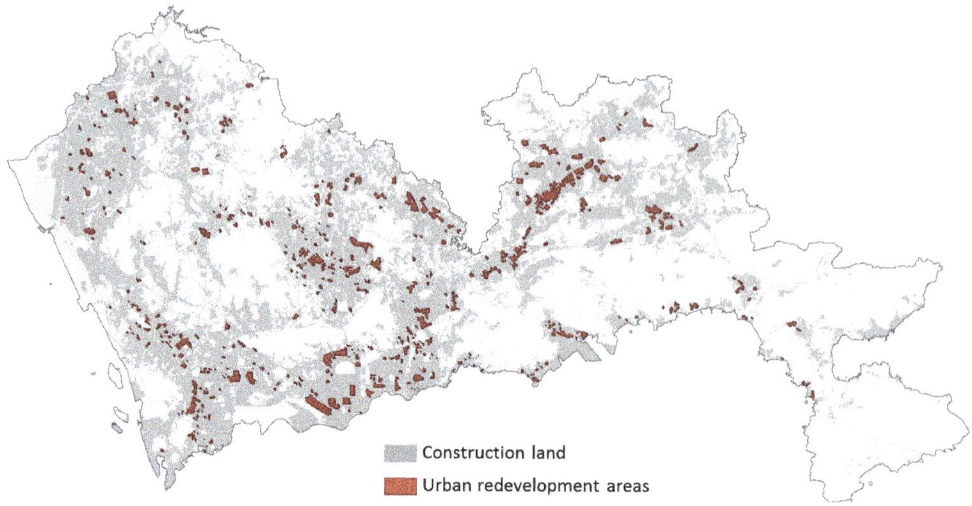

Figure 9. Urban redevelopment areas of Shenzhen.

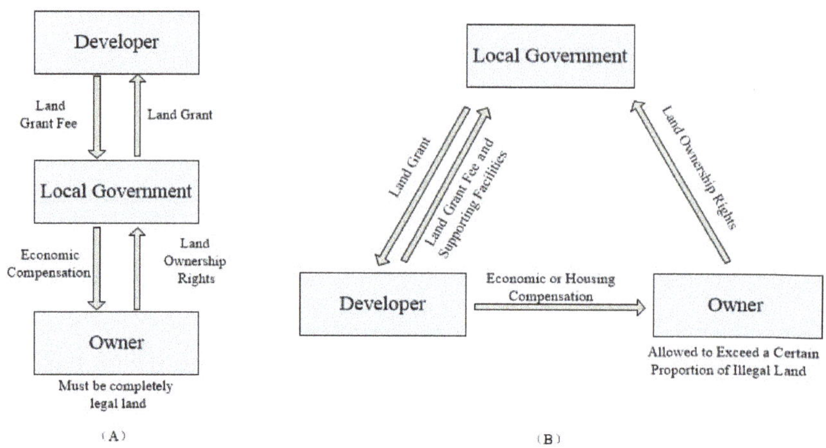

Figure 10. A before-and-after comparison of Shenzhen's urban redevelopment policy.

5.5. Investigation and Prosecution of Illegal Construction

As noted above, in the process of rapid urbanization in Shenzhen, the local farmers used informal and, strictly speaking, illegal methods to promote construction land expansion. They achieved two goals through wanton expansion or the construction of new homes. First, they obtained economic returns from leasing real estate. Second, they expected to obtain higher economic compensation in demolition talks with the government by means of this type of actual land occupation. Because these illegal buildings could improve the livelihood of local villagers and provide migrant workers with relatively inexpensive housing, the government adopted an ambiguous attitude toward these violations, leading to the development of large-scale illegal construction. Several times real estate market prosperity led to illegal building booms: in 1998, 2003, and 2008 (Figure 11). As of 2013, there were a total of 373,000 illegal buildings in Shenzhen. The building area was 428 million square meters, accounting for 43% of the total construction area. Furthermore, this was a unique situation countrywide [77,78].

In recent years, the Shenzhen government has realized that the proliferation of illegal construction has to be stopped, and demolition, confiscation, fines, and other measures against illegal land use have been used. Meanwhile, Remote Sensing Techniques, unmanned aerial vehicles, and other advanced technical means have been used to monitor violations, forming a highly effective control of illegal land use that has effectively inhibited local villagers' speculative behavior and luck-oriented mindset. Under a strict punishment mechanism, informal urban sprawl and expansion have effectively been controlled. Since 2008, the average annual rate of decline in the city's new illegal construction area has reached 12%. From 2010 to 2014, the total area of illegal land decreased by 213.75 hectares, an average annual decrease of 13.4% (Figure 11).

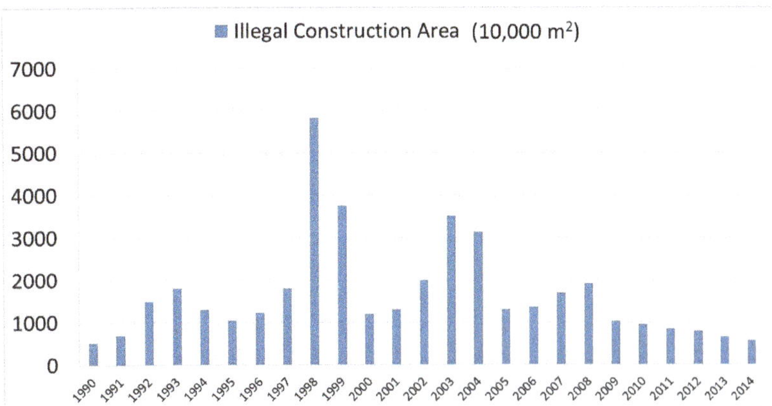

Figure 11. Shenzhen's illegal construction areas in 1990–2014. Note: several times of real estate market prosperity led to illegal building booms: in 1993, 1998, and 2003. Since 2008, the illegal construction area has declined year after year.

6. Conclusions

From 1979 to 2014, Shenzhen's urban land expansion has experienced three development stages. In the second to third stages of the transformation process, Shenzhen's sustainable land use policy has played an extremely important role.

Compared with other countries, Shenzhen policies have their own characteristics. The concept and role of the Shenzhen UGB are basically similar to the UGB in the United States and the Greenbelt in the United Kingdom. Slight differences mainly lie in the following aspects: first, the Shenzhen UGB not only establishes an artificial boundary or "greenbelt" in the city's outer ring but is also a boundary for ecological land protection; and second, due to the topography of Shenzhen, the shape of UGB is not a circle or a band but multiple closed sets of boundaries. As for the urban redevelopment policy, the main purpose in developed countries is to promote inner city revival and it is a process of city self-perfection. However, urban redevelopment policy in Shenzhen was proposed against a backdrop of high-speed urbanization; thus, the main purpose is to save land resources and enhance land use efficiency. Farmland protection policy, because of the different land property rights system, is mainly dominated by the market rules in developed countries, and the government only serves as a supervisor. In China, current arable land protection system is determined by the fertility and agricultural value of the land through a mandatory apportioning method that issues the country's arable land holding tasks level by level down to the local governments [79]. Such measures likely generate great location choice constraints on urban development, affecting the overall efficiency of land resources [80]. Especially in Shenzhen, the value added by agriculture accounts for less than 0.1% of GDP. The ecological function of the cultivated land has clearly gone beyond food production. Shenzhen successfully combined national control requirements with actual local realities. The government initiated the reclamation of arable land and established farmland protection areas, which both accomplished the arable land tasks issued by the central government and led to improvements in local ecological quality.

The United Nations Development Program predicts that China's urbanization rate will reach 70% by 2030 [81]. Some scholars, coming from an economic perspective, predict that the average year of maximum growth in China's construction land will be approximately 2047 [82]. The additional implication is that a considerable amount of urban land expansion will continue to occur. However, urbanization is an irreversible human activity [83]. Allowing market mechanisms to dominate urban spatial growth may undermine non-renewable resources and a pleasant ecological environment. If proper measures and methods can be employed at present to suppress urban land expansion, it may be possible to circumvent the enormous challenges confronting sustainable urban development and the enormous costs that will be produced. Shenzhen's policies can provide a reference and valuable lessons for other cities to explore sustainable land use.

First, Shenzhen government reduced the scale of rural-to-urban land conversion; it essentially changed the single-goal orientation in pursuit of economic interests and performance maximization into a composite orientation that also considered social, economic, and ecological benefits. Of course, a wide range of drivers was bound to exist in this transformation process, such as the scarcity of land resources, growing public ecological demands, *etc.* Shenzhen's experience can direct rulers to make scientific and rational urban development goals, control the speed and strategy of rural-to-urban land conversion, and reduce the profit-drives and blindness of urban land expansion.

Second, other cities can learn from Shenzhen's policies adopted in the fields of arable land and green space protection, urban redevelopment, illegal construction, and carrying out the planning and policy research according to the local situations. In fact, with the aid of experience in Shenzhen, the central government has demanded that 14 cities carry out the pilot work of UGB. However, there are huge differences between various cities in China. Each city is in a different development stage and faces different problems. An across-the-board management and control model cannot adjust to cities' diverse states. Therefore, local governments should formulate sustainable land use policy according to the local realities.

Finally, as Ding [84] has stated, even the most upright official will have a strong incentive to play the role of a land developer, thereby converting agricultural land into urban land. The problems of the local government's impulsivity, pursuit of profit, and blindness should be solved fundamentally. It is necessary to change the fiscal revenue and GDP-based assessment indicator and promotion mechanism for officials and establish a cadre performance evaluation mechanism that is based on guiding people's livelihood and sense of wellbeing.

Acknowledgments: This study was supported by the National Natural Science Foundation of China (Project No. 41271455 and 41371427). The authors would like to thank Ren Fu from Wuhan University for their valuable suggestions. We also thank the Urban Planning, Land & Resources Commission of Shenzhen Municipality for providing the data and materials.

Author Contributions: Jing Qian conceived and designed the study with the support of Qingyun Du. All the co-authors drafted and revised the article together. All authors read and approved the final manuscript.

Conflicts of Interest: The authors declare no conflict of interest.

References

1. China N.B.S. Statistical Communiqué of the People's Republic of China on the 2014 National Economic and Social Development. Available online: http://www.stats.gov.cn/tjsj/zxfb/201502/t20150226_685799.html (accessed on 26 February 2015).
2. The State Council. The New Urbanization Plan (2014–2020). Available online: http://www.gov.cn/gongbao/content/2014/content_2644805.htm (accessed on 9 December 2015).
3. Freedmann, J. Four theses in the study of China's urbanization. *Int. J. Urban Reg. Res.* **2006**, *30*, 440–451.

4. Gao, J.; Wei, Y.D.; Chen, W.; Yenneti, K. Urban Land Expansion and Structural Change in the Yangtze River Delta, China. *Sustainability* **2015**, *7*, 10281–10307.
5. Timberlake, M.; Wei, Y.D.; Ma, X.; Hao, J. Global cities with Chinese characteristics. *Cities* **2014**, *41*, 162–170.
6. Seto, K.C.; Kaufmann, R.K. Modeling the drivers of urban land use change in the Pearl River Delta, China: Integrating remote sensing with socioeconomic data. *Land Econ.* **2003**, *79*, 106–121.
7. Li, X.; Qu, F.; Zhang, S.; Zhang, S.; Wang, Y. Urban sprawl control in China based on international comparison and strategy selection. *Trans. Chin. Soc. Agric. Eng.* **2011**, *27*, 1–10.
8. Qian, W. Market failure, government shortcoming and excessive expanding of land-use scales in China's city. *Econ. Geogr.* **2001**, *21*, 456–460.
9. Nelson, A.C.; Moore, T. Assessing growth management policy implementation: Case study of the United States' leading growth management state. *Land Use Policy* **1996**, *13*, 241–259.
10. Bollens, S.A. State growth management: Intergovernmental frameworks and policy objectives. *J. Am. Plan. Assoc.* **1992**, *58*, 454–466.
11. Chinitz, B. Growth Management Good for the Town, Bad for the Nation? *J. Am. Plan. Assoc.* **1990**, *56*, 3–8.
12. Nelson, A.C. Comparing states with and without growth management analysis based on indicators with policy implications. *Land Use Policy* **1999**, *16*, 121–127.
13. Dempsey, J.A.; Plantinga, A.J. How well do urban growth boundaries contain development? Results for Oregon using a difference-in-difference estimator. *Reg. Sci. Urban Econ.* **2013**, *43*, 996–1007.
14. Bengston, D.N.; Fletcher, J.O.; Nelson, K.C. Public policies for managing urban growth and protecting open space: Policy instruments and lessons learned in the United States. *Landsc. Urban Plan.* **2004**, *69*, 271–286.
15. Knaap, G.J.; Hopkins, L.D. The inventory approach to urban growth boundaries. *J. Am. Plan. Assoc.* **2001**, *67*, 314–326.
16. Kühn, M. Greenbelt and Green Heart: Separating and integrating landscapes in European city regions. *Landsc. Urban Plan.* **2003**, *64*, 19–27.
17. Zhao, C.; Jin, G.; Zhou, P. Seoul's expansion and South Korea's urban development policy. *Urban Probl.* **2007**, *1*, 90–96. (In Chinese).
18. Hayashi, K. Land readjustment in Nagoya. In *Land Readjustment: A Different Approach to Financing Urbanization*; Doebele, W., Ed.; Lexington Books: Lexington, KY, USA, 1982.
19. Misra, B. Japanese experience in physical development and land management. In *Transferability of Development Experience (Regional Development Dialogue, Special Issue 1984)*; Okita, S., Ed.; United Nations Centre for Regional Development: Nagoya, Japan, 1984.
20. Henderson, S.; Bowlby, S.; Raco, M. Refashioning local government and inner-city regeneration: The Salford experience. *Urban Stud.* **2007**, *44*, 1441–1463.
21. Ponzini, D.; Rossi, U. Becoming a creative city: The entrepreneurial mayor, network politics and the promise of an urban renaissance. *Urban Stud.* **2010**, *47*, 1037–1057.

22. Duke, J.M.; Lynch, L. Farmland retention techniques: Property rights implications and comparative evaluation. *Land Econ.* **2006**, *82*, 189–213.
23. Nickerson, C.J.; Lynch, L. The effect of farmland preservation programs on farmland prices. *Am. J. Agric. Econ.* **2001**, *83*, 341–351.
24. Avin, U.; Bayer, M. Right-sizing Urban Growth Boundaries. *Planning* **2003**, *69*, 22–27.
25. Howell-Moroney, M. Studying the effects of the intensity of US state growth management approaches on land development outcomes. *Urban Stud.* **2007**, *44*, 2163–2178.
26. Carruthers, J.I. The Impacts of State Growth Management Programmers: A Comparative Analysis. *Urban Stud.* **2002**, *39*, 1959–1982.
27. Hall, P. The containment of urban England. *Geogr. J.* **1974**, *140*, 386–408.
28. Nelson, A.C. Preserving prime farmland in the face of urbanization: Lessons from Oregon. *J. Am. Plan. Assoc.* **1992**, *58*, 467–488.
29. Gennaio, M.P.; Hersperger, A.M.; Bürgi, M. Containing urban sprawl—evaluating effectiveness of urban growth boundaries set by the Swiss Land Use Plan. *Land Use Policy* **2009**, *26*, 224–232.
30. Zhu, Z.; Yao, S.; Xu, G. Urban expansion and its spatial growth management in Nanjing. *Hum. Geogr.* **2003**, *5*, 11–16. (In Chinese).
31. Chen, J.; Ren, L.; Yu, C.; Xu, X.; Li, X. Remarks on urban spatial growth management research. *Plan. Stud.* **2009**, *10*, 19–24. (In Chinese).
32. Feng, K.; Wu, C.; Wei, S.; Liu, Y. Urban Expansion Management: Review on the Theory of UGB and its Reference for China. *China Land Sci.* **2008**, *5*, 77–81. (In Chinese).
33. Liu, H.Y.; Zhang, P.G. Study on Application of Growth Management in Chinese Urban Planning. *Urban Plan. Int.* **2007**, *22*, 108–113. (In Chinese).
34. Yang, H. *Researches on the Law and Institutions for the Usage of Control of Land*; Law Press: Beijing, China, 2010; pp. 212–214.
35. Yu, B.; Lu, C. Change of cultivated land and its implications on food security in China. *Chin. Geogr. Sci.* **2006**, *16*, 299–305.
36. Tao, F.; Yokozawa, M.; Liu, J.; Zhang, Z. Climate change, land use change, and China's food security in the twenty-first century: An integrated perspective. *Clim. Chang.* **2009**, *93*, 433–445.
37. Zou, Y.; Zhao, W.; Mason, R. Marketization of collective-owned rural land: A breakthrough in Shenzhen, China. *Sustainability* **2014**, *6*, 9114–9123.
38. Shenzhen Evening News. Shenzhen Carrying 18 Million People Currently. Available online: http://wb.sznews.com/html/2014-01/25/content_2765782 (accessed on 25 January 2014).
39. Shenzhen N.B.S. Statistical Communiqué of Shenzhen on the 2014 National Economic and Social Development. Available online: http://www.sztj.gov.cn/xxgk/tjsj/tjgb/201504/t20150424_2862885.htm (accessed on 24 April 2015).
40. Futian Government Online. The Government Work Report 2005. Available online: http://www.southcn.com/news/dishi/shenzhen/ggl/200506060358.htm (accessed on 6 June 2005).

41. Shenzhen Statistical Yearbook (2014). Available online: http://www.sztj.gov.cn/nj2014/start2014.htm (accessed on 31 March 2015).
42. Beijing Statistical Yearbook (2014). Available online: http://www.bjstats.gov.cn/nj/main/2014-tjnj/CH/index.htm (accessed on 23 December 2015).
43. Shanghai Statistical Yearbook (2014). Available online: http://www.stats-sh.gov.cn/data/toTjnj.xhtml?y=2014 (accessed on 25 September 2015).
44. Guangzhou Statistical Yearbook (2014). Available online: http://data.gzstats.gov.cn/gzStat1/chaxun/njsj.jsp (accessed on 7 April 2015).
45. Ng, M.K. Shenzhen. *Cities* **2003**, *20*, 429–441.
46. Hin, L.L.; Xin, L. Redevelopment of urban villages in Shenzhen, China–an analysis of power relations and urban coalitions. *Habitat Int.* **2011**, *35*, 426–434.
47. Shi, Y.; Peng, Z.; Chen, H.; Chen, Y.; Fan, H. Features of land use change and impact factors of construction land of international metropolises and significances to Shanghai. *Urban Plan. Forum* **2008**, *6*, 32–39.
48. Huang, Z.; Wei, Y.D.; He, C.; Li, H. Urban land expansion under economic transition in China: A multi-level modeling analysis. *Habitat Int.* **2015**, *47*, 69–82.
49. Li, H.; Wei, Y.D.; Liao, F.H.; Huang, Z. Administrative hierarchy and urban land expansion in transitional China. *Appl. Geogr.* **2015**, *56*, 177–186.
50. Wei, Y.D. Restructuring for growth in urban China: Transitional institutions, urban development, and spatial transformation. *Habitat Int.* **2012**, *36*, 396–405.
51. Tang, B.; Wong, S.; Liu, S. Institutions, property taxation and local government finance in China. *Urban Stud.* **2011**, *48*, 847–875.
52. Qun, W.; Yongle, L.; Siqi, Y. The incentives of China's urban land finance. *Land Use Policy* **2015**, *42*, 432–442.
53. Ding, C. Policy and praxis of land acquisition in China. *Land Use Policy* **2007**, *24*, 1–13. (In Chinese).
54. Guo, S.; Liu, L.; Zhao, Y. The business cycle implications of land financing in China. *Econ. Model.* **2015**, *46*, 225–237.
55. Long, Y.; Han, H.; Lai, S.K.; Mao, Q. Urban growth boundaries of the Beijing Metropolitan Area: Comparison of simulation and artwork. *Cities* **2013**, *31*, 337–348.
56. Guo, G.C.; Wu, Q. On System Obstruction of Cultivated Land Protection in China from the Viewpoint of Principal- agent Theory. *China Land Sci.* **2008**, *4*, 49–55. (In Chinese).
57. Chen, H.; Wu, Y.; Ou, M. Theoretical Construction of Compensation Mechanism for Farmland Protection. *J. Nanjing Agric. Univ.* **2009**, *9*, 67–71. (In Chinese).
58. Shan, Y.; Yang, L.; Wang, J. Economic analysis of loss of cultivated land in China. *Syst. Sci. Compr. Stud. Agric.* **2002**, *18*, 300–303. (In Chinese).
59. Jin, X.; Shen, Z. The Cap-and-trade Model of Incremental Construction Land Quota: An Analog with the Emission Permits System. *China Popul. Resour. Environ.* **2010**, *20*, 86–91. (In Chinese).
60. Jin, X.M. Market-oriented Reform of Administration in Incremental Construction Land Quota. *China Land Sci.* **2009**, *23*, 19–23. (In Chinese).

61. Zheng, H.; Wang, X.; Cao, S. The land finance model jeopardizes China's sustainable development. *Habitat Int.* **2014**, *44*, 130–136.
62. Tian, L. Land use dynamics driven by rural industrialization and land finance in the peri-urban areas of China: "The examples of Jiangyin and Shunde". *Land Use Policy* **2015**, *45*, 117–127.
63. Shenzhen City Planning and Land Resources. The Comprehensive Planning of Shenzhen City (2010–2020). Available online: http://www.szpl.gov.cn/ (accessed on 29 September 2010).
64. Orea, L.; Perez, J.A.; Roibas, D. Evaluating the double effect of land fragmentation on technology choice and dairy farm productivity: A latent class model approach. *Land Use Policy* **2015**, *45*, 189–198.
65. Latruffe, L.; Piet, L. Does land fragmentation affect farm performance? A case study from Brittany, France. *Agric. Syst.* **2014**, *129*, 68–80.
66. Zheng, H.W.; Shen, G.Q.; Wang, H.; Hong, J. Simulating land use change in urban renewal areas: A case study in Hong Kong. *Habitat Int.* **2015**, *46*, 23–34.
67. Donaldson, R.; du Plessis, D. The urban renewal programmer as an area-based approach to renew townships: The experience from Khayelitsha's Central Business District, Cape Town. *Habitat Int.* **2013**, *39*, 295–301.
68. Tong, D.; Feng, C.C.; Li, G.C.; Chen, P. Categories and characteristics of urban villages by GIS-based analysis: A case study of the Shenzhen Special Economic Zone. *Sci. China Technol. Sci.* **2010**, *53*, 150–157.
69. De, T.; Changchun, F.; Jin, J.D. Spatial evolution and cause analysis of urban villages: A case study of Shenzhen Special Economic Zone. *Geogr. Res.* **2011**, *30*, 437–446.
70. Song, Y.; Zenou, Y. Urban villages and housing values in China. *Reg. Sci. Urban Econ.* **2012**, *42*, 495–505.
71. Hao, P.; Sliuzas, R.; Geertman, S. The development and redevelopment of urban villages in Shenzhen. *Habitat Int.* **2011**, *35*, 214–224.
72. Wang, Y.P.; Wang, Y.; Wu, J. Urbanization and informal development in China: Urban villages in Shenzhen. *Int. J. Urban Reg. Res.* **2009**, *33*, 957–973.
73. Liu, Y.; He, S.; Wu, F.; Webster, C. Urban villages under China's rapid urbanization: Unregulated assets and transitional neighborhoods. *Habitat Int.* **2010**, *34*, 135–144.
74. Lai, Y.; Peng, Y.; Li, B.; Lin, Y. Industrial land development in urban villages in China: A property rights perspective. *Habitat Int.* **2014**, *41*, 185–194.
75. Wang, Q.; Zhang, X.; Wu, Y.; Skitmore, M. Collective land system in China: Congenital flaw or acquired irrational weakness? *Habitat Int.* **2015**, *50*, 226–233.
76. Ding, C. Land policy reform in China: Assessment and prospects. *Land Use Policy* **2003**, *20*, 109–120.
77. Ye, W. The Analysis on the Illegal Building in Urban Village in Shenzhen. *China Bus. Mark.* **2011**, *25*, 69–72. (In Chinese).
78. Comprehensive Research Group, National School of Development, Peeking University. Market solutions to urban redevelopment—A study on shenzhen's land reform. *Int. Econ. Rev.* **2014**, *3*, 56–71.

79. Zhang, X.; Ou, M.; Li, J. Analysis on the Institution Change of Cultivated Land Protection and Its Performance in China. *Soc. Sci.* **2007**, *8*, 13–20. (In Chinese).
80. Tang, H.; Yun, W.; Guan, X.; Liang, M. Farmland Utilization in Big Cities must Be Based on National Spatial Service. *Nat. Resour. Econ. China* **2015**, *28*, 11–13. (In Chinese).
81. Sustainable and Livable Cities: Toward Ecological Civilization. Available online: http://rcid.cau.edu.cn/art/2013/12/2/art_7631_229156.html (accessed on 2 December 2013).
82. Li, X.; Zhang, S.; Wang, Y. Quantitative Study of Construction Land Increase Limit Year in the Economic Transition Stage in China. *J. Nat. Resour.* **2011**, *26*, 1085–1095. (In Chinese).
83. Seto, K.C.; Fragkias, M.; Güneralp, B.; Reilly, M.K. A meta-analysis of global urban land expansion. *PLoS ONE* **2011**, *6*, e23777.
84. Ding, C. *Urban Growth and Policy: International Perspective and China Development*; Higher Education Press: Beijing, China, 2009; pp. 283–290.

What is the Influence of the Planning Framework on the Land Use Change Trajectories? Photointerpretation Analysis in the 1958–2011 Period for a Medium/Small Sized City

Mário Monteiro and Alexandre Oliveira Tavares

Abstract: Medium/small sized cities create a polycentric urban system representing the backbone of their territory, characterized by profound changes on land use. The objective of this study is to evaluate the spatial and temporal dynamics of land use in a medium/small city in Portugal during the period 1958–2011, and the relation with the planning framework. The assessment involves land use changes recognition in the period 1958–2011, calculation of the stability grade indicator, and the losses and gains between classes. The rate of artificialization and its relation with the planning framework was also evaluated. The results for the city of Viseu showed a main decrease in annual crops and arable lands, with an increase of the continuous and discontinuous urban fabric. The changes are systematic transitions, marked by planning framework and its typology, objectives and scale enforcement. A plan addressing the city encouraged land use changes in the fringe, while a municipal Master Plan determined sequent transformation in all areas. The land use changes and the artificialization processes enabled the identification of three periods where planning framework was forthright. The study underlines the importance of a planning framework for medium/small sized cities for urban sprawl and artificialization processes.

Reprinted from *Sustainability*. Cite as: Monteiro, M.; Tavares, A.O. What is the Influence of the Planning Framework on the Land Use Change Trajectories? Photointerpretation Analysis in the 1958–2011 Period for a Medium/Small Sized City. *Sustainability* **2015**, *7*, 11727–11755.

1. Introduction

Land changes in urban areas, especially the conversion of cropland and forest land to urban uses, is one of the most important forms of global environmental changes [1]. Cities of small and medium scale are frequently characterized by profound changes in land use, a heterogeneous occupation, and a high level of infrastructure and equipment, where hazardous processes, natural and technological, have cumulative presence with economically and socially relevant damages and losses [2]. The relevance of these cities is also pointed out by some authors [3–6], presenting a complex social distribution and organization, with the increase of territorial vulnerability. However, as stated by the European Territorial Strategy and the Territorial Agenda of European Union [7,8], these cities create a polycentric urban system, where their interdependency forms are especially important for rural regions and represent the backbone of their territory because of the ability to generate jobs and permit the dissemination of certain urban functions [9].

Different studies consider land use and the urban expansion as a combined outcome of societal needs, supported by planning preparation actions [10–13]. Understanding long-term dynamics of the landscape and transitions of land use is fundamental to dimension the past and future effects and for the recognition of present landscape structure [14–18].

Some authors have investigated land-use changes through a comparative evaluation of land-cover image sources [13,19–23], using detection techniques involving different images and sources.

The use of satellite imagery and aerial photographs for interpretation has made it possible to integrate sets of images from several observation periods obtained from different sources and using different scales. The reconstitution of the transformation path using those images, in accordance with uniform classification principles and field validation capabilities, has been explored in other studies to reveal the dynamics of land-use transformation [20,24–27].

The analysis of changes on land use and occupation in the territory, with special focus in the processes from small and medium cities, allows the evaluation of consistency in urban expansion scenarios, the vulnerability of ecosystems, and the maintenance of the distinctive morphological and socio-cultural features [1,4,21,26,28–30].

Several authors present indicators to monitor urban expansion and temporal dynamics which are major triggering factors in land use change processes [31–33]. Different research groups point sets of indicators responsible for the characteristics of landscapes and land use change [34–37], namely related with the planning framework [13,38–40].

The main objective of this study is to evaluate the spatial and temporal dynamics of land use over time in an inland medium/small city of Portugal during the period 1958–2011, and the relation with the planning framework. The research questions we want to answer are:

What are the changes in land use in the period 1958–2011, using five moments of analysis?

What are the stability grade and the dynamics of losses and gains between classes of land use?

What is the zoning model for land use change and artificialization process?

What is the relation between land use trajectories and the planning framework?

2. Materials and Methods

2.1. Study Area

The study area is located in the Central Region of Portugal and it is a part of the municipal area of Viseu (Figure 1). The county of Viseu is a district capital and centre of the NUT III (Dão-Lafões), which presents one of the highest population consolidations in the inland of Portugal, polarizing several neighbouring municipalities.

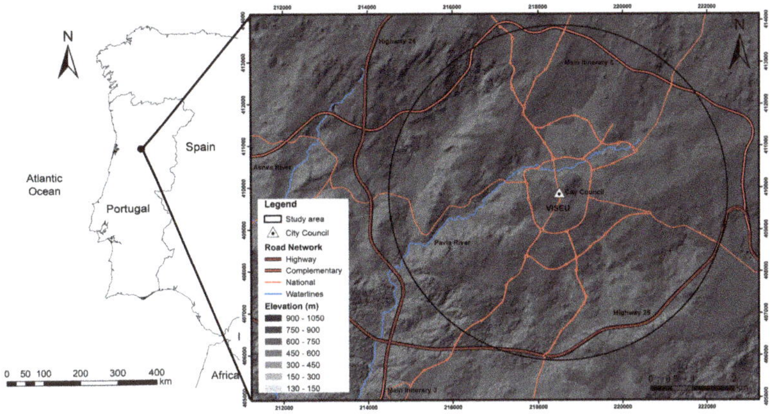

Figure 1. Location of the study area.

In morphological terms, the area is surrounded by mountains where the regional hypsometric values can reach 1043 m about 17 km away from the city centre. The study area is characterized by hills sometimes marked by river valleys, with altitudes values ranging from 300 to 600 m, approximately.

The geology of the area consists essentially of hercinic granite rocks, with small expressions of metasedimentary rocks and sedimentary and alluvial deposits preserved in tectonic depressions and related to the large rivers that cross the region [41].

The region has a Mediterranean climate with a well-defined dry season, although short (July and August). The average annual temperatures vary between 10 °C and 15 °C and the annual precipitation ranges from 400 to 2400 mm, with unstable winter atmospheric conditions, due to the influence of mountains, with long lasting periods of concentrated rainfall [42,43].

The land occupation is a further evidence of the heterogeneity of the municipality of Viseu. It is dominated by forest areas, essentially wild pine followed by agricultural areas, uncultivated areas, and artificial areas. The most distant parishes of the city centre have a higher forest area [42].

In biogeographical terms, the municipality of Viseu is located in the Eurosiberian Region. More precisely, the study area represents the Galician-Portuguese Sector, specifically in the Miniense Subsector, that is characterized by the existence of seasoned oak of *Rusco aculeate-Quercetum roboris quercetosum suberis* and furze (*Ulex europaeus* and *Ulex micranthus*) [44].

Regarding human occupation, the municipality of Viseu has 99,274 inhabitants according to the 2011 Census [45]. Figure 2 shows the evolution, in the period 1950–2011, of the resident population in the municipal area and in the parishes that are within the study area. It presents a small variation by the year 1970, with a sequent population increase (24%), where the growth of the municipal population is essentially in the study area.

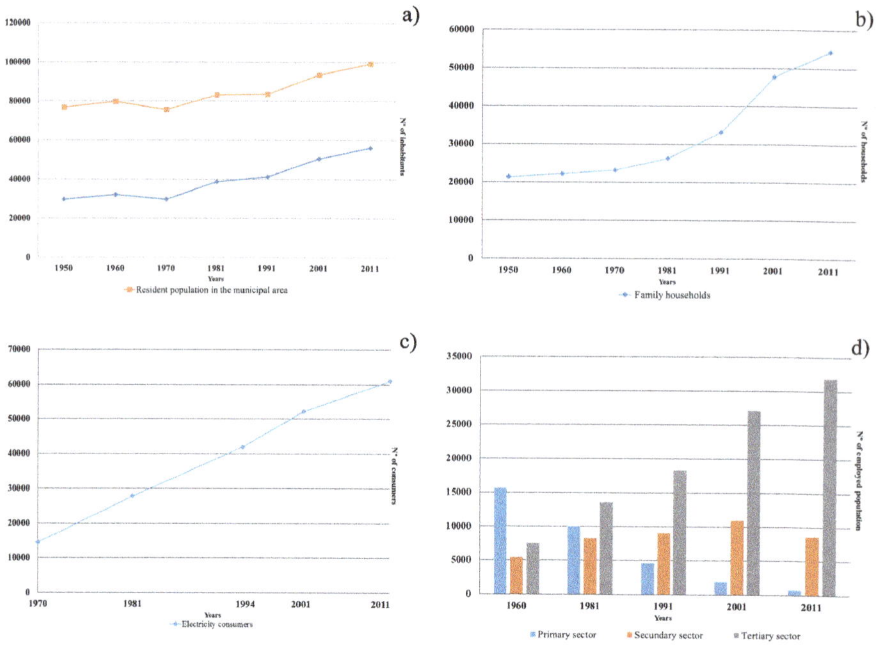

Figure 2. Demographic and socio-economic indicators: (**a**) municipal and study area resident population; (**b**) number of family households; (**c**) number of electricity consumers; (**d**) employed population by activity sectors, according the PORDATA database [46].

Figure 2 presents some municipal demographic and socio-economic indicators which demonstrate the progressive number of the resident population and the crescent number of family households. The population increase between 1970 and 1990 was influenced by the social integration of returned Portuguese, both from the former colonies and emigrants in Europe. The population and household increase after 1990 expresses the general improvement of well-being after joining the European community and concentration on the city from residents from surrounding rural municipalities. These socio-economic dynamics are expressed by the sharp increase in the number of electricity consumers, accompanying the increase of the population active in the tertiary sector at the same time reducing the importance of the primary sector. Industrialization, in the seventies, and the tertiary activities in the eighties and nineties, were influenced by road infrastructure construction, with the correspondent accessibility improvement.

Those aspects reinforced the increment of centrality and attractiveness of the town, which is expressed by a spatial urban growth, and achieving the characteristics of a medium-sized city in the national context [47], and a small-sized city according the EC definition [48].

To evaluate the land use and occupation changes, and the progressive urban expansion and infrastructure, which reflect the planning framework, a study area was selected, presented in Figure 1. It is centred in the City Council of Viseu, neighbour of the XV ancient town, and has a circular shape with 4 km radius, with 50.26 km^2 of area. The circle represents or partially covers nine parishes of the municipal territory, including the urban core and the peri-urban areas.

As mentioned by J. Almeida [47], the urban model of the city of Viseu is defined by a dominant and compact centre and develops in "stellar" form, growing the city quickly to the outside along the pre-existent road network, which provides the appearance of peri-urban areas of some size, later connected by concentric pathways [48,49]. For S. Almeida [49] the urban fabric development has been conditioned by urban structures, namely the encircling roads, but also the by collective equipment's location.

In the last 50 years two former master plans influenced urban expansion, the infrastructure network, and the constrained areas protection. A city urbanization plan was approved in 1952 and a Municipal master plan was published in 1995, which is currently being replaced. Between these two instruments there were partial plans, such as the Hidroprojecto Urban Plan (1971) and Macroplan Urban Plan (1983) that has never been approved, but with informal application for new urbanization areas, or the Polis Program (2000)—Urban Rehabilitation Program and Environmental Improvement of Cities. In Figure 3 some of the cartographic extracts of these plans are represented, emphasizing that by 1983 the plans were partial, focusing mainly in the urban core of the city, and in 1995 acquired the municipal level, which has been complemented by plans further detailed (layout plans, urbanization plans, and a rehabilitation plan), and also the 2013 Municipal Master Plan under approval.

2.2. Land Cover Data

The analysis of the land use and occupation change was supported by a set of five photographic mosaics from different periods, between 1958 and 2011, as described in Table 1. According to the initial characteristics of the images, they were digitalized, georeferenced, processed and analyzed using the ArcGIS 10.2 software (ESRI®). To adjust the different imagery scale three procedures were carried out (single operator, minimum unit of area, and backward analysis). Due to differing contrast levels, the 1958, 1974, and 1985 images had to be cross-referenced with other topographic representations. In order to integrate images from different sources consistently, a supervised analysis was performed using fieldwork as a control. A Datum_73_Hayford_Gauss_IGeoE was used as the geographic coordinate system in the process of georeferencing.

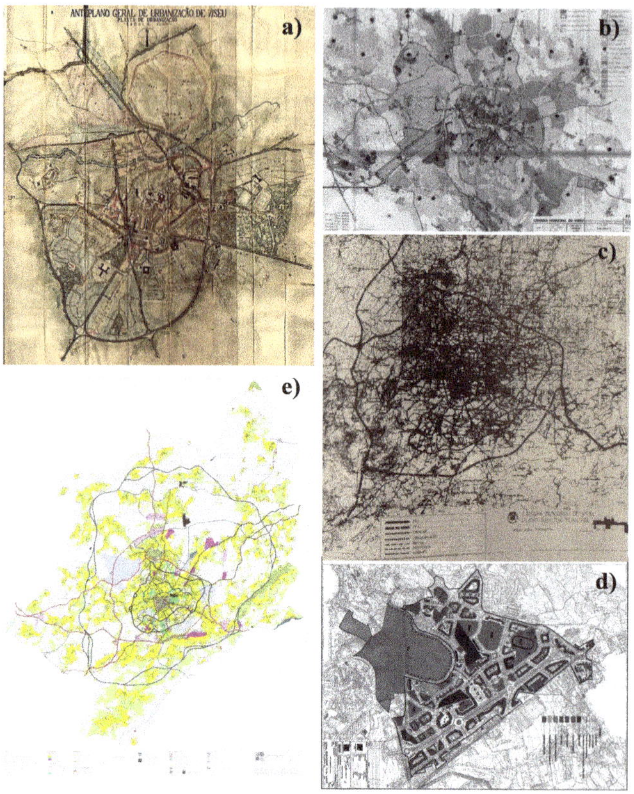

Figure 3. Examples of cartography associated to the planning framework of Viseu: (**a**) Urbanization Ante-Plan (1952); (**b**) Macroplan Urban Plan (1983); (**c**) Municipal Master Plan (1995); (**d**) Layout Plan (2008); (**e**) Spatial plan classification (Municipal Master Plan, 2013).

In another phase, based on the categories of land use and occupation of the Corine Land Cover [50], a classification system has been developed and identified by classes organized according to numerical codes (Table 2) [51,52].

Table 1. Data source characteristics.

Date	Format	Type image	Scale	Source
1958	Print	Black & White	1: 26,000	IGeoE
1974	Digital	Black & White	1: 15,000	IGP
1985	Digital	Black & White	1: 15,000	IGP
1995	Digital	False colour	1: 10,000	CNIG
2011	Digital	Colour	1: 10,000	IGP

IGeoE: Army Geographic Institute; IGP: Portuguese Geographic Institute; CNIG: National Center of Geographical Information.

Table 2. Classification terms.

<table>
<tr><th rowspan="2">Codes</th><th colspan="2">Corine Land Cover</th></tr>
<tr><th>Classes and Subclasses</th><th>Applied Classes</th></tr>
<tr><td>111</td><td>Continuous urban fabric</td><td></td></tr>
<tr><td>112</td><td>Discontinuous urban fabric</td><td></td></tr>
<tr><td>121</td><td>Industrial or commercial units</td><td></td></tr>
<tr><td>122</td><td>Road and rail networks and associated land</td><td>Road and rail networks</td></tr>
<tr><td>131</td><td>Mineral extraction areas</td><td></td></tr>
<tr><td>132</td><td>Dump site</td><td></td></tr>
<tr><td>133</td><td>Construction sites</td><td></td></tr>
<tr><td>141</td><td>Green urban areas</td><td></td></tr>
<tr><td>142</td><td>Sport and leisure facilities</td><td></td></tr>
<tr><td>211</td><td>Non-irrigated arable land</td><td>Arable land</td></tr>
<tr><td>221</td><td>Vineyards</td><td></td></tr>
<tr><td>222</td><td>Fruit trees and berry plantations</td><td rowspan="2">Fruit trees and olive groves</td></tr>
<tr><td>223</td><td>Olive groves</td></tr>
<tr><td>231</td><td>Pastures</td><td></td></tr>
<tr><td>241</td><td>Annual crops associated with permanent crops</td><td></td></tr>
<tr><td>242</td><td>Complex cultivation patterns</td><td></td></tr>
<tr><td>243</td><td>Land principally occupied by agriculture, with significant areas of natural vegetation</td><td rowspan="2">Agriculture with natural spaces and agro-forestry areas</td></tr>
<tr><td>244</td><td>Agro-forestry areas</td></tr>
<tr><td>311</td><td>Broad-leaved forest</td><td></td></tr>
<tr><td>312</td><td>Coniferous forest</td><td></td></tr>
<tr><td>313</td><td>Mixed forest</td><td></td></tr>
<tr><td>321</td><td>Natural grassland</td><td></td></tr>
<tr><td>322</td><td>Moors and heathland</td><td>Woods</td></tr>
<tr><td>323</td><td>Sclerophyllous vegetation</td><td></td></tr>
<tr><td>324</td><td>Transitional woodland/shrub</td><td>Degraded forest areas</td></tr>
<tr><td>332</td><td>Bare rock</td><td></td></tr>
<tr><td>333</td><td>Sparsely vegetated areas</td><td></td></tr>
<tr><td>334</td><td>Burnt areas</td><td></td></tr>
<tr><td>511</td><td>Water courses</td><td></td></tr>
<tr><td>512</td><td>Water bodies</td><td></td></tr>
</table>

Then, the classification of polygons of land use from each set of images was performed and incorporated into the geographic information system generated [17,18,53] wherein all polygons were classified using a minimum unit of analysis with 1 hectare and supported on a 1:10,000 scale.

2.3. *Spatial and Temporal Changes in Land Use and Annual Rate of Artificialization*

In order to evaluate the dynamics of land use and occupation a map to map comparison method was used involving the set of successive images, with cross-references to define the transitions of land use and land cover and also the measurement of the stability grade (SG) in each time period.

The maps created for each year were used to produce matrices for the periods 1958–1974, 1974–1985, 1985–1995, 1995–2011, and 1958–2011. In the Table 3, the calculation algorithms for the different indicators considered are presented.

Table 3. Equations of the calculated indicators.

Designation	Equation		Legend
Change in land cover	$CHi = (pi. - p.i)/p.i$	(1)	*CHi*—change in land cover for the period in column *i*
			pi.—the column total for the grid cells
			p.i—row total for the grid cells in the same category *i*
Conversion areas corresponding to gains and losses for a given year	$P(i).j = (p_{j,i} - p_{i,j})/(p_j - p_i) \times 100\, i \neq j$	(2)	$P_{(i),j}$—percentage by type *j* in the total conversion of category row *i*
			pj,i and *pi,j*—express the individual entry in the change matrix
Stability grade	$SG = [(C_{ix}A_{2\ldots5} - C_{ix}A_1)/TA] \times 100$	(3)	SG—indicator of the consistency and represent the global stability of the classes of land use in the year *A1* for the year $A_{2\ldots5}$
			$C_{ix}A_{2\ldots5}$—sum of the areas of the different classes (*ix*) at the subsequent time point
			$C_{ix}A_1$—sum of the areas of the different classes (*ix*) at time point 1
			TA—total area studied (km^2)
Annual rate of artificialization of surfaces	$AS = [UA_{n+i} - UA_i/nTA_{n+i}] \times 100\%$	(4)	AS—evaluastudy area
			TA_{n+i}—total study area to be calculated at the time point $i + n$
			UA_{n+i} and UA_i—surfaces with urban fabric and infrastrutures in the target unit at time $i + n$ and i
			n—number of years in each time period

The Equation (1), used in other studies [13,36,54] was applied in order to express the land use transition dynamics between two periods.

The conversion areas corresponding to gains and losses for a given year, relative to the year under comparison were calculated in relation to the total land-use type using Equation (2).

To express the total area that has not experienced a possible transition to a different category of land use and occupation, the indicator stability grade (SG) was defined as shown in Equation (3).

An annual rate of artificialization (AS) was calculated using Equation (4), expressing the unceasing increase in urbanization, namely related with peri-urbanization. This indicator combines urban transformation processes that express soil sealing and vegetation cover change, from the continuous and discontinuous urban fabric, industrial or commercial units, road and rail networks, mineral extraction areas, dump site, construction sites, green urban areas, and sport and leisure facilities.

In order to understand if the major changes in land use and occupation in the study area occur in the central area or in the periphery, and according to other studies [22,55,56], it was decided to divide the area into 4 different sections with increasing distances from the center point (city hall), as shown in Figure 4.

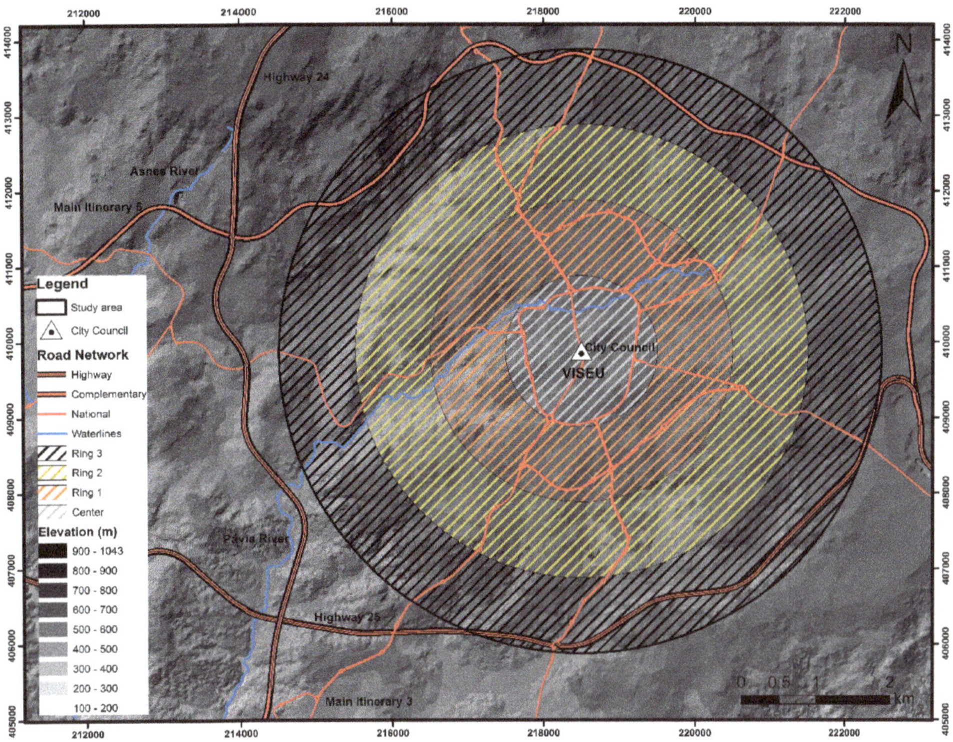

Figure 4. Subdivision of the study area in four different sections to calculate the stability grade in each one.

For this purpose, four buffers were defined with distances of 1, 2, 3, and 4 km respectively, relative to city council point. As a result, a central circle and three rings with increasing distances from the central point were originated. The focus of this analysis was to establish a model of land use change dynamics according with the spatial planning framework which presents different scales and spatial enforcement. Thereafter the stability grade was calculated for the four sections in the five different periods of analysis.

To analyze the relationship between land use change and planning processes in the city of Viseu, different instruments of planning were compiled, between 1952 and 2008. For each of them there was an analysis of the objectives, the type, and area of enforcement.

For these plans, a content analysis of the planning documents and associated reports was made, supported by methods [57,58]. The special protection legal regimes which present constraints to planning and land use transformations were also identified.

Table 4 presents the planning framework for the period of analysis, and also indicates the previous plans and the recent master plan.

Table 4. Planning framework.

Date of Approval	Planning Framework Description	Enforcement Area
1952	Urbanization Ante-Plan, with the major purpose to resolve the intra-urban circulation and regional connectivity.	City
1971	Hidroprojecto Urban Plan for encircling road definition, urban requalification and new residential neighborhoods proposal.	City
1983	Macroplan Urban Plan projecting new residential neighborhoods, health and educational equipment's location, and encircling road consolidation.	City
1992	RAN—Agricultural Protection Regime, for soils with agricultural potential.	Municipal non urban
1988 to 1993	17 Layout Plans with local occupation proposals and infrastructure implementation rules.	Local
1995	Municipal Master Plan, establishing a development strategy, a spatial plan model and defining urban policies and regulatory guidelines for the municipal territory.	Municipal
1996	REN—Ecologic Regime Protection, for sensitive ecological and natural risk susceptibility areas.	Municipal non urban
1996 to 98	4 Layout Plans with local occupation proposals and infrastructure implementation rules.	Local
2000	POLIS Programme—Urban Infrastructure Planning and Urban Rehabilitation providing urban interventions in disqualified areas and develop green urban areas and leisure facilities.	Local
2003 to 2013	7 Layout Plans with local occupation proposals and infrastructure implementation rules.	Local
2013	Municipal Master Plan (Revision), with development strategy and an urban consolidation proposal.	Municipal

As can be observed, during the period of analysis the first initial plans had a central city enforcement, and were responsible for creating new residential areas, infrastructures and an encircling road network. These plans were supported by local plans for urban occupation proposal. The plans with municipal characteristics and enforcement were approved in 1995, and supported by a regulatory agricultural and ecological protection regime for non-urban areas. Yet, in addition to these plans, some urbanization plans, layout plans and a rehabilitation plan emerged.

3. Results

3.1. Temporal Patterns for Land Use Change in the Study Area

Using the methods described a systematic analysis and classification of land use and occupation was produced for the five moments (1958, 1974, 1985, 1995, 2011) in the study area.

Table 5 shows the classes applied to the use and occupation of land, the number of polygons of each class and the respective area and percentage. Twenty six different classes of land use and occupation have been identified in the circle area under the analysis (Figure 5). The images only presented five classes with episodic representation (mineral extraction areas, dump site, construction sites, burnt areas, and water bodies). As observable, over the analysis period there was a steady increase in the urban area, namely on the discontinuous urban fabric. This process was still associated with increasing complex cultivation patterns. Since 1985 there has been a considerable decrease in areas with annual crops associated with permanent crop and pasture use. The display also shows the maintenance of forest areas, with constant values above 30% of the study area.

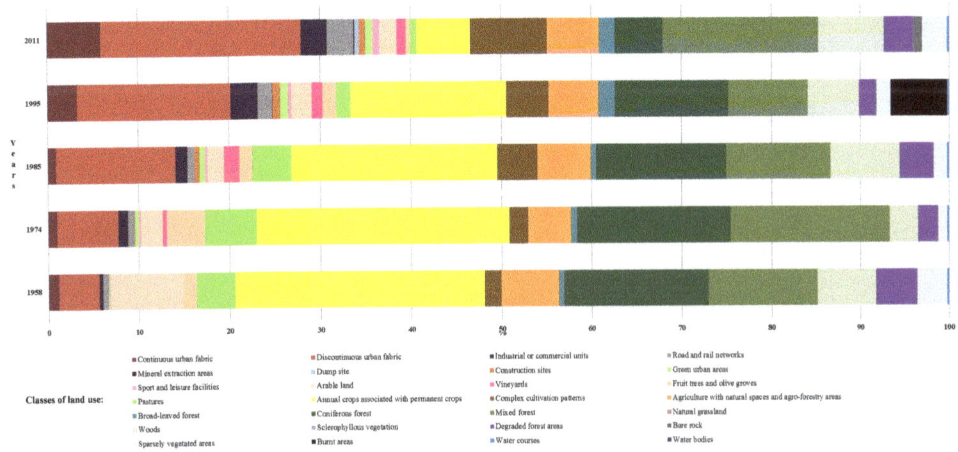

Figure 5. Representations of subclasses of land use from 1958 to 2011 in the study area.

Table 5. Number of patches, areas, and percentages for each land use and occupation in the study area, according to the Corine Land Cover classification.

Classes Applied	Patch (no)					Area (ha)					Percentage				
	Year					Year					Years				
	1958	1974	1985	1995	2011	1958	1974	1985	1995	2011	1958	1974	1985	1995	2011
Continuous urban fabric	1	1	1	4	10	59.85	50.17	46.78	165.22	299.31	1.19	1.00	0.93	3.29	5.95
Discontinuous urban fabric	30	50	74	92	94	225.15	342.33	664.74	851.49	1104.77	4.48	6.81	13.23	16.94	21.98
Industrial or commercial units	2	11	17	36	31	18.18	51.23	63.90	143.19	142.95	0.36	1.02	1.27	2.85	284
Road and rail networks	2	1	1	2	2	31.28	36.66	39.26	78.83	146.22	0.62	0.73	0.78	1.57	2.91
Mineral extraction areas				1	2				1.79	5.89				0.04	0.12
Dump site				1	2				1.12	23.17				0.02	0.46
Construction sites		1	6	10	7		1.59	27.00	42.53	36.63		0.03	0.54	0.85	0.73
Green urban areas	1	2	2	6	4	6.64	22.72	34.00	45.75	41.47	0.13	0.45	0.68	0.91	0.83
Sport and leisure facilities	1	2	4	5	9	4.48	9.59	11.17	19.29	39.28	0.09	0.19	0.22	0.38	0.78
Arable land	20	11	12	16	25	403.45	121.75	91.74	113.3	93.71	8.03	2.42	1.83	2.25	1.86
Vineyards		4	18	15	16		23.22	81.56	60.57	50.27		0.46	1.62	1.21	1.00
Fruit trees and olive groves	12	38	22	18	9	73.05	212.07	69.47	73.85	23.90	1.45	4.22	1.38	1.47	0.48
Pastures	26	47	43	24	13	213.12	280.87	218.69	80.53	35.39	4.24	5.59	4.35	1.60	0.70
Annual crops associated with permanent crops	38	65	73	91	57	1381.42	1401.64	1139.36	862.04	296.87	27.48	27.89	22.67	17.15	5.91
Complex cultivation patterns	13	27	57	59	121	92.03	103.69	223.73	236.81	428.84	1.83	2.06	4.45	4.71	8.53
Agriculture with natural spaces and agro-forestry areas	45	47	63	67	83	322.89	242.37	300.87	278.81	291.44	6.42	4.82	5.99	5.55	5.80
Broad-leaved forest	7	11	8	21	24	27.96	32.51	28.18	90.81	86.48	0.56	0.65	0.56	1.81	1.72
Coniferous forest	38	35	47	64	36	808.75	860.43	729.02	635.46	267.40	16.09	17.12	14.50	12.64	5.32
Mixed forest	25	39	49	54	63	611.20	894.55	584.78	445.33	871.79	12.16	17.80	11.63	8.86	17.34
Woods	27	36	47	49	60	331.73	161.03	388.73	289.58	371.89	6.60	3.20	7.73	5.76	7.40
Degraded forest areas	16	21	30	26	47	233.74	116.64	196.92	93.16	165.71	4.65	2.32	3.92	1.85	3.30
Bare rock				1	1			6.03	50.37			0.12	1.00		
Sparsely vegetated areas	12	2	10	17	25	169.22	50.77	75.01	78.37	141.02	3.37	1.01	1.49	1.56	2.81
Burnt areas				10				320.25				6.37			
Water courses	1	2	2	3	3	12.24	10.55	11.47	10.97	10.27	0.24	0.21	0.23	0.22	0.20
Water bodies				1	1				1.3	1.34				0.03	0.03
Total	317	453	586	693	745	5026.38	5026.38	5026.38	5026.38	5026.38	100.00	100.00	100.00	100.00	100.00

The cartographic outputs present in Figure 6 show differentiated changes in the different sectors of the area with a variety of trajectories. The analysis indicates some systematic transitions involving an increase in urban areas and a decrease in agricultural areas, as well as the transformation of forest types and the episodic occurrence of wildfires. The analysis also reveals a continuous increase in the number of polygons between 1958 and 2011, which demonstrates the fragmentation of the landscape, representing a fast expansion process with new corridors [4] with impacts on natural systems and landscape homogeneity [59,60].

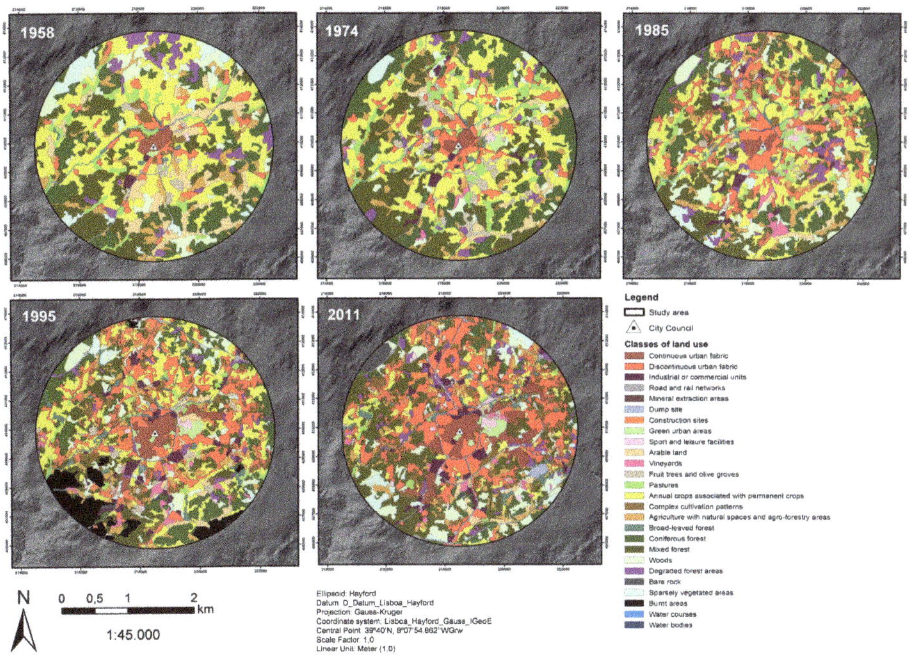

Figure 6. Land use and occupation in the area studied, based on image classifications by year.

In the study area, and having as focal point the city center—center of the considered circle, two different transformations can be observed. Firstly, there is an increase of the urban areas with a continuous fabric, discontinuous fabric, industrial or commercial units, road and rail networks, green urban areas, and sport and leisure facilities, as well as complex cultivation patterns. Secondly, a systematic decrease in the annual crops associated with permanent crops, and a routing on the forest occupation.

The analysis of land use change shows some temporal transitional processes (e.g., forest and semi-natural areas, forest and burnt areas) representing casuistic losses of forest space.

The results also show the growth of pattern complexity and reveal the general increase of the artificial areas from the central part of the study area to the periphery, contrasting with a continuous decrease of the agricultural areas. These results illustrate the peri-urbanization and rururbanization processes with the increase of classes of urban fabric and complex cultivation patterns.

3.2. Dynamic Analysis of Changes in Land Use in the Study Area

The transformation matrices for 1958–1974, 1974–1985, 1985–1995, 1995–2011, and 1958–2011 allowed for a detailed study of the dynamics of land use and occupation in five periods of analysis. For each period of analysis a transformation matrix was generated, and the respective stability grade calculated.

The results obtained in the matrices allowed the creation of a scheme (Figure 7), illustrating the dynamics of land use in the different periods and highlighting the visual trajectory of transformation between classes. The scale used to achieve the results presented in the figure, is comprised of three ranges (0.2 to 0.5 km^2; 0.5 to 1 km^2, and > 1km^2) that represent the higher or lower amount of area transferred between classes from period to period. The results show both the dominant dynamics and the differentiated processes for the various periods analyzed, highlighting specific gains and losses within the overall transformation and indicates a continuous trajectory of urbanization for the overall period (1958–2011), namely supported by the annual crops. The scheme also demonstrated a dynamic involving agricultural areas, with a loss in the annual crops and pastures and a gain in the complex cultivation patterns and agriculture with natural spaces and agro-forestry areas. The period between 1958–1974 shows an increase in the forest, with consequent evolution to woods and degraded forest areas, with the exception of the casuistic transition associated with the burnt areas.

When analyzing the dynamics of land use with the stability grade indicator, it reveals distinctive frames for particular periods. The global stability indicator presents contrasting values ranging from 36.87% (in the period 1995–2011) to 46.54% (in the period 1958–1974). The data shows a discontinuous sequence in which the stability grade increases from 1958 to 1985 and decreases from 1985 to 2011. In the periods 1958–1974 and 1974–1985, the SG values are quite similar (46.54% and 49.20%), but from 1985 to 2011 the SG decreases by approximately 12%, meaning that the largest transitions in land use in the study area took place from 1985 onwards.

In order to express the percentage of conversions in relation to the total land cover type, an internal transition matrix was created, for the period 1958–2011. Table 6 detailed the gains and losses for the period under analysis (1958–2011) for the 26 classes of land use and occupation. This matrix was obtained by cross-referencing the maps of 1958 and 2011 and the results show the units of change for each land use class and indicate the (%) rate of change. The results in general show an increase in the urban fabric and infra-structures associated, supported by the arable lands, fruit trees and olive groves, pastures, annual crops and agriculture with natural spaces and agro-forestry areas. In the agricultural areas, it is possible to verify an increase in the complex cultivation patterns (360.22%), opposed to the rest of the classes that present losses. Gains in broad-leaved, mixed forest and woods should also be noted. The results also reflect the decrease in coniferous forest, degraded forest areas, and sparsely vegetated areas. The matrix demonstrates the low stability grade (SG) for the period 1958–2011 (20.61%) reflecting the changes and dynamics between the different land use classes. This means that 79.39% of the land cover has experienced changes, showing intensive land use dynamics during the study period.

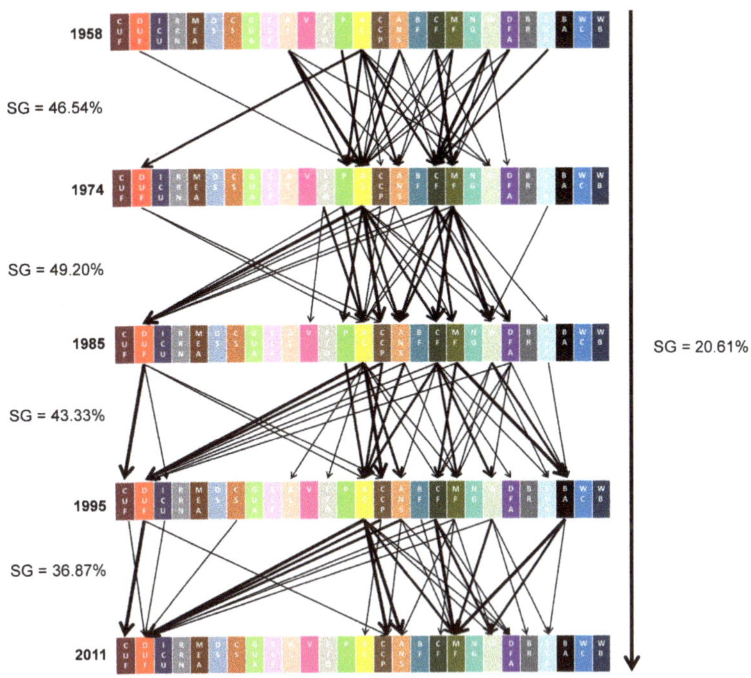

Figure 7. Dynamic transformations of land use for the study area and the stability grade (SG) for each period. CUF: Continuous urban fabric; DUF: Discontinuous urban fabric; ICU: Industrial or commercial units; RRN: Road and rail networks; MEA: Mineral extraction areas; DS: Dump site; CS: Construction sites; GUA: Green urban areas; SLF: Sport and leisure facilities; AL: Arable land; V: Vineyards; FT/OL: Fruit trees and olive groves; P: Pastures; AC: Annual crops associated with permanent crops; CCP: Complex cultivation patterns; ANS/AF: Agriculture with natural spaces and agro-forestry areas ; BF: Broad-leaved forest; CF: Coniferous forest; MF: Mixed forest; W: Woods; DFA: Degraded forest areas; BR: Bare rock; SVA: Sparsely vegetated areas; BA: Burnt areas; WC: Water courses; WB: Water bodies.

Table 6. Transformations in land use and occupation in the study area (%).

Land cover	Loss or gain in 2011	Type (1)	%	Type (2)	%	Type (3)	%	Type (4)	%	Type (5)	%	Type (6)	%	Type (7)	%	Type (8)	%
CUF	406.78	AC	30.10	DUF	17.35	MF	14.81	AL	8.97	ANS/AF	7.40	DFA	5.68	CF	4.38	CCP	4.18
DUF	393.75	AC	43.53	CF	15.39	W	10.00	ANS/AF	8.01	MF	7.75	DFA	6.87	FT/OL	3.52	CCP	2.23
ICU	741.18	AL	16.20	MF	13.00	AC	12.74	CF	12.28	P	11.83	ANS/AF	9.40	DFA	5.83	W	4.99
RRN	377.42	AC	24.75	AL	15.29	DFA	10.13	CF	10.08	ANS/AF	8.03	W	7.78	MF	7.19	P	6.58
MEA		CF	66.45	AC	22.30	W	11.24										
DS		MF	35.69	ANS/AF	24.60	AL	21.09	AC	12.05	CF	6.58						
CS		CF	38.07	AC	27.43	P	16.33	DFA	7.50	MF	7.49	AL	1.71	FT/OL	1.24	ANS/AF	0.23
GUA	566.67	MF	49.83	AC	22.18	BF	12.82	CCP	7.64	DUF	6.71	CUF	−2.85	AL	3.55	RRN	0.48
SLF	850.00	AL	32.14	CF	30.16	CCP	17.14	AC	12.66	MF	5.10	DFA	4.72	WC	0.62	W	0.30
AL	−76.73	AC	7.10	W	1.82	CF	2.91	ANS/AF	14.19	P	1.82	DFA	1.11	DUF	8.19	MF	13.31
V		AL	34.30	AC	32.19	CF	12.25	MF	9.45	ANS/AF	9.35	W	0.78	DUF	0.55	FT/OL	0.47
FT/OL	−67.12	AC	−15.04	AL	−16.10	SLF	−1.99	MF	7.91	DUF	63.08	CF	−0.28	W	5.94	RRN	4.43
P	-83.10	AC	12.16	AL	−3.16	CCP	6.87	CF	177.73	SVA	−0.34	ANS/AF	8.04	MF	22.54	BF	5.90
AC	−78.44	AL	−2.03	P	−1.99	ANS/AF	9.49	CF	1.39	DUF	35.30	CCP	21.99	W	2.71	MF	11.53
CCP	360.22	AC	70.80	ANS/AF	12.81	AL	9.52	CF	5.83	DFA	2.72	P	3.62	W	3.27	DUF	−5.84
ANS/AF	−10.46	AC	−327.34	AL	−139.70	CF	−44.80	MF	116.87	W	0.34	P	−45.43	DFA	24.41	FT/OL	−16.61
BF	230.77	AC	44.24	AL	25.29	CF	21.95	P	17.92	ANS/AF	5.31	MF	3.20	CCP	3.28	W	1.20
CF	−66.96	DFA	−0.93	W	19.38	MF	35.32	AC	−2.79	SVA	1.52	AL	−1.67	ANS/AF	2.60	CCP	3.63
MF	42.25	CF	73.38	AC	47.99	W	2.34	ANS/AF	14.11	AL	15.82	P	15.37	DFA	6.12	SVA	−9.95
W	12.35	CF	261.31	MF	−15.18	AC	73.12	SVA	16.18	DFA	17.38	ANS/AF	0.27	AL	14.07	P	17.08
DFA	−30.21	AC	−60.97	CF	7.43	W	10.26	P	−21.10	ANS/AF	−11.29	MF	23.43	SVA	3.68	AL	−5.07
BR		SVA	81.13	CF	18.87												
SVA	−17.06	MF	−91.97	CF	−29.12	W	23.05	DFA	−8.88	ANS/AF	−9.22	AL	−10.81	BF	−5.71	AC	−2.72
BA																	
WC	−9.09	P	−70.42	AC	3.96	AL	−6.62	DUF	26.71	CCP	25.30	CF	−14.12	CUF	4.08	MF	10.66
WB		AL	83.29	CF	16.71												

Land cover	Loss or gain in 2011	Type (9)	%	Type (10)	%	Type (11)	%	Type (12)	%	Type (13)	%	Type (14)	%	Type (15)	%	Type (16)	%	Type (17)	%
CUF	406.78	P	3.33	FT/OL	2.80	GUA	0.41	W	0.42	RRN	0.34	ICU	−0.23	WC	0.03	BF	0.02		
DUF	393.75	AL	2.88	P	3.08	SVA	1.40	BF	0.72	RRN	0.12	ICU	−0.55	WC	0.06	CUF	−4.72	GUA	−0.27
ICU	741.18	DUF	3.86	SVA	4.49	RRN	1.63	CCP	1.40	BF	0.91	CUF	0.44	FT/OL	0.45	WC	0.39	GUA	0.15
RRN	377.42	CCP	6.01	SVA	5.71	DUF	−0.89	FT/OL	1.89	WC	0.08	BF	0.01						
MEA																			
DS																			
CS																			
GUA	566.67	SLF	0.10	FT/OL	0.09														
SLF	850.00	BF	0.22	P	0.04	ANS/AF	0.01												
AL	−76.73	WC	0.04	RRN	5.67	FT/OL	2.55												
V		DFA	0.47	SLF	0.13	SVA	0.04												
FT/OL	−67.12	CCP	20.44																
P	-83.10	WC	0.78																
AC	−78.44	DFA	3.82	FT/OL	0.68	WC	−0.01	RRN	2.62										
CCP	360.22	MF	2.68	FT/OL	2.98	WC	0.15	SVA	0.12	BF	−0.57	RRN	−2.05						
ANS/AF	−10.46	DUF	224.09	BF	9.87	SVA	8.26	WC	−3.50	CCP	137.21	RRN	29.36						
BF	230.77	DFA	0.58	WC	0.38	DUF	−10.80												
CF	−66.96	P	0.12	DUF	25.00	ICU	2.83	RRN	2.14										
MF	42.25	FT/OL	1.49	DUF	−26.17	BF	−0.72	CCP	−3.47	RRN	−3.17	WC	0.08						
W	12.35	FT/OL	7.27	DUF	−219.16	CCP	−27.42	BF	−1.74	RRN	−22.28								
DFA	−30.21	CCP	13.48	FT/OL	−3.41	DUF	88.83	WC	−0.38										
BR																			
SVA	−17.06	P	2.13	RRN	23.27														
BA																			
WC	−9.09	BF	11.22	ANS/AF	55.99														
WB																			

The final results show the general gains and losses for the discontinuous urban fabric from annual crops and coniferous forest (positive values), generating continuous urban fabric, industrial and commercial units and green urban areas (negative values). The conversion areas involving agriculture with natural spaces and agro-forestry areas display substantial losses to discontinuous urban fabric, complex cultivation patterns, and mixed forest.

3.3. Model for Land Use Change and Artificialization Processes

Expressing the land use change dynamics in the study area, Table 7 presents the aggregation of five classes of land use, scaling up the classes determined over time the process of artificialization. The results point, for each period, to the gain of artificial areas with the percentage that remains artificial and the one added from other previous land use. This expresses soil sealing, vegetation cover change, and antrophic actions with soil consummation for urban fabric, industrial or commercial units, road and rail networks, mineral extraction areas, dump sites, construction sites, green urban areas and sport and leisure facilities. As can be observed in Figure 6 and Table 7, the biggest transformations happened from the agricultural and shrub vegetation areas. The forest conversion to artificial area is successive, with a major rate in the period 1995–2010.

When analyzing the evolution for the center zone and the three rings it is noted that the artificialization of agricultural areas decreases from the center to the periphery, the decrease from 1985 in center area and on ring 1 being especially evident. The artificialization of forest areas occurred rapidly since 1974 in the center area, verifying that there is a progressive use on the different rings, but more noticeably from 1995 onwards.

The use of shrub areas and sparse vegetation for artificialization processes has a tendency of gradual use over time, more evident in the center than at the periphery, but also expressing temporal and spatial peak usage.

This process of artificialization in general involves a sequence of land use change and occupation from: (1) agricultural areas to forest; (2) forest to shrub or sparsely vegetation areas; (3) shrub or sparse vegetation areas to artificial areas, is recognized as common dynamic in the peri-urbanization processes.

The results also emphasize the changes to artificialized areas with an urban fabric and infrastructures associated, namely during the periods of 1985–1995 and 1995–2011.

In order to realize the major changes in the land use and occupation and the relation with the spatial variation from the central area or in the periphery, the partial stability grade (SG) was calculated for the 4 sections considered, in the five different periods of analysis.

Looking to the stability grade study into the four sectors, the dynamics of land use also reveal distinctive frames, as we can see in Figure 8, which shows the results of the stability grade of each sector in the different periods of time analyzed.

Table 7. Transformations in land use and occupation in the study area (%).

	1958–1974	1974–1985	1985–1995	1995–2011	1958–2011
Study area			**Artificial Areas (%)**		
Artificial areas	78.55	81.27	85.31	89.77	90.91
Agricultural areas	6.72	13.07	18.44	19.91	36.91
Forest areas	2.62	6.61	6.94	13.57	26.12
Shrub vegetation areas	5.48	11.55	17.35	25.46	36.51
Sparsely vegetated areas	2.94	2.04	8.00	8.15	15.29
Center			**Artificial Areas (%)**		
Artificial areas	91.18	94.08	98.51	97.07	98.55
Agricultural areas	28.08	29.31	66.27	56.52	89.80
Forest areas	8.70	30.43	64.29	57.14	82.61
Shrub vegetation areas	0.00	0.00	100.00	100.00	0.00
Sparsely vegetated areas	0.00	0.00	0.00	0.00	0.00
Ring 1			**Artificial Areas (%)**		
Artificial areas	72.84	80.31	85.58	89.76	86.08
Agricultural areas	7.09	13.25	21.08	25.31	46.10
Forest areas	13.33	10.85	13.33	21.31	45.03
Shrub vegetation areas	9.09	15.22	31.48	45.76	50.00
Sparsely vegetated areas	33.33	33.33	26.67	46.15	40.00
Ring 2			**Artificial Areas (%)**		
Artificial areas	70.91	75.89	84.47	89.66	87.72
Agricultural areas	5.06	13.83	14.64	19.02	32.30
Forest areas	1.19	9.96	7.01	16.31	33.13
Shrub vegetation areas	13.66	15.85	23.58	27.27	55.43
Sparsely vegetated areas	6.25	50.00	40.00	17.74	47.06
Ring 3			**Artificial Areas (%)**		
Artificial areas	70.42	69.90	76.62	83.28	80.28
Agricultural areas	4.01	9.57	13.99	17.14	24.88
Forest areas	1.30	3.75	5.14	9.85	15.84
Shrub vegetation areas	1.12	8.78	7.22	11.18	26.05
Sparsely vegetated areas	1.36	0.00	1.82	4.24	8.90

The Figure 8 analysis shows a similar trend in the periods 1958–1974 and 1974–1985 for stability grade, where the SG decreases in the following order: central sector, ring 3, ring 1 and ring 2. This means that larger changes occur in the intermediate sectors (ring 1 and ring 2) in relation to the proximity to the city center, but also expressing the city planning framework. This made a tight transformation process for the excluded plan areas in the fringe of the central area possible, and potentiated the peri-urbanization and rururbanization processes. In the period 1985–1995 the trend of SG is different, presenting a continuous decrease from ring 3 to the central sector, showing that the dynamic of land use is more intensive in the urban core of the study area with the influence of the Hidroprojecto Urban Plan (1971) and Macroplan Urban Plan (1983) that increased the artificialization. In the last period (1995–2011), with a municipal master plan that includes all the sectors in the planning framework, the trend is the inverse of the period 1985–1995, with a continuous decrease from the center to the ring 3. In this period the high value of the SG in the central sector (63.06%) is tangible compared with the rings 1, 2 and 3, which means that larger changes occurred in the peripheral zones of urban core. The stabilization of the urban core and consequent intensive dynamics of land use in peripheral zones are in accordance with the directives of the Master Plan approved in 1995.

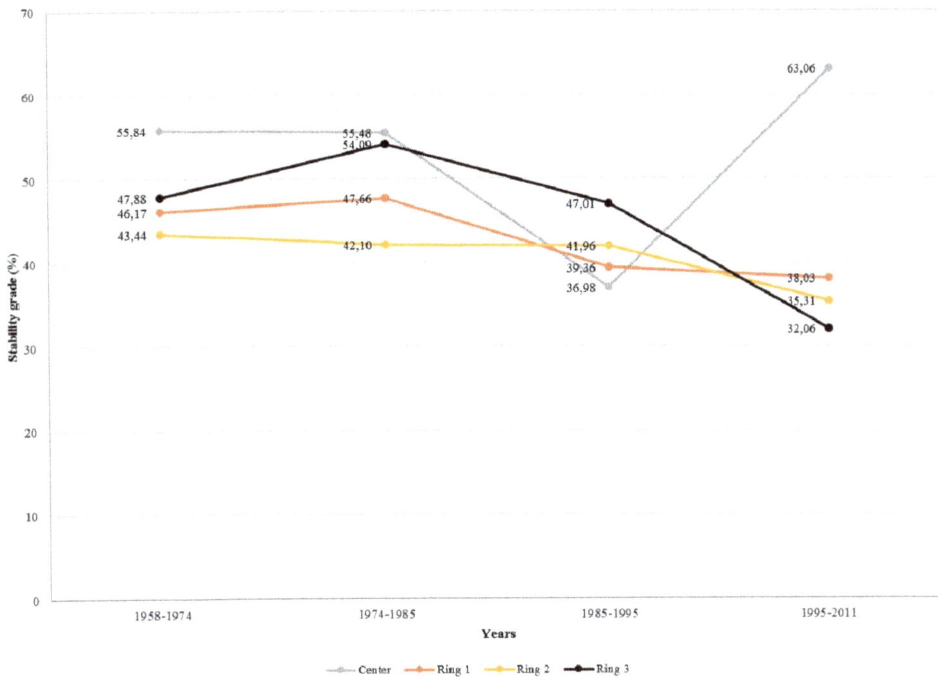

Figure 8. Stability grade for the 4 sections of the study area in the periods of analysis.

When analyzing the overall trend for the period 1958–2011, it can be observed that the periurbanization and rururbanization processes that occurred outside the boundaries of the planning framework trace the whole evolution of land use and occupation, corresponding to both ring 1 and ring 2 low stability grades (18.60% and 17.22%, respectively).

3.4. *Land Use Trajectories and Planning Framework*

The focus of this analysis is to deepen the model of land use change dynamics according to the spatial planning framework which presents different scales and spatial enforcement.

The graphic present in Figure 9 also show a relationship between the planning process and the rate of artificialization.

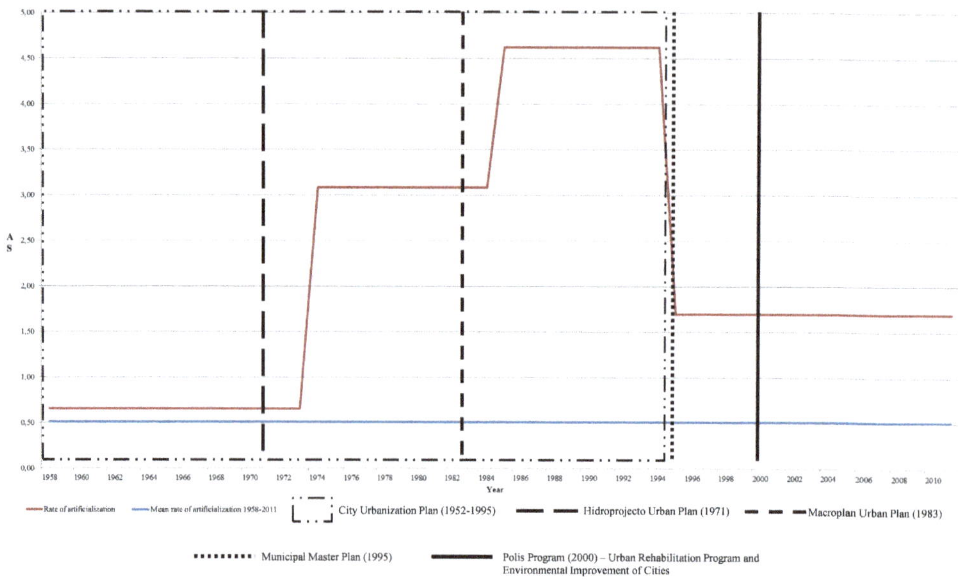

Figure 9. Annual rate of the artificialization of surfaces (AS) and the respective mean for the period 1958–2011 and the relation with the planning framework.

The results show that the informal partial Plans (Hidroprojecto Urban Plan and Macroplan Urban Plan) influence the increase of the artificialization rate, as they respond to pressures to the occupation of the territory for new residential or service areas. These plans are directly related to the peri-urbanization processes and the definition of new road infrastructures. With the approval of the Master Plan in 1995 there has been a decrease in the artificialization rate. This plan fits all land use and occupation processes for the whole county and is a regulatory plan supported by two restriction regimes (agriculture-RAN and ecological-REN), leading to a decrease in the annual rate of artificialization, despite the increase in continuous and discontinuous urban areas.

The increasing distances from the center point (city hall), as shown in Figure 4, and the different planning framework typology, objectives, and enforcement in the study area produced during time different dynamics on land use change.

A zoning model for land use change and the artificialization processes can be summarised as:

(a) In the period 1958–1985 the major transformation occurred in the fringe of the central area (ring 1 and ring 2) which are outer areas for the central urban planning enforcement. The agricultural areas are the principal source of the artificialization process, with the subsequent use of the forest areas;
(b) For the period 1985–1995 the stability grade is directly related to the distance for the city council central point, which reflects the influence of the non-formalised central urban plan, and the encircling road consolidation. This period also reflects the intensification of the artificialization processes namely using the shrub vegetation areas and sparsely vegetated areas, which represent abandoned areas or areas with reserve urban qualification.

The period 1995–2011 expresses the influence of a master plan with municipal enforcement. The degree of stability intensely downloaded to the central area now operates in reverse to the different rings. So, there is a major change towards the peripheral ring, which expresses a process of municipal urban sprawl with periurbanization and rururbanization processes. Observing the artificialization areas, the gains are mainly from the agricultural areas, forest areas, shrub vegetation areas, and sparsely vegetated areas.

These processes namely represent a deep transformation on rings 1 and 2, supported by partial plans and the road infrastructure design and construction, which often outweighs the planning processes by economic or political decision propose. Those artificialized areas present considerable gains from annual crops, arable lands, and pastures, defining those classes as one of the most important in the trajectories of land use and occupation in the study area, but associated with systematic transitions involving the forest, shrub vegetation, and sparsely vegetated areas. The casuistic transformation with land use classes resulting from forest fires, the appearance of mineral extraction areas, and waste disposal areas do not reflect the planning framework, in opposition to the modification on the stream waters margins related with legal constrains or rehabilitation plan.

4. Discussion and Conclusions

The five cartographic outputs for changes in land use and occupation underline the main decrease in annual crops and arable lands, the increase in artificial areas, namely the continuous and discontinuous urban fabric, the industrial and commercial units and the road network as well as the complex cultivation patterns and the forest routing involving different typologies.

The data representation demonstrates that urban occupation has increased gradually since 1958. The increase of artificial areas since the 70s results in the follow-up, at first, of industrial growth and afterwards, tertiary sector growth with the overall improvement of well-being [47].

The transformation of the agricultural land to urban areas is evident in this study case, as well in other contexts [5,61,62], and reflecting similar changes in Portugal [13,63]. In the area, an increase in landscape fragmentation, accompanied by increased complex cultivation patterns, translates the evolution of the rural areas [64,65] with peri-urbanization and rururbanization processes.

In general terms, there has been a continuous increase in the urban area (from 5.7% to 27.9% of the area), and the stability of the area occupied by agriculture with natural spaces and agro-forestry areas (around 6%). It was also possible to observe a decrease in the annual crops (from 27.5% to 5.9% of the area) and in the coniferous forests (from 16.1% to 5.3% of the area), especially underlined in the 1995 and 2011 images. This represents the general transition of the rural areas, with loss of agricultural areas and forest degradation, and can be clearly associated with small and medium sized city urban expansion.

The results for the study area confirm processes that can be classified as systematic transitions, contrasting with some land use changes marked by casuistic trajectories in the case of land use classes resulting from forest fires or the appearance of mineral extraction areas and waste disposal areas.

In a broad sense there has been a continuous increase in the artificial areas, with a general transition of the rural areas that can be clearly associated with the small and medium sized cities urban expansion. For this study area, but also referred to other urban contexts [13,56,66,67] a general sequence of land use change and occupation from: (1) agricultural areas to forest; (2) forest to shrub or sparsely vegetated areas; (3) shrub or sparsely vegetated areas to artificial areas, is recognized as common dynamic in the artificialization processes.

This study stressed the importance of monitoring the stability grade to identify the sequences of major changes in land use. This indicator points to a low level of stabilization, 20.61%, which reflects the intensive dynamics of the changing classes in the area. The temporal dynamics between the 26 classes identified also reveals distinct transformation phases with loss and gain conversions, justifiable by territorial planning framework [47], which is also noted in other geographical contexts [13,63,68].

Some of the systematic transitions are clearly marked by the planning process, involving partial Plans (city urban, urbanization or layout plans), Master Plans, and Regulatory Protection Regimes.

This does point out that a plan that addresses only the city encourages the land use change in the fringe, while the municipal Master Plan determines the transformation processes in all areas, revealing systematic transformations.

The results demonstrated that the transformation processes within the study area are supported by a history of urban occupation, including the discontinuous fabric and encircling road network as well as the loss of agricultural activities to the emergence of new housing residences and equipments. The results have also shown the urban planning framework as a driver on land use and occupation changes, differencing and marking the territory and its systematic transformations, according to the type, objectives and scale, as pointed by [40].

The Figure 10 summarized the two forces that act on the land use change: the territorial dynamics and the planning framework.

The zoning model approach for land use change and the artificialization processes enables the identification of three periods where the planning framework was forthright. The results show different dynamics on land use change, to the central areas and the fringe, directly related with planning scope and the scale enforcement.

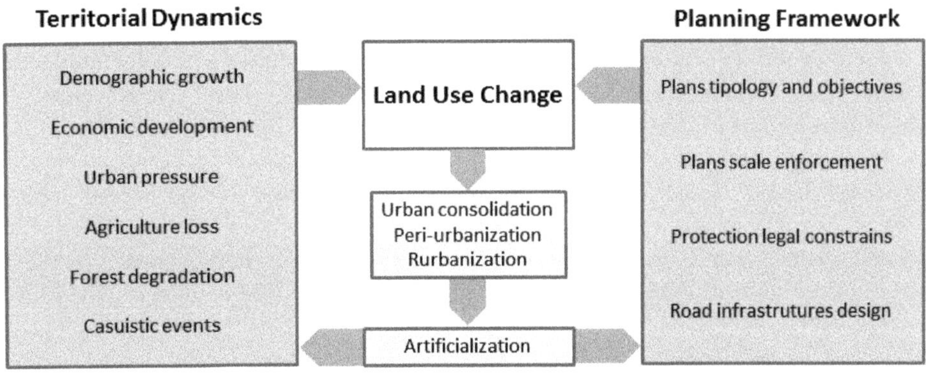

Figure 10. Forcers acting on land use and occupation changes, promoting the artificialization processes.

This conclusion highlighted the importance of the planning framework for medium/small sized cities on processes of urban sprawl and the periurbanization and rururbanization processes. The study also underlined the relevance of road infrastructure design and construction on zoning land use changes and enabled territorial dynamics with systematic transformations. According to Antrop [4] the evolution can translate an expanding changing pattern for the period 1958–1985, an exploding changing for the period 1985–1995 corresponding to a several urbanization and layout plans on the central area and the close rings, and a broad ring development for the period 1995–2011 promoted by the municipal Master Plan.

The results obtained show temporal and spatial processes of land use and occupation changes that affect the medium and small sized cities as recognized from different authors [1,3,24,69–72], and underlining the national cycles of evolution on the processes of urbanization and infrastructures construction that spread the artificialization in different ways through space [31,40,66,73,74].

These achievements support the discussion about the influence of the national planning system in relation to the spatial inequalities and as drivers on the land use change and management processes [75,76].

Land use change studies of medium/small cities and there specific patterns and processes of artificialization are needed. Planning the territory in a grounded and creative way requires a deep knowledge about the systematic and casuistic processes of land use and occupation and the different forcers, stressing the goals pointed out by different authors [33,77,78].

Giving consistent attention to historical planning framework ascendance on the land use change processes could support new strategies and regulatory regimes to manage and maintain the landscape heterogeneity and the multifunctional land use, avoiding conflicts and promoting recognition from actors.

Reshaping or rebuilding the medium/small sized cities is a constant challenge for planner and local inhabitants, balancing new functionalities and infrastructures with diversity values and ecological corridors where the land use dynamics are drivers to create new functions and offer a basis for planning and decision making.

Acknowledgments: The authors would like to thank the support provided by the FORLAND project.

Author Contributions: Mário Monteiro author designed the research and analyzed the data and wrote the paper. Alexandre Oliveira Tavares designed the research, analyzed the data and wrote the paper. All authors read and approved the final manuscript.

Conflicts of Interest: The authors declare no conflict of interest.

References

1. Rounsevell, M.D.A.; Reginster, I.; Araújo, M.B.; Carter, T.R.; Dendoncker, N.; Ewert, F.; House, J.I.; Kankaanpää, S.; Leemans, R.; Metzger, M.J.; *et al.* A coherent set of future land use change scenarios for Europe. *Agric. Ecosyst. Environ.* **2006**, *114*, 57–68.
2. Turok, I.; Mykhnenko, V. The trajectories of European cities, 1960–2005. *Cities* **2007**, *24*, 165–182.
3. Antrop, M. Changing patterns in the urbanized countryside of Western Europe. *Landsc. Ecol.* **2000**, *15*, 257–270.
4. Antrop, M. Landscape change and the urbanization process in Europe. *Landsc. Urban Plan.* **2004**, *67*, 9–26.
5. Kasanko, M.; Barredo, J.I.; Lavalle, C.; McCormick, N.; Demicheli, L.; Sagris, V.; Brezger, A. Are European cities becoming dispersed? A comparative analysis of 15 European urban areas. *Landsc. Urban Plan.* **2006**, *77*, 111–130.
6. Schwarz, N. Urban form revisited—Selecting indicators for characterising Europe cities. *Landsc. Urban Plan.* **2010**, *96*, 29–47.

7. Euroepan Comission (EC). *ESDP European Spatial Development Perspective. Towards Balanced and Sustainable Development of the Territory of the European Union*; Euroepan Comission: Luxembourg, 1999.
8. European Union. Territorial Agenda of the European Union. In Proceedings of the Agreed on the Informal Ministerial Meeting on Urban Development and Territorial Cohesion, Leipzig, Germany, 24–25 May 2007.
9. European Spatial Planning Observatory Network (ESPON). *The Role of Small and Medium-Sized Towns in Europe*; European Spatial Planning Observatory Network: Viena, Austria, 2006.
10. Fahmi, F.Z.; Hudalah, D.; Rahayu, P.; Woltjer, J. Extended urbanization in small and medium-sized cities: The case of Cirebon, Indonesia. *Habitat Int.* **2014**, *42*, 1–10.
11. Han, S.S. Urban expansion in contemporary China: What can we learn from a small town? *Land Use Policy* **2010**, *27*, 780–787.
12. Solon, J. Spatial context of urbanization: Landscape pattern and changes between 1950 and 1990 in the Warsaw metropolitan area, Poland. *Landsc. Urban Plan.* **2009**, *93*, 250–261.
13. Tavares, A.O.; Pato, R.L.; Magalhães, M.C. Spatial and temporal land use change and occupation over the last half century in a peri-urban area. *Appl. Geogr.* **2012**, *34*, 432–444.
14. Acosta, A.; Carranza, M.L.; Giancola, M. Landscape change and ecosystem classification in a municipal district of a small city (Isernia, Central Italy). *Environ. Monit. Assess.* **2005**, *108*, 323–335.
15. Shahraki, S.Z.; Sauri, D.; Serra, P.; Modugno, S.; Seifolddini, F.; Pourahmad, A. Urban sprawl pattern and land-use change detection in Yazd, Iran. *Habitat Int.* **2011**, *35*, 521–528.
16. Siciliano, G. Urbanization strategies, rural development and land use changes in China: A multiple-level integrated assessment. *Land Use Policy.* **2012**, *29*, 165–178.
17. Skokanová, H.; Havlíček, M.; Borovec, R.; Demek, J.; Eremiášová, R.; Chrudina, Z.; Mackověin, P.; Rysková, R.; Slavík, P.; Stránská, T.; *et al.* Development of land use and main land use change processes in the period 1836–2006: Case study in the Czech Republic. *J. Maps* **2012**, *8*, 88–96.
18. Tavares, A.O.; Monteiro, M.; Vargas, M.A.; Pato, R.L.; Serra, R. Land use change and forest routing in a rural context: The relevance of the community-based management and planning framework. *Appl. Geogr.* **2014**, *52*, 153–171.
19. Aspinall, R. Modelling land use change with generalized linear models—A multi-model analysis of change between 1860 and 2000 in Gallatin Valley, Montana. *J. Environ. Manag.* **2004**, *72*, 91–103.
20. Braimoh, A.K. Random and systematic land-cover transitions in northern Ghana. *Agric. Ecosyst. Environ.* **2006**, *113*, 254–263.
21. Díaz-Pacheco, J.; García-Palomares, J.C. A highly detailed land-use vector map for Madrid region based on photo-interpretation. *J. Maps* **2014**, *10*, 424–433.
22. Díaz-Palacios-Sisternes, S.; Ayuga, F.; García, A.I. A method for detecting and describing land use transformations: An examination of Madrid's southern urban-rural gradient between 1990 and 2006. *Cities* **2014**, *40*, 99–110.

23. Sönmez, N.K.; Onur, I.; Sari, M.; Maktav, D. Monitoring changes in land cover/use by CORINE methodology using aerial photographs and IKONOS satellite images: A case study for Kemer, Antalya, Turkey. *Int. J. Remote Sens.* **2009**, *30*, 1771–1778.
24. Catalan, B.; Sauri, D.; Serra, P. Urban sprawl in the mediterranean?: Patterns of growth and change in the Barcelona Metropolitan Region 1993–2000. *Landsc. Urban Plan.* **2008**, *85*, 174–184.
25. Deng, J.S.; Wang, K.; Hong, Y.; Qi, J.G. Spatio-temporal dynamics and evolution of land use change and landscape pattern in response to rapid urbanization. *Landsc. Urban Plan.* **2009**, *92*, 187–198.
26. Liu, T.; Yang, X. Monitoring land changes in an urban area using satellite imagery, GIS and landscape metrics. *Appl. Geogr.* **2015**, *56*, 42–54.
27. Tapiador, F.J.; Casanova, J.L. Land use mapping methodology using remote sensing for the regional planning directives in Segovia, Spain. *Landsc. Urban Plan.* **2003**, *62*, 103–115.
28. Henríquez, C.; Azócar, G.; Romero, H. Monitoring and modeling the urban growth of two mid-sized Chilean cities. *Habitat Int.* **2006**, *30*, 945–964.
29. Verburg, P.H.; van Berkel, D.B.; van Doorn, A.M.; van Eupen, M.; van den Heiligenberg, H.A.R.M. Trajectories of land use change in Europe: A model-based exploration of rural futures. *Landsc. Ecol.* **2010**, *25*, 217–232.
30. Wu, K.; Zhang, H. Land use dynamics, built-up land expansion patterns, and driving forces analysis of the fast-growing Hangzhou metropolitan area, eastern China (1978–2008). *Appl. Geogr.* **2012**, *34*, 137–145.
31. Diogo, V.; Koomen, E. Land Use Change in Portugal, 1990–2006: Main Processes and Underlying Factors. *Cartographica* **2012**, *47*, 237–249.
32. Li, X.; Zhou, W.; Ouyang, Z. Forty years of urban expansion in Beijing: What is the relative importance of physical, socioeconomic and neighborhood factors? *Appl. Geogr.* **2013**, *38*, 1–10.
33. Verburg, P.; Schot, P.; Dijst, M.; Veldkamp, A. Land use change modelling: Current practice and research priorities. *GeoJournal* **2004**, *61*, 309–324.
34. Hietel, E.; Waldhardt, R.; Otte, A. Linking socio-economic factors, environment and land cover in the German Highlands, 1945–1999. *J. Environ. Manag.* **2005**, *75*, 133–143.
35. Hietel, E.; Waldhardt, R.; Otte, A. Statistical modelling of land-cover changes based on key socio-economic indicators. *Ecol. Econ.* **2007**, *62*, 496–507.
36. Long, H.; Tang, G.; Li, X.; Heilig, G.K. Socio-economic driving forces of land-use change in Kunshan, the Yangtze River Delta economic area of China. *J. Environ. Manag.* **2007**, *83*, 352–364.
37. Serra, P.; Pons, X.; Saurí, D. Land-cover and land use change in a Mediterranean landscape: A spatial analysis of driving forces integrating biophysical and human factors. *Appl. Geogr.* **2008**, *28*, 189–209.
38. Long, Y.; Gub, Y.; Han, H. Spatiotemporal heterogeneity of urban planning implementation effectiveness: Evidence from five urban Master Plans of Beijing. *Landsc. Urban Plan.* **2012**, *108*, 103–111.

39. Parcerisas, L.; Marull, J.; Pino, J.; Tello, E.; Coll, F.; Basnou, C. Land use changes, landscape ecology and their socioeconomic driving forces in the Spanish Mediterranean coast (El Maresme County, 1850–2005). *Environ. Sci. Policy* **2012**, *23*, 120–132.
40. Pato, R.; Castro, P.; Tavares, A. The relevance of physical forces on land-use change and planning process. *J. Environ. Plan. Man.* **2015**.
41. Ferreira, N.; Godinho, M.M.; Neves, L.; Pereira, A.; Sequeira, A.; Castro, P.; Bento dos Santos, T. *Explanatory Notice of 17-A Sheet of the Geological Map of Portugal (1:50,000)*; National Laboratory for Energy and Geology: Lisbon, Portugal, 2010.
42. CMV. *Municipal Plan for the Defense of Forest Fire*; Câmara Municipal de Viseu: Viseu, Portugal, 2007.
43. IGP. *Forest Fire Risk Cartography—Viseu District Report*; Portuguese Geographic Institute: Lisbon, Portugal, 2004.
44. Costa, J.C.; Aguiar, C.; Capelo, J.; Lousã, M.; Neto, C. Biogeography of Continental Portugal. Available online: http://www.researchgate.net/profile/Jorge_Capelo/publication/228540962_Biogeografia_de_Portugal_continental/links/0912f50bdcbe96ec29000000.pdf (accessed on 24 August 2015).
45. INE. *2011 Census Definitive Results—Portugal*; National Institute of Statistics: Lisbon, Portugal, 2012.
46. PORDATA—Database Contemporary Portugal. Available online: http://www.pordata.pt/ (accessed on 10 October 2015).
47. Almeida, J.R. The Urban Dynamic of Viseu in the Second Half of the 20th Century. Master's Thesis, University of Porto, Porto, Portugal, 2000.
48. Dijkstra, L.; Poelman, H. *Cities in Europe: The New OECD-EC Definition*; European Commission: Brussels, Belgium, 2012.
49. Almeida, S. Anchors of Development. The Collective Equipment in Urban Design: The Case of Viseu of the Twentieth Century to the Most Recent Expansion. Master's Thesis, University of Coimbra, Coimbra, Portugal, 2015.
50. EEA. *Corine Land Cover (CLC1990) 100 m—Version 12/2000*; European Environmental Agency: Copenhagen, Denmark, 2000.
51. Bossard, M.; Feranec, J.; Otahel, J. *Corine Land Cover Technical Guide—Addendum*; Technical Report No 40; European Environmental Agency: Copenhagen, Denmark, 2000.
52. Nery, F. *CORINE Land Cover Nomenclature: Commented Portuguese Version*; Portuguese Geographic Institute: Lisbon, Portugal, 2007.
53. Piwowar, J.M. Digital image analysis. In *Remote Sensing*; Aronoff, S., Ed.; ESRI Press: Redlands, CA, USA, 2005; pp. 287–335.
54. Castanheira, J.M.; Aranha, J. Study of the change in land cover in the Alto Tâmega valley. Multi-temporal study with Landsat-5 TM e Landsat-7 ETM^+ images. In Proceedings of the VIII Meeting of Geographic Information Users, Oeiras, Portugal, 2–4 June 2004.
55. Freire, S.; Santos, T.; Tenedório, J.A. Recent urbanization and land use/land cover change in Portugal—The influence of coastline and coastal urban centers. *J. Coast. Res.* **2009**, *56*, 1499–1503.

56. Freiria, S.; Tavares, A. Towards the acknowledgment of the urban-rural interface as a spatial category. *IJEE* **2011**, *5*, 292–300.
57. Elo, S.; Kyngäs, H. The qualitative content analysis process. *J. Adv. Nurs.* **2008**, *62*, 107–115.
58. GAO. *Content Analysis: A Methodology for Structuring and Analyzing Written Material*; Program Evaluation and Methodology Division, United States General Accounting Office: Washington, DC, USA, 1996.
59. Giulio, M.; Holderegger, R.; Tobias, S. Effects of habitat and landscape fragmentation on humans and biodiversity in densely populated landscapes. *J. Environ. Manag.* **2009**, *90*, 2959–2968.
60. Pătru-Stupariu, I.; Stupariu, M.; Tudor, C.; Grădinaru, S.; Gavrilidis, A.; Kienast, F.; Hersperger, A. Landscape fragmentation in Romania's Southern Carpathians: Testing a European assessment with local data. *Landsc. Urban Plan.* **2015**, *143*, 1–8.
61. Russo, P.; Tomaselli, G.; Pappalardo, G. Marginal periurban agricultural areas: A support method for landscape planning. *Land Use Policy* **2014**, *41*, 97–109.
62. Su, S.; Jiang, Z.; Zhang, Q.; Zhang, Y. Transformations of agricultural landscapes under rapid urbanization: A threat to sustainability in Hang-Jia-Hu region, China. *Appl. Geogr.* **2011**, *31*, 439–499.
63. Jones, N.; de Graaff, J.; Rodrigo, I.; Duarte, F. Historical review of land use changes in Portugal (before and after EU integration in 1986) and their implications for land degradation and conservation, with a focus on Centro and Alentejo regions. *Appl. Geogr.* **2011**, *31*, 1036–1048.
64. Bodesmo, N.; Pacicco, L.; Romano, B.; Ranfa, A. The role of environmental and socio-demographic indicators in the analysis of land use changes in a protected area of the Natura 2000 Network: The case study of Lake Trasimeno, Umbria, Central Italy. *Environ. Monit. Assess.* **2012**, *184*, 831–843.
65. Hasse, J.E.; Lathrop, R.G. Land resource impact indicators of urban sprawl. *Appl. Geogr.* **2003**, *23*, 159–175.
66. Araya, Y.; Cabral, P. Analysis and modeling of urban land cover change in Setubal and Sesimbra, Portugal. *Remote Sens.* **2010**, *2*, 1549–1563.
67. De Noronha Vaz, E.; Nijkamp, P.; Painho, M.; Caetano, M. A multi-scenario forecast of urban change: A study on urban growth in the Algarve. *Landsc. Urban Plan.* **2012**, *104*, 201–211.
68. Gong, J.; Chen, W.; Liu, Y.; Wang, J. The intensity change of urban development land: Implications for the city master plan of Guangzhou, China. *Land Use Policy* **2014**, *40*, 91–100.
69. Aguilera, F.; Valenzuela, L.; Botequilha-Leitão, A. Landscape metrics in the analysis of urban land use patterns: A case study in a Spanish metropolitan area. *Landsc. Urban Plan.* **2011**, *99*, 226–238.
70. Romano, B.; Zullo, F. Models of urban land use in Europe: Assessment tools and criticalities. *Int. J. Agric. Environ. Inform. Syst.* **2013**, *4*, 80–97.

71. Roose, A.; Kull, A.; Gauk, M.; Tali, T. Land use policy shocks in the post-communist urban fringe: A case study of Estonia. *Land Use Policy* **2013**, *30*, 76–83.
72. Salvati, L.; Sateriano, A.; Bajocco, S. To grow or to sprawl? Land cover relationships in a Mediterranean City Region and implications for land use management. *Cities* **2013**, *30*, 113–121.
73. Caetano, M.; Araújo, A.; Nunes, A.; Nunes, V.; Pereira, M. *Accuracy Assessment of the CORINE Land Cover 2006 Map of Continental Portugal. Technical Report*; Geographic Portuguese Institute: Lisbon, Portugal, 2009.
74. Petrov, L.; Lavalle, C.; Kasanko, M. Urban land use scenarios for a tourist region in Europe: Applying the MOLAND model to Algarve, Portugal. *Landsc. Urban Plan.* **2009**, *92*, 10–23.
75. Cardoso, R.; Breda-Vázquez, I. Social justice as a guide to planning theory and practice: Analyzing the Portuguese planning system. *Int. J. Urban Reg. Res.* **2007**, *31*, 384–400.
76. Pires, A.R. The fragile foundations of European spatial planning in Portugal. *Eur. Plan. Stud.* **2005**, *13*, 237–252.
77. Claessens, L.; Schoorl, J.; Verburg, P.; Geraedts, L.; Veldkamp, A. Modelling interactions and feedback mechanisms between land use change and landscape processes. *Agric. Ecosyst. Environ.* **2009**, *129*, 157–170.
78. Turcu, C. Re-thinking sustainability indicators: Local perspectives of urban sustainability. *J. Environ. Plan. Manag.* **2013**, *56*, 695–719.

Section II:
New Dimensions of Urban Land Use and Urban Space

The Making of a Sustainable Wireless City? Mapping Public Wi-Fi Access in Shanghai

Mingfeng Wang, Felix Haifeng Liao, Juan Lin, Li Huang, Chengcheng Gu and Yehua Dennis Wei

Abstract: In the context of the global information economy, ready access to the Internet is critical to a city's competitiveness, which has prompted a number of cities to launch plans to establish wireless networks. Most literature on the development of wireless cities focuses on cities in Western countries, and few have discussed how Chinese cities have adopted wireless technologies in their urban infrastructure development efforts. This paper examines recent development and spatial distribution of public Wi-Fi access in Shanghai, a leading business hub in China. We mapped Wi-Fi hotspots through the government sponsored "i-Shanghai" project and China Mobile Communications Corporation (CMCC). We find that while telecommunication providers have been proactively deploying WLAN (wireless local area network, a proxy of public Wi-Fi or wireless access) hotspots in Shanghai, neither government sponsored WLAN hotspots nor facilities established by CMCC could cover the old traditional neighborhoods in the central city and sub-districts in remote rural areas. We also address the development of a more sustainable wireless city in Shanghai with a particular focus on digital divide and social equity issues.

Reprinted from *Sustainability*. Cite as: Wang, M.; Liao, F.H.; Lin, J.; Huang, L.; Gu, C.; Wei, Y.D. The Making of a Sustainable Wireless City? Mapping Public Wi-Fi Access in Shanghai. *Sustainability* **2016**, *8*, 111.

1. Introduction

Wireless communication technologies have emerged as a major force underlying the recent development and change of the global economy [1]. Under such notions as *digital cities* [2], *intelligent cities* [3], *mobile cities* [4], *wireless cities* [5], *ubiquitous cities* [6], and *smart cities* [7,8], new planning strategies that emphasize the adoption and adaptation of information and communication technologies (ICTs) have attracted considerable attention from scholars and policy makers. In the context of the global information economy, the usage of wireless technologies has become a key indicator of the competitiveness of a city [5,9–11]. A number of cities, such as Singapore and Taipei in Asia, Philadelphia, San Francisco and Boston in the United States, and Perth in Australia, have either expressed an intention to establish a wireless network or launched specific plans to develop wireless cities [5,12,13]. While most literature on wireless cities and the digital divide focuses on cities in Western countries, few have discussed how Chinese cities have adopted wireless technologies in infrastructure development, and even fewer have addressed the issue of digital divide in China's urban landscapes.

By the end of 2014, China's mobile Internet users had reached 649 million, as compared to 22.5 million in 2000, accounting for 19% of the total number in the world [14], among which the number of smart phone users increased to 557 million by the end of 2014 [15]. Internet development, especially a dramatic increase in the number of smart phone users, has resulted in a booming demand for wireless access in Chinese cities [16]. In collaboration with major telecommunication operators, including China Mobile Communications Corporation (CMCC), China Unicom, and China Telecom, local governments in Chinese cities have launched plans for public Wi-Fi network access and the development of "wireless" cities. For example, since 2012, the Shanghai Municipal Government has launched a project named "i-Shanghai" (or *loving Shanghai*). The project aims to deploy over 4000 Wi-Fi hotspots by 2020, making Shanghai a large-scale wireless city.

This paper examines the development of public Wi-Fi access in Shanghai and compares the WLAN hotspots established by CMCC and those sponsored by the Shanghai Municipal Government through the "i-Shanghai" project. Notably, the total number of Internet users in Shanghai has increased by 500% during 2000–2014, among which the share of Internet users through smart phone or other mobile devices has increased from 64% in 2009 to 79% in 2013 [17,18]. We analyze the spatial distribution of Wi-Fi access at multiple scales (*i.e.*, district and sub-district levels) and in different areas (e.g., central city and suburban areas). By applying exploratory spatial data analysis (ESDA) methods, such as concentric analysis and spatial hot spot analysis, we underscore the proactive role played by telecommunication companies in shaping the Wi-Fi geographies in Chinese cities. The remainder of the paper is organized as follows: the next section briefly reviews the concept of wireless city and related literature on wireless city development and the digital divide. This is followed by the spatial distribution analysis of public Wi-Fi access in Shanghai provided by the "i-Shanghai" project and CMCC. The last section summarizes the major findings and discusses the characteristics of Shanghai's pathway towards a sustainable wireless city.

2. Background and Literature Review

Wireless cities are essentially cities completely covered by high-speed broadband and public Wi-Fi access, where the Internet can be openly accessed and used by their citizens. Wireless technologies are an important part of infrastructure development through which the efficiency and equity of governmental service can be improved [5]. Wireless network coverage is also a key step to extend fiber broadband to the public, and the access to Wi-Fi, and more broadly the Internet, is regarded as "the city's fifth major infrastructure" next to water, electricity, gas, and roads [19–21]. The impact of wireless city development is significant in many aspects, such as providing mobility values [22], changing travel behaviors [23], altering the perception of community and forming social networks in urban space [24–26]. Previous literature found that in wireless cities, transaction costs can be reduced [27] and social well-being of citizens is better off through closer social interactions [22,28,29].

Given the benefits of developing wireless networks or ready access to the Internet, a great number of cities are deploying or have plans to establish wireless broadband networks [30]. Previous literature has also documented a number of factors that determine the locations of public Wi-Fi access mostly in Western cities. Customer choice is regarded as a basic driving force behind the development of the wireless broadband market, and the layout of wireless facilities is obviously influenced by local social and economic factors [31]. Oyana [32] studied the distribution of wireless facilities in Southern Illinois, USA and identified three main factors that influence the distribution of wireless facilities: the higher educational population rate, age-specific group, and average family income. Driskella and Wang [33] examined the spatial layout of Wi-Fi access in Louisiana, USA, suggesting that the determinants of public Wi-Fi hotspot location are residents' socio-economic disadvantages in neighborhoods and their household characteristics. Grubesic and Murray [31] analyzed the distribution of Wi-Fi hotspots in four neighborhood communities in Cincinnati, USA, and reported that the degree of network access in each community was closely related to social, economic, demographic, and spatial factors, but evident differences exist between rich and poor blocks' Wi-Fi access; the network access in the inner city is concentrated in areas around commercial and office buildings.

Urban planning and government policies also play an important role in the making of a wireless city [34]. For example, through a comparative study of Singapore and Taipei, Hu *et al.* [5] suggested that the development of the wireless city in Taipei is mainly driven by top-down government policy while such development depends more on the market-driven participation of customers in Singapore. The development of wireless city is also driven by the innovation and widespread use of smartphone technologies. In a case study of Salt Lake City, USA, Torrens [35] identified that Wi-Fi access has provided a possible solution to "last-mile" problems in the city, and Evans-Cowley [36] also addressed the influences of mobile phones on urban life.

Given the nature of Internet access as a public utility, the role of government in providing Wi-Fi infrastructure is of particular concern in the literature [30,37,38]. Ballon *et al.* [30] reviewed the public-private partnership in shaping wireless city networks in both EU and the U.S. and found that public authorities such as governments will strive for the optimal trade-off between minimizing their inputs for directly being involved in the establishment and operation of wireless city networks and maximizing the leverage for the purpose of specific policy goals such as narrowing the digital divide and so on. Girth [37] explored the variation in approaches and examined the structural factors that give rise to public-private partnerships. Some literature also have examined the role of the Asian government, China especially, in telecommunication industry development and infrastructure construction. For instance, Soh and Yu [39] analyzed the development of 3G networks in China and explained how regulatory impacts from government and domestic and foreign network operators are interdependent to each other.

Despite the positive impacts of the wireless city development, the spatial and social inequalities of information technology and its access has also drawn a lot of attention [40,41], derived from the traditional thoughts of urban sociology [42], with a particular focus on the digital divide. Moreover, this type of restructuring is often associated with a new round of social differentiation in different urban spaces [31,43]. From this perspective, the term digital divide, which refers to the fact that certain segments of the population and specific social groups may be excluded from access to the Internet, has gained considerable scholarly attention [5,7,38,44,45]. As Castells [46] pointed out, the gap between network haves and have-nots increases sources of spatial and social inequality. Combined with the historical socioeconomic divides in the city, the strong commercial bias of Internet access may strengthen an uneven geography of public Wi-Fi access [47], which can be manifested at different scales, from global to regional ones, and to neighborhoods and communities [43,48–50]. Popularized by Gray Andrew Pole, a New York Times journalist, the notion of digital divide has called for more attention in regards to equitable access to public Wi-Fi and the uneven distribution of access to the Internet within cities [35,48]. For example, Prieger [50] investigated the gaps in broadband usage for minorities and found that fewer fixed broadband options were available to Blacks and Hispanics, but they tend to have more mobile broadband providers available. These digital disparities are closely associated with social polarization and inequalities, imposing new challenges for the sustainability of families, communities, and cities in the context of the new information economy [51].

In short, despite an emerging body of literature on wireless cities and digital divide in Western countries, Chinese cities have largely escaped from scholarly attention and few work has been done to address potential digital divide in the course of the Chinese wireless city development [52]. Specifically, to the best of our knowledge, most of the previous literature on the digital divide or Internet development in China has focused on disparities or spatial distribution at the provincial, city or regional levels [53,54], the development of wireless city and digital divide at the intra-city level have rarely been researched.

3. Data and Study Region

The data used in this study was mainly obtained from the official websites of Shanghai CMCC [55] and the "i-Shanghai" project [56]. The raw data contains the attributes of each WLAN hotspot established by CMCC and the "i-Shanghai" project, including the name, address, type, located sub-district, and also the same attributes of Wi-Fi hotspots. WLAN is one of the most widely used broadband wireless technologies that provides public Wi-Fi access in China. By July 2013, the number of WLAN hotspots provided by Shanghai CMCC had increased to 6800, as compared to less than 703 in 2009. In this study, the locations of all the CMCC and "i-Shanghai" WLAN hotspots in 2013 were geocoded. In addition, as shown in Figure 1, our study area included 18 districts and counties: Huangpu, Jingan, Xuhui, Changning, Putuo, Zhabei, Hongku, Yangpu, Pudong, Minhang, Baoshan, Songjiang, Jiading, Qingpu, Jinshan, Fengxian, Nanhui and Chongming, and 208 sub-districts/towns, which are the smallest administrative unit in China. Following Wei *et al.* [57], we divided the 18 districts and 208 sub-districts into four areas: traditional city property area (TCPA, which is composed of Huangpu and Jingan), expanded central city area (ECCA, which includes Xuhui, Changning, Putuo, Zhabei, Hongkou and Yangpu), inner-suburban area (ISA, which consists of Pudong, Minhang and Boshan), and outer suburban area (OSA, which includes Chongming, Jiading, Songjiang, Qingpu, Jinshan, Fengxian and Nanhui).

As the largest business hub and financial center in China, Shanghai has played a leading role in developing wireless cities in China. As illustrated in Figure 2, the Internet penetration, as measured by the total number of Internet users divided by total population, increased from less than 30% before 2005 to over 70% in 2013. The number of Internet users through smartphones in Shanghai also increased from 7.47 million to 12.6 million during 2009–2013, and its shares of Internet users rose from 63.8% to 78.5% [17,58]. On par with Beijing, Shanghai is currently one of the two Chinese cities with the highest rates of Internet penetration. In fact, as early as the 1990s, Internet development was one of the items that were prioritized in the agenda of Shanghai municipal government. The theme of the 2010 Shanghai World Expo, "Better City, Better Life", reiterated the importance of developing Shanghai into a wireless city. In 2011, the Shanghai Municipal Government issued "The Shanghai's Promotion Plan for the Development of Smart City (2011–2013)", which focused on the construction of wireless city. In collaboration with major corporations in the Chinese telecommunication industry, especially CMCC, one of the goals in the plan was to build a wireless broadband network that covers over 80% of the city, with the speed of Internet of 20 Mbps.

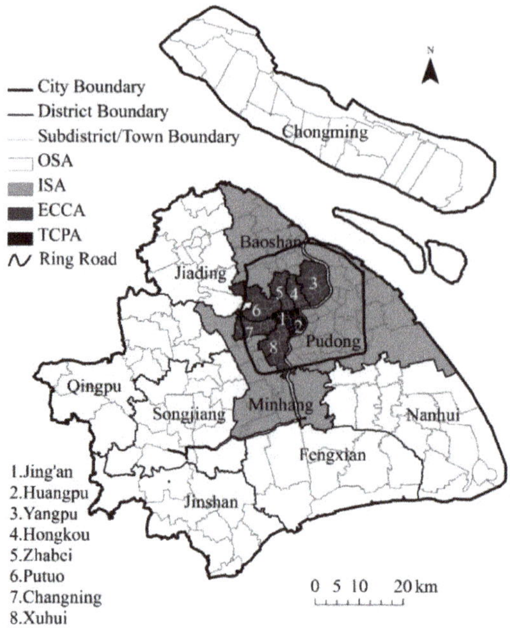

Figure 1. Administrative division and geographic structure of Shanghai. Note: TCPA = traditional city proper area; ECCA = expanded central city area; ISA = inner-suburban area; OSA = outer-suburban area.

Figure 2. Internet penetration in Shanghai, as compared to China and other centrally administrated municipalities; Source: China Internet Network Information Center, 2004–2013.

4. Mapping Public Wi-Fi Access in Shanghai

This section comprehensively analyzes the geographies of Wi-Fi access in Shanghai, with an emphasis on those hotspots provided by the Shanghai Municipality Government through the "i-Shanghai" project and those established by the CMCC. CMCC has contributed significantly to the wireless city development in Shanghai. In 2013, CMCC had established more than 6000 hotspots, and the majority of these Wi-Fi hotspots are not free to access. The "i-Shanghai" project was launched in 2012, and by the end of 2013, 450 WLAN hotspots were deployed to provide free Wi-Fi access to the public. As these WLAN hotspots are free to access, a lot of benefits have been brought to those who could not afford Internet access, although the free access could only last for 2 hours.

Figure 3 describes different locations of CMCC and "i-Shanghai" WLAN hotspots in Shanghai, indicated by the type of these WLAN hotspots. CMCC WLAN hotspots are more likely to be located in commercial districts and office buildings, which accounted for 26% and 31% of all CMCC hotspots, respectively, in 2013. In comparison with CMCC hotspots, the majority of hotspots provided by the "i-Shanghai" project were placed in those areas associated with governmental and public services, especially places of interest, parks, exhibition and sports centers, and hospitals (Figure 3).

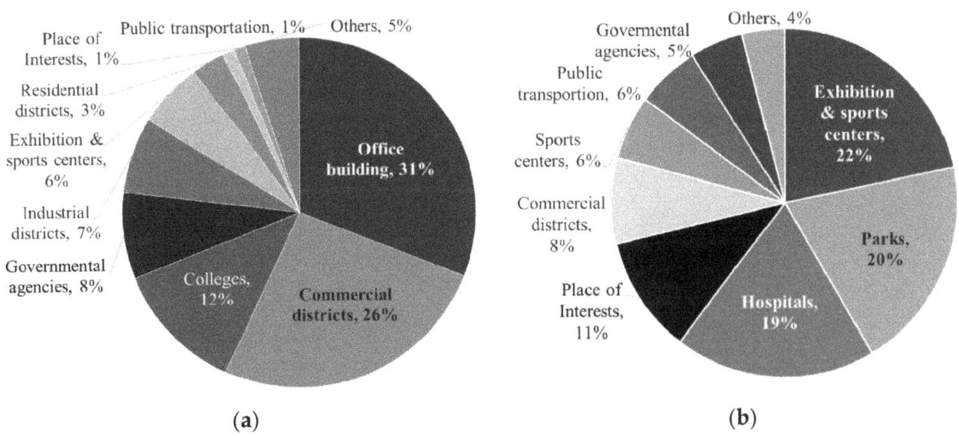

Figure 3. Types of CMCC and "i-Shanghai" WLAN hotspots, 2013. Source: CMCC and "i-Shanghai". (**a**) CMCC; (**b**) "i-Shanghai".

Geographically, both CMCC and "i-Shanghai" hotspots were concentrated in the areas within the outer Ring Road in Shanghai or 15 kilometer to the people's square or the city center (Figure 1). In general, "i-Shanghai" Wi-Fi hotspots were more likely to locate close to the city center (Figure 4b), and the CMCC Wi-Fi hotspots spread out more extensively to suburban districts. It should be noted that, in 2013, only eight out of 206 sub-districts in Shanghai had no CMCC WLAN hotspots and most of these sub-districts/towns were located in outer suburbs like Chongming Island, Jinshan, and Nanhui (Figure 4a). Regardless of CMCC or "i-Shanghai" WLAN hotspots, a few sub-districts/towns with a very small number of WLAN hotspots, say less than five, were located in the traditional city area of Shanghai characterized by geographical concentrations of work force communities and those neighborhoods in remote rural areas (Figure 4).

(a) (b)

Figure 4. Spatial Distribution of CMCC and "i-Shanghai" WLAN hotspots, 2013. (**a**) CMCC; (**b**) "i-Shanghai".

Table 1 presents the top ten sub-districts/towns with the largest number of CMCC and "i-Shanghai" WLAN hotspots. Some interesting findings emerge: first, in comparison with the list of "i-Shanghai" hotspots, more sub-districts in the suburban areas were found as the top ranking sub-district/towns with the largest number of CMCC WLAN hotspots. Seven out of the top-ten ranking sub-districts were in suburban areas, such as Fangsong in Songjiang district and Jiangchuan Road in Minhang district. Second, the total number of CMCC WLAN hotspots in most of the top-ranking sub districts/towns were installed around a number of office buildings and higher education and R & D institutes, such as Xujiahui in Xuhui district and Zhangjiang in Pudong district. By contrast, the top two sub-districts with the largest number of the "i-Shanghai" hotspots, *i.e.*, Lujiazui and Huapu, were located in the inner-suburban area (ISA). Most of these places are important places of interest and locations where governmental buildings are concentrated. Overall, sub-districts with agglomerations of higher education institutes and R & D centers tend to have more CMCC WLAN hotspots, but the "i-Shanghai" hotspots tend to agglomerate in those landmarks in Shanghai, such as the Bund, Huangpu, and Lujiazui financial district, *etc.* (Table 1).

Table 1. Ranking of China Mobile Communications Corporation (CMCC) and "i-Shanghai" WLAN (wireless local area network) hotspots in Shanghai's sub-districts/towns in 2013 (Top 10).

CMCC				"i-Shanghai"			
Rank	**Subdistrict**	**Location**	**N**	**Rank**	**Subdistrict**	**Location**	**N**
1	Fangsong	OSA	395	1	Lujiazui	ISA	16
2	Nanqiao	OSA	171	2	Huangpu	ISA	15
3	Huamu	ECCA	158	3	The Bund	TCPA	11
4	Xiayang	OSA	131	4	Chuansha	ISA	10
5	Zhangjiang	ISA	124	5	Nanqiao	OSA	9
6	Jiangchuan	ISA	123	-	Huaihai Zhong Road	TCPA	9
7	The Bund	TCPA	120	7	Jiangsu Road	ECCA	8
8	Xujiahui	ECCA	113	-	Ou-yang Road	ECCA	8
9	Huinan	ISA	102	-	Youyi Road	ISA	8
10	Zhelin	OSA	100	10	East Nanjing Road	TCPA	7
					Fenglin Road	ECCA	7
					Tianping Road	ECCA	7
					Zhoujiadu	ISA	7

TCPA = traditional city proper area; ECCA = expanded central city area; ISA = inner-suburban area; OSA = outer-suburban area.

In order to further explore the Wi-Fi geographies in Shanghai represented by the spatial distribution of CMCC and "i-Shanghai" WLAN hotspots, concentric analysis based on a 5-km radius interval and ESDA methods were used, following Wang and Gu [52]. Results of concentric analysis demonstrate that the share of CMCC WLAN hotspots within 5 kilometers from the city center (Figure 5), which is defined as the location of People's Square, was less than 20%, while its counterpart under the "i-Shanghai" project was over 35%. There were also nearly 40% of the CMCC hotspots located 20 kilometers or further from the center (*i.e.*, outer suburbs), whereas the share of "i-Shanghai" hotspots was just above 25% in the same region (Figure 5).

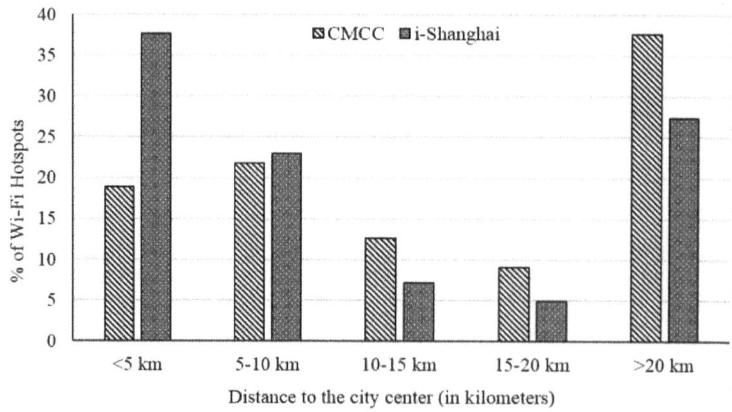

Figure 5. Spatial distribution of "i-Shanghai" and CMCC hotspots based on concentric analysis.

Spatial hot spot analysis (Getis-Ord Gi*), which measures the spatial association of a variable and identifies the characteristics of patterns [59], was used to further detect the clusters of WLAN hotspots at the sub-district/town level. This method detects the statistically significant spatial clusters of high values, or sub-districts/towns with large number of hotspots in this study, and identifies them as "hot spots", and the statistically significant spatial clusters of low values, or sub-districts/towns with little number of hotspots, and identifies them as "cold spots". The Getis-Ord Gi* statistic is widely used in analyzing the distribution of geographically referenced data. The Gi* of each unit *i* could be calculated as:

$$G_i^* = \frac{\sum_{j=1}^{n} w_{i,j}x_j - \overline{X}\sum_{j=1}^{n} w_{i,j}}{S\sqrt{\frac{\left[n\sum_{j=1}^{n} w_{i,j}^2 - \left(\sum_{j=1}^{n} w_{i,j}\right)^2\right]}{n-1}}}; \quad (1)$$

where x_j is the number of hotspots in administrative unit j, $w_{i,j}$ is the spatial weight matrix, n is the number of administrative units, and S is the standard deviation of the observations. The results of hot spot analysis is essentially a z-score for each spatial unit, measuring the spatial clustering of high/low values. A statistically significant (e.g., $p < 0.05$) positive z-score implies the presence of a hot spot. By contrast, a statistically significant (e.g., $p < 0.05$) negative z-score implies the presence of a cold spot [59,60].

As shown in Figure 6b, a typical "core-periphery" structure of WLAN hotspots was found with respect to "i-Shanghai" Wi-Fi hotspots. The hot spots within the urban area (inside of the Shanghai Outer Ring Expressway) constituted the core area, indicating the clustering of "i-Shanghai" WLAN hotspots. These sub-districts/towns are mostly important commercial districts in Shanghai, such as Xujiahui, the Bund, and Lujiazui financial district. In contrast to hot spots, the cold spots were located in remote suburban sub-districts/towns, favoring less developed districts, such as Chongming Island. In comparison with the spatial pattern of "i-Shanghai" hotspots, the spatial distribution of CMCC Wi-Fi hotspots displayed a bi-center distribution. The downtown area (within five kilometers from the People's Square) remained as the core area, and what is more interesting is that another core area has emerged in the suburbs to the southwestern part of Shanghai. The new core area is where new development zones and new higher education institutes are located, including Songjiang New City, Qingpu New city, Jiangchuan subdistrict, and Nanqiao New Town. A lot of sub-districts are characterized by geographical concentrations of higher education institutes such as Songjiang University town in Songjiang New Town, Haiwan University town in Tuolin town, and Shanghai Jiao Tong University and Shanghai University of Electric Power in Jiangchuan sub-district.

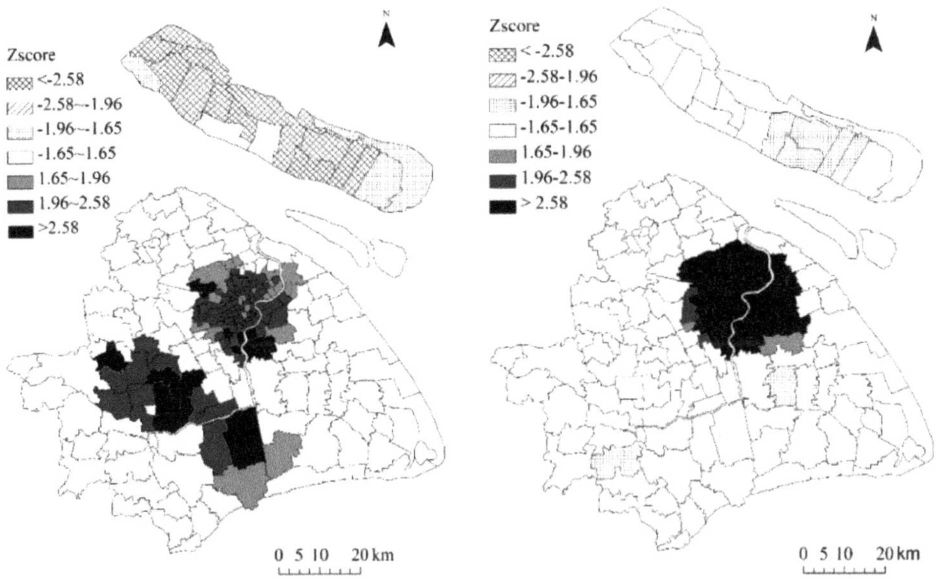

(**a**) (**b**)

Figure 6. Hotspot analysis of CMCC and "i-Shanghai" hotspots in Shanghai, 2013. (**a**) CMCC; (**b**) "i-Shanghai".

In comparison with the spatial layout of "i-Shanghai" hotspots, CMCC tends to be more proactive in deploying WLAN hotspots. As shown in Figure 6a, the bi-center distribution of hot spots of CMCC Wi-Fi hotspots is greatly driven by a rapid process of suburbanization in Shanghai in recent years [61,62]. In this process, large-scale industrial enterprises extend to suburban areas first, followed by the growth of population and new housing development [63]. In addition, given the skyrocketing land price and restricted land supply in the downtown area, many business institutions and government agencies have recently moved to the suburbs in Shanghai. As a result, the rapid population growth and increased demand for Wi-Fi access have prompted CMCC to develop new cyber infrastructure like WLAN hotspots in these sub-districts (Figure 6a).

5. Micro-Scale Analysis of Three Sub-Districts

The mapping above has revealed the spatial distribution of WLAN hotspots or public Wi-Fi access in Shanghai. Contrasting Wi-Fi geographies shaped by CMCC and "i-Shanghai" WLAN hotspots were found, which highlight the proactive role of telecommunication operators in Shanghai's wireless city development. The findings motivate the concern regarding how key stakeholders, such as government, telecommunication operators, and communities, can play a role in Shanghai's wireless city development. The following section will focus on three specific sub-districts, including Nanqiao New Town, Songjiang University Town, and Pengpu Xincun, given their unique features regarding the wireless development at the community level (Figure 7).

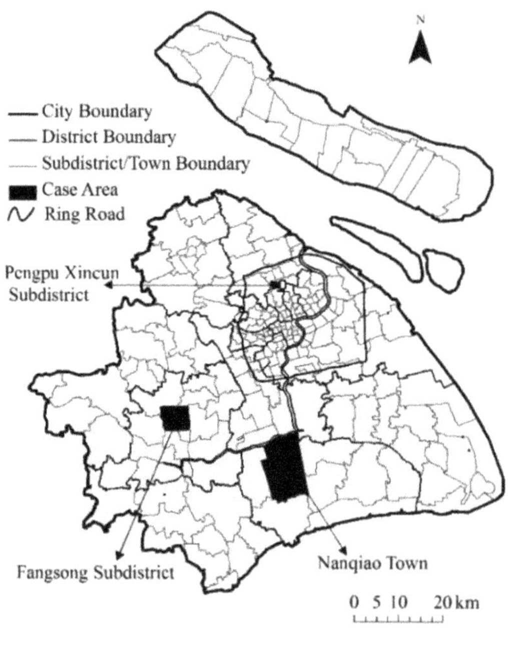

Figure 7. Locations of case studies.

5.1. Nanqiao New Town

Nanqiao is one of the top-ten ranking sub-districts that has experienced a rapid wireless development driven by the deployment of both CMCC and "i-Shanghai" WLAN hotspots (Table 1). In 2013, the number of CMCC hotspots increased to approximately 170, ranking as the second largest one in Shanghai. As shown in Figure 8a, there have been five clusters of CMCC hotspots: 1. Fengpu Commercial Street, especially those within the Fengpu Mansion, which is a commercial office building; 2. Bailian shopping mall; 3. the historical community (old town); 4. Government of the Fengxian District; 5. Nanqiao Township Government. In contrast to the spatial layout of CMCC WLAN hotspots, the distribution of public Wi-Fi access provided by "i-Shanghai" was centered on government agencies, including the Fengxian District Conference Center and Nanqiao Township Government, and they were also located around large-scale businesses companies or the Shopping malls, namely RT-MART and LOTUS (Figure 8b).

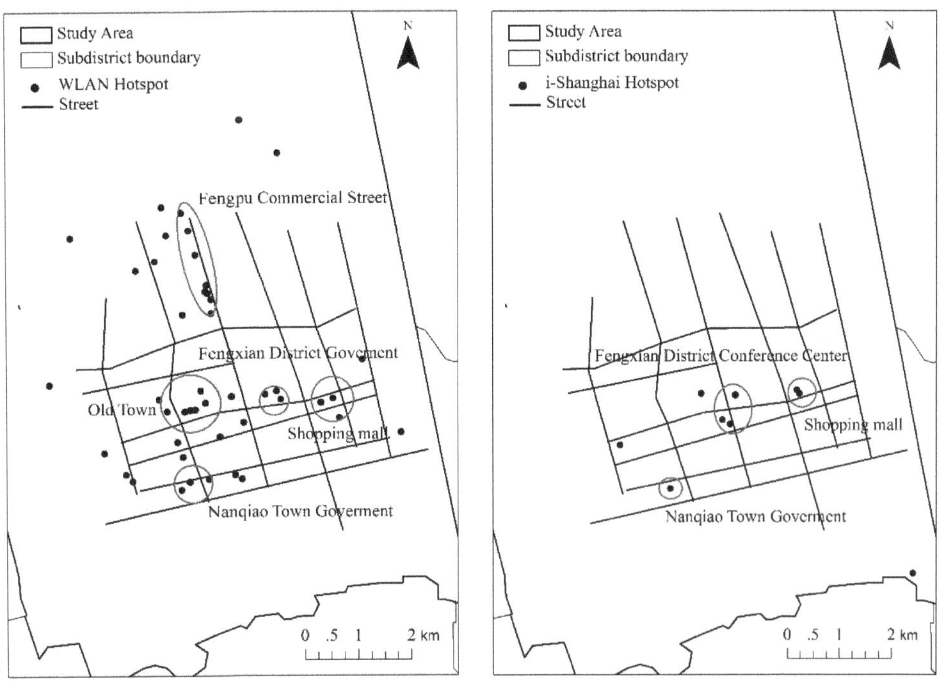

(**a**) (**b**)

Figure 8. Spatial distribution of CMCC and "i-Shanghai" hotspots in Nanqiao New Town, 2013. (**a**) CMCC; (**b**) "i-Shanghai".

In the case of Nanqiao, the government policy and urban planning have played an important role in the expansion of public wireless access. Nanqiao New Town is the political, economic, and cultural center in Fengxian district of Shanghai, and it is also planned to be a core area providing comprehensive services for the north coast of Hangzhou Bay. As one of the three major new towns in Shanghai, the development of Nanqiao was emphasized in Shanghai's 12th Five-Year Plan [64]. Based on its development plan, new ICTs or wireless technologies were the key instrument to integrate various functions within this sub-district. These government investments aimed to provide wireless access especially in the eastern part of the Nanqiao New Town (Figure 8b). It was also designed to provide favorable development conditions for enterprises and high quality living for residents, and to promote a more competitive local business environment, with great potential demand for Internet access.

5.2. Songjiang University Town

Fangsong sub-district in Songjiang is another typical case in terms of wireless development. In comparison with Nanqiao New Town, Fangsong sub-district has achieved a rapid growth of CMCC WLAN hotspots. There have been 395 CMCC hotspots in this area whereas the number of "i-Shanghai" hotspots was four at the end of 2013. These hotspots have provided public Wi-Fi access to thousands of people especially college students. Different from Nanqiao, the wireless development was heavily driven by the establishment of Songjiang University Town. Among the 395 CMCC WLAN hotspots in 2013, about 320 were located within Songjiang University Town. In this sub-district, there are a large number of college students and faculty in the seven higher education institutions, including Shanghai International Studies University, University of International Business and Economics, Shanghai Lixin University of Commerce, Donghua University, East China University of Political Science and Law, and Shanghai University of Engineering Science. The case of Songjiang University Town marks a more proactive role of telecommunication company, *i.e.*, CMCC, in recent wireless development in Shanghai, by particularly tracing the new demand from college students in university towns [65].

5.3. Pengpu Xincun Sub-District

Pengpu Xincun sub-district serves as a case where specific communities have been overlooked in the course of wireless city development in Shanghai. As shown in Figure 7, Pengpu is located in the downtown area. However, there has been very limited public Wi-Fi access either through CMCC WLAN hotspots or the Wi-Fi access provided by the "i-Shanghai" project. In 2013, only five CMCC WLAN hotspots, as compared to over one hundred CMCC WLAN hotspots in other nearby sub-districts in the downtown of Shanghai. Similarly, there was even no "i-Shanghai" WLAN hotspot located in this sub-district. The lagging development of wireless access in Pengpu has its own historical background. The dwelling houses in the sub-district were built for low-paid workers in previous state owned enterprises. Like other shanty towns in Asian cities [66], most buildings in Pengpu were characterized by old and small living spaces, even though the residential population in Pengpu was relatively large, about 160,000 residents in 2013. The development of Pengpu community has slowed down in recent years because of the limited space, and the unfavorable socio-economic conditions in the community have resulted in the backward wireless development in these areas. In short, the case of Pengpu reflects that even though the WLAN hotspots have been widely built up due to the efforts made by either telecommunication companies or the government, there have been many new "blind spots" in Shanghai's communities where the gap between information haves and have-nots is still evident [67].

6. Policy Implications

Our mapping of public Wi-Fi access in Shanghai has important policy implications in developing wireless cities and addressing the digital divide issue in the context of China. First, as shown in this study, the planning in China is more focused on physical dimensions and deemphasizes issues related to social equity [68], as illustrated by the digital disparities at the community and neighborhood levels in this study. More efforts should be made to address the social dimension of wireless city development in Shanghai and other Chinese cities. Second, Chinese cities are transitional cities. Local governments and market forces are collectively shaping the Wi-Fi geographies especially through the establishment of new development zones and towns in their suburban areas [69]. Hence, how to develop a mutually beneficial collaborative relationship between the government and market participants should be underscored in the future infrastructure planning. Third, although the study has been focused on the spatial distribution of public Wi-Fi access in Shanghai, the development of wireless cities should address not only the construction of hardware facilities, but also the education and supporting system that improves access to the Internet. Therefore, not only the quantity of wireless access, but also the quality of public access to the digital world, deserves attention from policy makers.

7. Conclusions

Wireless technologies play an increasingly critical element in the Chinese urban and infrastructure developments. Local governments and major network operators have increasingly embraced the notion of wireless city, as the access to Internet through public Wi-Fi has become an integral part of a city's economic competitiveness [70,71]. This research traces the recent wireless city development in Shanghai using data gathered from both the government and the telecommunication providers and several interesting findings are uncovered. First, a substantial amount of public Wi-Fi facilities has been deployed and built up, shaping Wi-Fi geographies in Shanghai at multiple spatial scales [69], and a mix of market participants and governmental agencies has been involved in the process of wireless city development in Shanghai. Second, by mapping public Wi-Fi at both district and sub-district levels, we find that both local governments and major telecommunication companies, such as CMCC, have significantly contributed to the widespread Wi-Fi access to the Internet in Shanghai. Results also suggest that telecommunication companies, such as CMCC, have played a more proactive role in the deployment of WLAN hotspots. These findings are similar to the recent wireless development in other cities in Asia, such as Singapore and Taipei [5], while the results also pinpoint the unique characteristics of the Chinese government in the pursuit of wireless city development under rapid urban growth and economic transition. As shown in our micro-scale analysis, Naoqiao New Town, as a major new business community in Shanghai, has fueled the new wireless development by having both governmental agencies and large-size business corporations involved. Songjiang University Town presents a typical case that telecommunication operators, such as CMCC, have played a proactive role in shaping Wi-Fi geographies. In contrast, Pengpu Xingchu has exhibited a sluggish wireless development despite its advantageous location in the central area of Shanghai. Third, from the perspective of social inequalities and sustainability, despite the rapid expansion of both CMCC and "i-Shanghai" WLAN hotspots, the digital divide is still evident especially at the community level. Deployment of WLAN hotspots is obviously biased towards business centers and university campuses as well as their surrounding areas in Shanghai. Public Wi-Fi access was rarely found in poor residential areas in the inner city and remote rural areas of Shanghai. This polarized structure indicates that the problem of digital divide is far from being solved while new digital divide has been created, imposing profound challenges for making Shanghai a sustainable wireless city in the future. Finally, although we have mapped the public Wi-Fi access in Shanghai, studies of digital divide are also promising from a user or customer's perspective, especially when this type of data becomes available. In addition, as China is a country characterized by its huge size and regional differentials, more research is needed to compare the development of wireless cities in different geographical regions or across urban hierarchy in China [72].

Acknowledgments: The authors would like to acknowledge the funding of the National Natural Science Foundation of China (41371175, 41329001), and the Ford Foundation (0155-0883). This study was also supported by the Key Project of Key Research Institute of Humanities and Social Sciences at Universities (13JJD840010) funded by the Ministry of Education of China. Felix Liao would like to acknowledge partial support from the Idaho EPSCoR (Experimental Program to Stimulate Competitive Research) Program and by the National Science Foundation under award number IIA-1301792.

Author Contributions: Mingfeng Wang, Felix Liao, Juan Lin, and Chengcheng Gu collected and analyzed the data. Felix Liao, Mingfeng Wang, and Li Huang wrote the manuscript. Dennis Wei edited and revised the paper. All authors read and approved the manuscript.

Conflicts of Interest: The authors declare no conflict of interest.

References

1. Castells, M.; Fernandez-Ardevol, M.; Qiu, J.L.; Sey, A. *Mobile Communication and Society: A Global Perspective*; Mit Press: Cambridge, MS, USA, 2009.
2. Ishida, T.; Isbister, K. *Digital Cities: Technologies, Experiences, and Future Perspectives*; Springer Science & Business Media: Berlin, Germany, 2000.
3. Komninos, N. *Intelligent Cities and Globalisation of Innovation Networks*; Routledge: London, UK, 2008; Volume 37, pp. 1–307.
4. Sheller, M.; Urry, J. *Mobile Technologies of the City*; Routledge: London, UK, 2006.
5. Hu, M.C. Antecedents for the adoption of new technology in emerging wireless cities: Comparisons between singapore and taipei. *Reg. Stud.* **2014**, *48*, 665–679.
6. Shin, D.-H. Ubiquitous city: Urban technologies, urban infrastructure and urban informatics. *J. Inf. Sci.* **2009**, *35*, 515–526.
7. Caragliu, A.; del Bo, C.; Nijkamp, P. Smart cities in europe. *J. Urban Technol.* **2011**, *18*, 65–82.
8. Gil-Castineira, F.; Costa-Montenegro, E.; Gonzalez-Castano, F.J.; Lopez-Bravo, C.; Ojala, T.; Bose, R. Experiences inside the ubiquitous oulu smart city. *Computer* **2011**, *44*, 48–55.
9. Cieslik, A.; Kaniewsk, M. Telecommunications infrastructure and regional economic development: The case of poland. *Reg. Stud.* **2004**, *38*, 713–725.
10. Pearson, I. The role of future ict in city development. *Foresight J. Futur. Stud. Strateg. Think. Policy* **2006**, *8*, 3–16.
11. Alfirevic, N.; Pavicic, J.; Znidar, K. *Knowledge and Information Technology and Urban (Regional) Development*; World Scientific and Engineering Acad and Soc: Athens, Greece, 2009; pp. 112–116.
12. Gunasekaran, V.; Harmantzis, F.C. Emerging wireless technologies for developing countries. *Technol. Soc.* **2007**, *29*, 23–42.
13. Gaspar, J.; Glaeser, E.L. Information technology and the future of cities. *J. Urban Econ.* **1998**, *43*, 136–156.
14. CNN. China's Online Users more than Double Entire U.S. Population. Available online: http://www. cnn.com/2015/02/03/world/china-internet-growth-2014/ (accessed on 2 October 2015).
15. China Internet Network Information Center. *Statistical Report on Internet Development in China*; CNNIC: Beijing, China, 2014.
16. CNNIC. *The 33rd Statistical Report on Internet Development in China*; China Internet Network Information Center: Beijing, China, 2014.
17. CNNIC. *Development Reports of Shanghai Internet, 2012*; China Internet Network Information Center: Shanghai, China, 2012.

18. CNNIC. *Development Reports of Shanghai Internet, 2009*; China Internet Network Information Center: Shanghai, China, 2009.
19. Piro, G.; Cianci, I.; Grieco, L.A.; Boggia, G.; Camarda, P. Information centric services in smart cities. *J. Syst. Softw.* **2014**, *88*, 169–188.
20. Steenbruggen, J.; Tranos, E.; Nijkamp, P. Data from mobile phone operators: A tool for smarter cities? *Telecommun. Policy* **2015**, *39*, 335–346.
21. Qi, L.; Zeng, J.; Zhang, Y. Study on the construction and operation in wireless city. In *Electronic Information and Electrical Engineering*; Chen, B.J., Hu, V., Kong, D., Eds.; Atlantis Press: Paris, France, 2012; Volume 19, pp. 70–74.
22. Le, A.N.H.; Wang, M.C.H.; Cheng, J.M.S.; Huang, S.C.F. The mobility value of internet services in a wireless city: The case of taipei, taiwan. *Asian J. Technol. Innov.* **2012**, *20*, 51–65.
23. Berry, M.; Hamilton, M. Changing urban spaces: Mobile phones on trains. *Mobilities* **2010**, *5*, 111–129.
24. Wilken, R. A community of strangers? Mobile media, art, tactility and urban encounters with the other. *Mobilities* **2010**, *5*, 449–468.
25. Rettie, R. Mobile phones as network capital: Facilitating connections. *Mobilities* **2008**, *3*, 291–311.
26. Choi, J.H.-J. The city is connections: Seoul as an urban network. *Multimedia Syst.* **2010**, *16*, 75–84.
27. Sandvig, C. An initial assessment of cooperative action in Wi-Fi networking. *Telecommun. Policy* **2004**, *28*, 579–602.
28. Sedgley, N.; Elmslie, B. Do we still need cities? Evidence on rates of innovation from count data models of metropolitan statistical area patents. *Am. J. Econ. Sociol.* **2011**, *70*, 86–108.
29. Helgason, O.; Kouyoumdjieva, S.T.; Karlsson, G. Does mobility matter? In Proceedings of the Wons 2010: Seventh International Conference on Wireless on-Demand Network Systems and Services, Kranjska Gora, Slovenia, 3–5 February 2010; pp. 9–16.
30. Ballon, P.; van Audenhove, L.; Poel, M.; Staelens, T. Business models for wireless city networks in the eu and the us: Public inputs and public leverage. In *Telecommunication Markets: Drivers and Impediments*; Curwen, P., Preissl, B., Haucap, J., Eds.; Springer: New York, NY, USA, 2009; pp. 325–340.
31. Grubesic, T.; Murray, A. "Where" matters: Location and Wi-Fi access. *J. Urban Technol.* **2004**, *11*, 1–28.
32. Oyana, T.J. Exploring geographic disparities in broadband access and use in rural southern illinois: Who's being left behind? *Gov. Inf. Q.* **2011**, *28*, 252–261.
33. Driskell, L.; Wang, F. Mapping digital divide in neighborhoods: Wi-Fi access in baton rouge, louisiana. *Ann. GIS* **2009**, *15*, 35–46.
34. Ford, G.S.; Koutsky, T.M. Broadband and economic development: A municipal case study from florida. *Rev. Urban Reg. Dev. Stud.* **2005**, *17*, 216–229.
35. Torrens, P.M. Wi-Fi geographies. *Ann. Assoc. Am. Geogr.* **2008**, *98*, 59–84.

36. Evans-Cowley, J. Planning in the real-time city: The future of mobile technology. *J. Plan. Lit.* **2010**, *25*, 136–149.
37. Girth, A.M. What drives the partnership decision? Examining structural factors influencing public-private partnerships for municipal wireless broadband. *Int. Public Manag. J.* **2014**, *17*, 344–364.
38. Park, N.; Lee, K.M. Wireless cities: Local governments' involvement in the shaping of Wi-Fi networks. *J. Broadcast. Electron. Media* **2010**, *54*, 425–442.
39. Soh, P.-H.; Yu, J. Institutional environment and complementary assets: Business strategy in china's 3g development. *Asia Pac. J. Manag.* **2010**, *27*, 647–675.
40. Castells, M. European cities, the informational society, and the global economy. *Tijdschr. Voor Econ. En Soc. Geogr.* **1993**, *84*, 247–257.
41. Cartier, C.; Castells, M.; Qiu, J.L.C. The information have-less: Inequality, mobility, and translocal networks in chinese cities. *Stud. Comp. Int. Dev.* **2005**, *40*, 9–34.
42. Harvey, D. Social-justice, postmodernism and the city. *Int. J. Urban Reg. Res.* **1992**, *16*, 588–601.
43. Grubesic, T.H.; Murray, A.T. Waiting for broadband: Local competition and the spatial distribution of advanced telecommunication services in the united states. *Growth Chang.* **2004**, *35*, 139–165.
44. Kellerman, A. Mobile broadband services and the availability of instant access to cyberspace. *Environ. Plan. A* **2010**, *42*, 2990–3005.
45. Perrons, D. Understanding social and spatial divisions in the new economy: New media clusters and the digital divide. *Econ. Geogr.* **2004**, *80*, 45–61.
46. Castells, M. *The Internet Galaxy. Reflections on the Internet Business and Society*; Oxford University Press: Oxford, UK, 2001.
47. Fuentes-Bautista, M.; Inagaki, N. Reconfiguring public internet access in austin, tx: Wi-fi's promise and broadband divides. *Gov. Inf. Q.* **2006**, *23*, 404–434.
48. Sahiner, T.; Ozbakir, A.; Kurt, G.K. Intra-city digital divide measurements through clustering. In Proceedings of 2013 ITU Kaleidoscope: Building Sustainable Communities (K-2013), Kyoto, Japan, 22–24 April 2013; pp. 233–238.
49. Strover, S.; Mun, S.-H. Wireless broadband, communities, and the shape of things to come. *Gov. Inf. Q.* **2006**, *23*, 348–358.
50. Prieger, J.E. The broadband digital divide and the benefits of mobile broadband for minorities. *J. Econ. Inequal.* **2015**, *13*, 373–400.
51. Myeong, S.; Kwon, Y.; Seo, H. Sustainable e-governance: The relationship among trust, digital divide, and e-government. *Sustainability* **2014**, *6*, 6049.
52. Mingfeng, W.; Chengcheng, G. Building a smart city: A spatial analysis and review of wlan hotspots in metropolitan shanghai. *Prog. Geogr.* **2015**, *34*, 438–447.
53. Li, R.; Shiu, A. Internet diffusion in china: A dynamic panel data analysis. *Telecommun. Policy* **2012**, *36*, 872–887.
54. Feng, G.C. Factors affecting internet diffusion in china: A multivariate time series analysis. *Telemat. Inf.* **2015**, *32*, 681–693.
55. Shanghai CMCC. Available online: www.10086.cn/sh/ (accessed on 18 January 2016).

56. The "i-Shanghai" Project. Available online: http://i-shanghai.sheitc.gov.cn (accessed on 18 January 2016).
57. Wei, Y.D.; Bi, X.; Wang, M.; Ning, Y. Globalization, economic restructuring, and locational trajectories of software firms in shanghai. *Prof. Geogr.* **2015**, 1–16.
58. Liu, G.Z.; Zhou, Q. An empirical analysis on industrial structural transition and regional inequality in china. In Proceedings of the 2009 International Conference on Management Science and Engineering, Moscow, Russia, 14–16 September 2009; Lan, H., Ed.; IEEE: New York, NY, USA, 2009; pp. 1044–1052.
59. Getis, A.; Ord, J.K. The analysis of spatial association by use of distance statistics. *Geogr. Anal.* **1992**, *24*, 189–206.
60. Ord, J.K.; Getis, A. Local spatial autocorrelation statistics: Distributional issues and an application. *Geogr. Anal.* **1995**, *27*, 286–306.
61. Shen, J.; Wu, F. Moving to the suburbs: Demand-side driving forces of suburban growth in china. *Environ. Plan. A* **2013**, *45*, 1823–1844.
62. Walcott, S.M.; Pannell, C.W. Metropolitan spatial dynamics: Shanghai. *Habitat Int.* **2006**, *30*, 199–211.
63. He, D.; Ning, Y. Evolution of population structure and spatial distribution in shanghai since 2000. In *Urban Development Challenges, Risks and Resilience in Asian Mega Cities*; Springer: Berlin, Germany, 2015; pp. 227–249.
64. Chen, L.; Song, S.; Wang, L. Low-Carbon City and Complex Community: From Planning Ideas to Planning Practice—A Case Study of Shanghai Nanqiao New City. *Urban Plan. Forum* **2011**, *4*, 1–8.
65. Wang, R. Planning and design of wireless city based on wlan. *Telecommun. Sci.* **2011**, *6*, 21–27. (In Chinese)
66. Li, Z.; Wu, F. Socio-spatial differentiation and residential inequalities in shanghai: A case study of three neighbourhoods. *Hous. Stud.* **2006**, *21*, 695–717.
67. Cheng, H.; Wang, M. Analysis on factors affecting floating population's using of internet at home: Insights from a survey of urban villages in shanghai. *China Urban Stud.* **2012**, *5*, 81–93. (In Chinese)
68. Wei, Y.D. Planning chinese cities: The limits of transitional institutions. *Urban Geogr.* **2005**, *26*, 200–221.
69. Wei, Y.D. Zone fever, project fever: Development policy, economic transition, and urban expansion in china. *Geogr. Rev.* **2015**, *105*, 156–177.
70. Macagnano, E.V. Intelligent urban environments: Towards e-inclusion of the disabled and the aged in the design of a sustainable city of the future. A south african example. In *Sustainable City V: Urban Regeneration and Sustainability*; Gospodini, A., Brebbia, C.A., Eds.; Wit Press/Computational Mechanics Publications: Southampton, UK, 2009; Volume 117, pp. 537–547.
71. Roper, S.; Grimes, S. Wireless valley, silicon wadi and digital island - helsinki, tel aviv and dublin and the ict global production network. *Geoforum* **2005**, *36*, 297–313.
72. Mack, E.A. Variations in the broadband-business connection across the urban hierarchy. *Growth Chang.* **2015**, *46*, 400–423.

Monocentric or Polycentric? The Urban Spatial Structure of Employment in Beijing

Daquan Huang, Zhen Liu and Xingshuo Zhao

Abstract: The spatial structure of Beijing has changed dramatically since the reforms of the late 1970s. It is not clear, however, whether these changes have been sufficient to transform the city's monocentric spatial structure into a polycentric one. This paper uses 2010 enterprise registered data to investigate the spatial distribution of employment in Beijing. Using a customized grid to increase the spatial resolution of our results, we explore the city's employment density distribution and investigate potential employment subcenters. This leads to several findings. First, Beijing still has strong monocentric characteristics; second, the city has a very large employment center rather than a small central business district; third, five subcenters are identified, including four in the suburbs; and fourth, a polycentric model that includes these subcenters possesses more explanatory power than a simple monocentric model, but by only four percentage. We conclude that the spatial structure of Beijing is still quite monocentric, but may be in transition to a polycentric pattern.

Reprinted from *Sustainability*. Cite as: Huang, D.; Liu, Z.; Zhao, X. Monocentric or Polycentric? The Urban Spatial Structure of Employment in Beijing. *Sustainability* **2015**, *7*, 11632–11656.

1. Introduction

China is urbanizing at an unprecedented rate, with many cities experiencing rapid population expansion [1]. Beijing, the country's capital, has been one of its fastest-growing cities over the past few decades [2–4]. According to census data, Beijing's population grew from 13.6 million in 2000 to more than 19.6 million in 2010, an increase of 44.5%. The aggregation of population in the city center has caused many problems, including traffic congestion [5,6], air pollution [7–9] and a lack of affordable housing [10,11].

These problems are often blamed on the capital's spatial structure, which critics argue is overly monocentric and should be decentralized. In fact, Beijing has been making efforts to establish a polycentric urban spatial structure by implementing a series of urban plans and related policies, targeting the decentralization of both employment and population [12,13]. Since the economic reforms of the late 1970s, Beijing has issued three comprehensive city plans, with planning periods of 1981–2000, 1991–2010 and 2004–2020 [14,15]. Each of these plans has included the formation of a multi-center development pattern among its goals. For example, the 1991–2010 Beijing Comprehensive Plan outlined a polycentric development pattern that it called "decentralized cluster style (*Fensan Zutuan Shi*)". In this proposal, the urban area would consist of the existing old city and ten edge clusters, each surrounded by a greenbelt (the first greenbelt) that would control urban sprawl and improve environmental quality. With the development of urban land and housing markets since the early 1990s, however, Beijing's urban area expanded quickly, encroaching on the greenbelt. By the early 2000s, the greenbelt had all been developed, and the old city merged completely with the edge clusters [16]. The 2004–2020 Beijing Comprehensive Plan followed its predecessor's principle of decentralized development, proposing a three-tier development structure composed of the central city, the new towns and the common towns. The central city referred to the old city and the ten edge clusters from the 1991–2010 plan; the new towns included eleven towns where suburban district and county governments were located [17]; and the common towns referred to other designated towns. Between the central city and the new towns, a second greenbelt was planned with the same purpose as the first.

The city's spatial structure and these efforts to control it have attracted broad interest from researchers and policy makers. Many studies have focused on the spatial distribution of Beijing's residential population, concluding that, in general, the city has shown a trend toward polycentricity and suburbanization [18–23]. However, just because a city's residential population has suburbanized does not necessarily mean that its urban spatial structure has also changed to a polycentric form, especially if employment is still concentrated in the city center. The spatial distribution of employment plays an important role in the structuring of urban spaces. Indeed, some urban economic theories begin from the spatial distribution of employment and determine population distribution from distance to the employment center, transport costs and other factors [24–27]. Thus, a polycentric city is usually defined by the presence of one or more employment subcenters outside of the central business district [28].

Unfortunately, few studies have examined the spatial distribution of employment in Chinese cities, probably due to the difficulty of data collection. China has conducted six population censuses, in 1953, 1964, 1982, 1990, 2000 and 2010, and thanks to the country's *hukou* (household registration) system, annual population data are available for most cities. In contrast, the country has conducted only two economic surveys of limited scope, in 2004 and 2008. What is more, data collected in these surveys are of doubtful quality due to the lack of unified classification standards among administrative departments at different levels [29].

These few studies have yielded no consensus. Some researchers suggest that Beijing is still dominated by a monocentric urban spatial structure [30,31], while others conclude that Beijing has already entered a polycentric era [31,32]. For example, Sun *et al.* [32] use employment data from the secondary and tertiary industries to identify five employment subcenters: *Yangfangdian* Street, *Zhongguancun* Street and *Shangdi* Street in *Haidian* District, *Heping* Street in *Chaoyang* District and *Yingfeng* Street in *Fangshan* District. Due to the limitations of existing employment data, these studies were mainly based at the town level, which leads to relatively coarse results [33].

This paper seeks to deepen the understanding of Beijing's urban spatial structure by using a better set of employment data: the 2010 enterprise registered data from the Beijing Industry and Commerce Bureau. We customize a 1.5 km $\times$ 1.5 km grid as the research unit and follow the two-stage method proposed by McMillen [28] to identify employment subcenters. Finally, we establish a polycentric model to explain the employment density distribution. Our findings will supplement the current understanding of the city's spatial structure and lay the foundation for better urban development, planning and policy-making in the future.

The paper is organized as follows: Section 2 describes the research area, data and method, while Section 3 provides a general description of employment density in Beijing in 2010. Section 4 evaluates a monocentric model, identifies Beijing's employment subcenters and examines the subcenters' ability to explain the entire employment density distribution of the city. Conclusions are presented in Section 5.

2. Research Area, Data and Methods

2.1. Research Area

Beijing has a total administrative area of approximately 16,410 km^2. The city's northwest is a primarily mountainous area of about 10,072 km^2 and is protected for ecological reasons, while the southeast is made up of plains of about 6338 km^2 (Figure 1). The main urban area is located on this plain, with the Ming and Qing Dynasty Imperial Palace (the Forbidden City) at its center. Since 1949, a system of ring roads has been developed around the Forbidden City, with the city's expansion requiring the construction of the 2nd, 3rd, 4th, 5th and 6th Ring Roads in turn [34]. With Tiananmen as the center, the average radius of these five ring roads is about 4 km, 7 km, 10 km, 15 km and 25 km, respectively. Eight radial expressways have also been constructed to link the central city and suburban areas (Figure 2). The city's subway system is a similar ring-and-radial network; Subway Lines 2 and 10 are loop lines, while Lines 13 and 4, as well as the *Changping, Fangshan, Yizhuang, Shunyi, Batong* and Airport lines are all radial lines.

We define a research area of 4429 km^2, which includes the area within the 6th Ring Road and a 10-km buffer zone beyond it (Figure 2). We selected this research area based on several considerations. First, in view of the subway system's extent and assumptions about the commuting range of private cars, this area represents a reasonable commuting scope and can be regarded as an integrated labor market.

Indeed, according to bus-pass data, the majority of transportation flow occurs within this area [35]. Second, the research area includes not only the central city, but also seven new towns, which have closely integrated with it [36] (Figure 3). Third, this area accommodates most of Beijing's population and firms, accounting for about 80.9% and 74.5%, respectively. Fourth, if the research area were expanded further, many mountainous regions with weak links to the central city would be included.

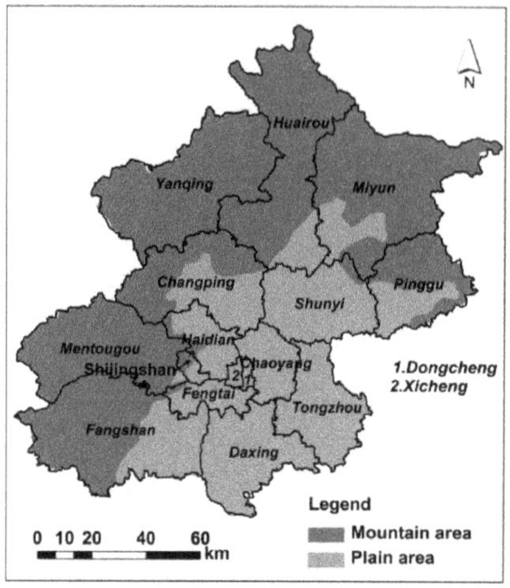

Figure 1. Topographic map of Beijing.

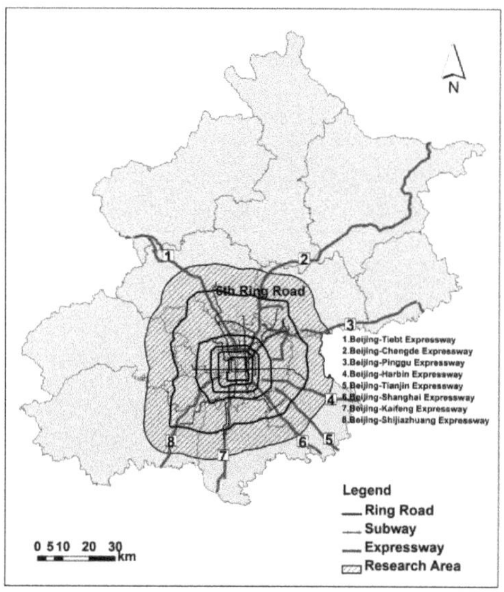

Figure 2. Research area of the study.

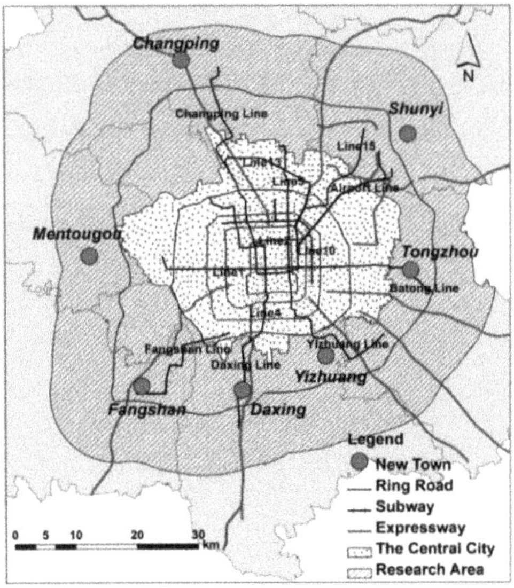

Figure 3. The Central city and planned new towns.

2.2. Data

The employment data are from the Beijing Industry and Commerce Bureau's 2010 list of registered enterprises and institutions. The registered enterprises include state-owned, private, collective, foreign-invested, foreign-owned and other types of enterprise, while the institutions include government agencies, institutions and social organizations. Collected information includes enterprise and institutional names, addresses, turnover, registered capital, the number of their employees, and so on. We use ArcGIS and Google Maps to locate the enterprise address, giving us the spatial location of all 411,300 enterprises [37].

The Beijing Industry and Commerce Bureau's dataset includes 14.91 million employees, the sectoral composition of which is in general accordance with the published data of the *Beijing Statistical Yearbook* [38] (Table 1). This dataset covers a wide range of industries; business services, technological exchange and promotion services and wholesale trade are the top three by employment, accounting for 6.3%, 5.9% and 5.8% of the total employment, respectively (Table 2).

One shortcoming of our data is that the employment figures for some large enterprises, such as the China National Tobacco Corporation, the China Network Communications Group Corporation and Sony (China) Limited, include employees working not only in Beijing, but also elsewhere in China. If these companies are included in the study, they might bias the spatial distribution of employment in the research area. Therefore, we eliminated 560 companies of this kind from the dataset, representing 0.14% of the total number of enterprises in the research area [39]. Due to their relatively small numbers and lack of spatial concentration, excluding these enterprises has little effect on our results. Because of these removals, the dataset used in this study contains 410,800 enterprises and 13.51 million employees.

Table 1. Composition of employment by sector.

Data Source	Primary Industry	Secondary Industry	Tertiary Industry
Beijing Statistical Yearbook	6.0%	19.5%	74.4%
Beijing Industry and Commerce Bureau	1.6%	22.3%	76.1%

Table 2. Top ten industries by number of employees.

Top Ten Industries	Employment	Percentage of Total Employment
Business services	942,957	6.3%
Technological exchange and promotion services	878,809	5.9%
Wholesale trade	857,681	5.8%
Education	575,360	3.9%
Construction	545,589	3.7%
Retail Trade	527,912	3.5%
Real estate	458,434	3.1%
Computer service industry	348,745	2.3%
Catering trade	340,799	2.3%
Health	315,535	2.1%
Top ten total	5,791,821	38.8%

2.3. Methods

2.3.1. Defining the Research Unit

The research area includes 195 towns (townships or streets), with an average administrative area of about 22 km^2. Towns are, therefore, too large to make an effective research unit by the standards established in the literature. For example, McMillen *et al.* [40] use 0.5 square miles (about 1.29 km^2) of a statistical sample as their research unit; McDonald *et al.* [41] use 1 square mile (about 2.56 km^2); and Giuliano *et al.* [42] use an average of 3 square miles as their research unit. In light of these scales, we defined a customized grid as our research unit.

Choosing the appropriate grid size is critical and can affect the calculated employment density. If the grid is too small, there will be many units with zero employment density, such as parks or residential blocks, and many others with extremely high employment density, such as blocks of high-rise office buildings. If the grid is too large, employment concentrations may be offset by neighboring parks, and centers of employment may be hard to identify [28].

We tested grids with cell sizes of 1 km × 1 km, 1.5 km × 1.5 km and 2 km × 2 km [43]. As shown in Table 3, the research area includes 4266 cells of 1 km × 1 km with 24.9% of them being null, 1883 cells of 1.5 km× 1.5 km with 9.2% of them being null and 1030 cells of 2 km× 2 km with 6.5% of them being null (Table 3). Considering both the number of samples and the percentage of null-value units, we chose to use the 1.5 km × 1.5 km grid as our research unit.

Table 3. Grid sizes compared.

Grid Cell Size	Study Unit Number	Null Value Unit		Employment Density (Person/km^2)		
		Number	Percentage	Max	Min	Mean
1 km × 1 km	4266	1062	24.9%	114,850	2	4767
1.5 km × 1.5 km	1883	173	9.2%	98,746	2	3822
2 km × 2 km	1030	67	6.5%	93,538	2	3670

2.3.2. Identifying Subcenters

A number of methods have been used to identify subcenters in metropolitan areas. Among them, the most direct relies on the researcher's knowledge and observations, but this is highly subjective. Building on observations and experience, some researchers use a threshold value-setting method for identifying employment subcenters. For example, Giuliano *et al.* [42] define a subcenter as a group of contiguous tracts that each have at least 10 employees per acre and that together have at least 10,000 employees. This method is also subjective, since the identification of subcenters is sensitive to the choice of the threshold. For example, McMillen *et al.* [44] use a threshold of 20 people/acre, with total employees exceeding 20,000.

McDonald [45] proposed a more objective method for identifying employment subcenters, defining a subcenter as a cluster of significant positive residuals from a simple regression of employment density on the distance from the central business district. McDonald *et al.* [41] subsequently used the same method to identify employment subcenters in Chicago, with reasonable results. McMillen [28] argued that a subcenter should also have a significant effect on overall employment in the city and proposed a two-stage method, which first identifies candidate subcenters and then estimates their explanatory power. This two-stage method is better at eliminating subjective influences and can easily be employed by other researchers [28,40].

We adopt a similar two-stage method. In the first stage, we follow McDonald's [46] method of identifying potential subcenters. First, we establish a monocentric model. The classical monocentric model assumes that all of a city's jobs are concentrated in its central business district [24,26,27]. However, this assumption cannot fit reality. When investigating the spatial distribution of employment, many studies assume that employment distribution is similar to population distribution, but more concentrated in the city center [47]. To capture this dynamic, a negative exponential model has been widely used.

$$D(s) = \alpha e^{\beta s + u} \tag{1}$$

D(s) is employment density; s is the distance to Tiananmen, often regarded as Beijing's center [48]; α is employment density at distance zero and β ($\beta < 0$) is density gradient; u is the error term.

Second, since the negative exponential model is nonlinear, we use its logarithmic form to estimate it by the ordinary least squares method (OLS).

$$LnD(s) = \beta s + c + u \tag{2}$$

Here, $c = ln\alpha$, and other symbols are the same as in model Equation (1).

Using this monocentric model, the residuals between real and predicted values can be calculated, and those observations with residuals significant at the 95 percent confidence level are chosen as potential subcenters.

Third, to deal with the potential problem that a unit surrounded by very low employment density units could be falsely identified as a candidate subcenter, we follow the threshold method of Muñiz *et al.* [49], which uses the average employment density of the research area as the threshold for determining potential employment subcenters.

In the second stage, we build a polycentric model to test whether our candidate subcenters help explain the research area's overall employment density distribution. Heikkila *et al.* [50] propose three different hypotheses about the role of subcenters, corresponding to three different polycentric models. As potential subcenters have different industries and different functions [51], the hypothesis that benefits from all subcenters being completely complementary is most suitable for our study. Based on an exponential function, a polycentric model that added control variables can be expressed as:

$$D(s, s_{sub_i}, X) = \alpha_0 e^{\beta s} e^{\sum_{i=1}^{n} \alpha_i s_{sub_i}} e^{\delta X} e^{u} \tag{3}$$

where $D(s, s_{sub_i}, X)$ is the density of a research unit, s and s_{sub_i} are the distance to Tiananmen and the potential employment subcenter *i*, respectively; *X* is control variables, including the distance to the nearest subway station, the highway and the airport; a_0, $a_i (i = 1, 2 \ldots)$ and δ are parameters to be estimated; u is an error term.

A log-transformed version of model Equation (3) is also adopted here, for the same reason. Thus, our polycentric model is as follows:

$$LnD(s, s_{sub_i}, X) = \beta s + \sum_{i}^{n} \alpha_i s_{sub_i} + \delta X + c + u \tag{4}$$

Here, $c = \ln\alpha_0$ and other symbols are the same with model Equation (3).

However, many studies suggest that the employment density declines more precipitously when moving away from subcenters than from the main city center; therefore, the inverse distances to the employment subcenters are often employed as interpretation variables [40,46], as shown in model Equation (5). This can also solve the problem of potential collinearity between the distance variables. In our research, model Equation (5) is adopted to estimate the effect of subcenters on overall employment density distribution.

$$LnD(s, s_{sub_i}, X) = \beta s + \sum_{i}^{n} \alpha_i s_{sub_i}{}^{-1} + \delta X + c + u \tag{5}$$

2.3.3. Effect of the Potential Subcenter on Local Employment Density

To examine the effect of potential subcenters on local employment density, we estimate regressions with the distance to Tiananmen and the distances to each potential subcenter as the explanative variables and confine the observations to certain scopes (5, 10 and 15 km radius) to the potential subcenters shown in model Equation (6).

$$LnD(s, s_{sub_i}) = \beta s + \alpha s_{sub_i} + c + u \tag{6}$$

Control variables are excluded in the above model Equation (6) since distances to subway stations and highways are in general highly correlated with the distance to the subcenters, and the distance to the airport has little effect on the employment density when the observations are limited to a local scope.

3. Employment Density in Beijing

The average employment density in the research area is 3125 employees/km^2. There is a great deal of variation around that mean, however, from just two employees/km^2 up to a high of 98,746 employees/km^2 in one grid cell on *Jianguomen* Street. The spatial distribution of employment displays four major characteristics as follows.

3.1. Density Declines from the Center to the Suburb

The research area is divided by the five ring roads into six concentric zones, numbered from I, in the center, to VI, on the periphery (Figure 4). Employment density varies among these ring zones, decreasing dramatically as one moves outward from the center. The employment densities of the six zones, from center to periphery, are 29,248 employee/km^2, 31,547 employee/km^2, 22,571 employee/km^2, 6897 employee/km^2, 1447 employee/km^2 and 563 employee/km^2 (Table 4). From the innermost to the outermost ring zone, the employment density drops by about 98%.

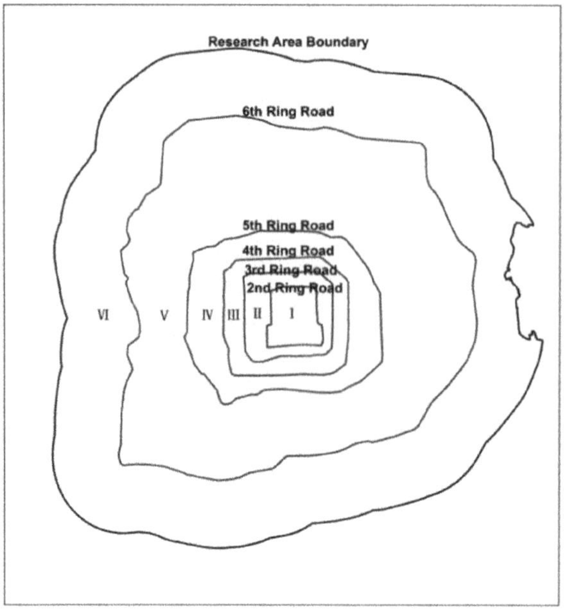

Figure 4. The six ring zones.

The highest employment density is not observed in the innermost ring zone, but in the next zone out, Zone II, which is not predicted by the classic monocentric model. Several factors may contribute. First, historic-site preservation has prevented economic concentration. The 2nd Ring Road was built at the site of the old city wall and contains within it the city's major sites of historical and cultural importance, including the Ming and Qing Dynasty Imperial Palace, Tiananmen (which together cover about 1.2 km^2), the Temple of Heaven and historic quadrangles and lanes (*hutong*). Preservation of the old city has long been an important part of city planning, and many important sites have been protected from modern development as a result. Second, development and construction within the 2nd Ring Road are subject to strict planning regulations, such as building height caps and density limits, which have influenced employment density. Third, the city has relocated populations and manufacturing from the center to suburban areas to reduce pressure on the central city.

Table 4. Employment density by ring zone.

Ring Zones	Area (km^2)	Area Share (Percentage)	Enterprise Number Share (Percentage)	Enterprise Density (Per km^2)	Employment Share (Percentage)	Employment Density (Per km^2)
I (within 2nd Ring Road)	62.0	1.4%	11.7%	775	13.1%	29,248
II (between 2nd and 3rd Ring Roads)	97.1	2.2%	20.8%	878	22.1%	31,547
III (between 3rd and 4th Ring Roads)	141.6	3.2%	23.8%	691	23.1%	22,571
IV (between 4th and 5th Ring Roads)	311.7	7.0%	15.6%	206	15.5%	6897
V (between 5th and 6th Ring Roads)	1664.6	37.6%	18.3%	45	17.4%	1447
VI (outside the 6th Ring Road)	2152.5	48.6%	9.8%	19	8.8%	563

Table 5. Cumulative employment density by ring zone.

Zones	Area (km^2)	Area Share (Percentage)	Enterprise Share (Percentage)	Enterprise Density (Per km^2)	Employment Share (Percentage)	Employment Density (Per km^2)
I (within 2nd Ring)	62.0	1.4%	11.7%	775	13.1%	29,248
I + II (within 3rd Ring)	159.1	3.6%	32.5%	838	35.2%	30,651
I + II + III (within 4th Ring)	300.7	6.8%	56.3%	769	58.3%	26,847
I + II + III + IV (within 5th Ring)	612.5	13.8%	71.9%	482	73.8%	16,693
I + II + III + IV + V (within 6th Ring)	2269.0	51.2%	90.2%	163	91.2%	5567
Research area	4429.6	100.0%	100.0%	93	100.0%	3125

3.2. *A Vast Employment Center*

It is also worth noting that the 4th Ring Road seems to be a boundary outside which the employment density plummets, while the employment densities of the areas within the 4th Ring Road (Zones I, II and III) are relatively similar. What is more, the area within the 4th Ring Road accounts for only 6.8% of the research area, but it hosts almost 60% of its employment (Table 5). It may be more appropriate to think of a large employment center of 300 km^2 than a small central business district, as assumed in theory (Figure 5).

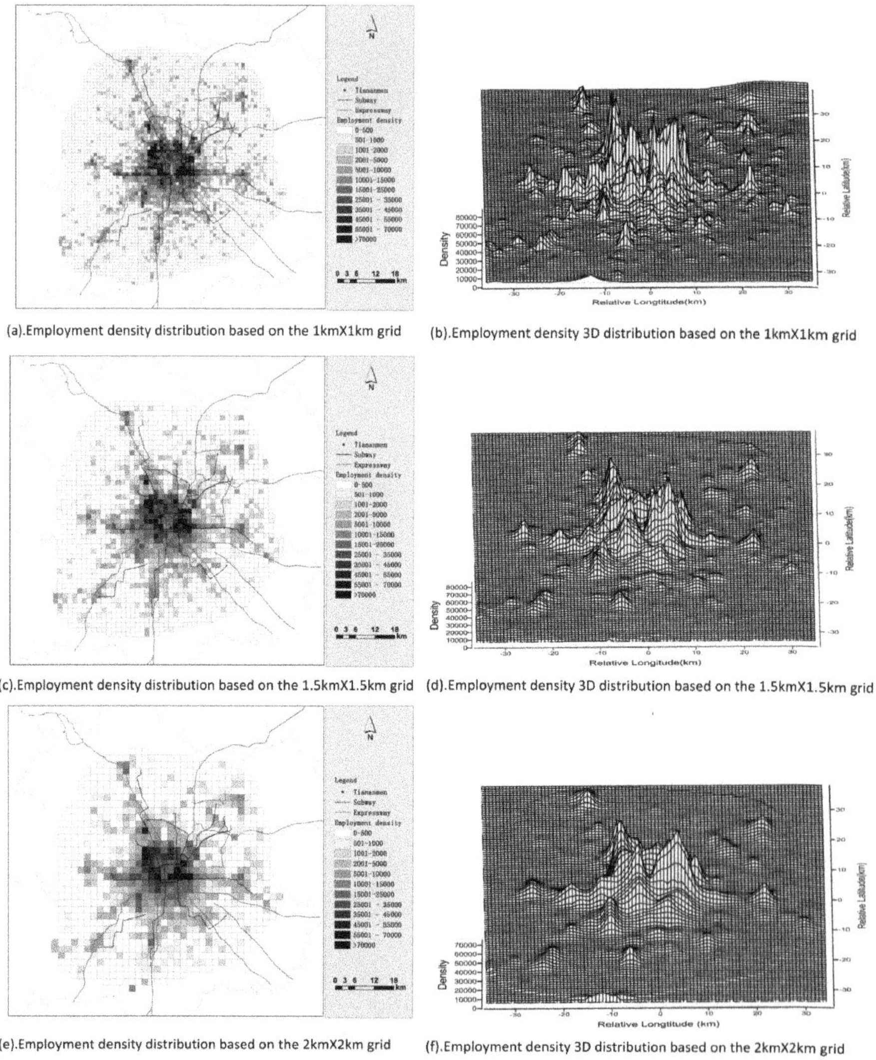

(a).Employment density distribution based on the 1kmX1km grid

(b).Employment density 3D distribution based on the 1kmX1km grid

(c).Employment density distribution based on the 1.5kmX1.5km grid

(d).Employment density 3D distribution based on the 1.5kmX1.5km grid

(e).Employment density distribution based on the 2kmX2km grid

(f).Employment density 3D distribution based on the 2kmX2km grid

Figure 5. Beijing's employment density distribution at different scales.

3.3. The Influence of Transportation

When the city's major roads are added to the employment density map, it appears that employment density is closely associated with major transit roads, extending radially along the expressways (Figure 5). For example, higher employment density can be observed along the east-west Subway Line 1 (Figures 1 and 5), the Beijing-Tibetan expressway to the northwest, the Beijing-Shijiazhuang expressway to the southwest and the Beijing-Kaifeng expressway to the south (Figures 2 and 5). We established a simple OLS regression model to further examine the effects of the highways and subway on employment density:

$$LnD(s, s_{highway}, s_{subway}) = \beta s + \gamma s_{highway} + \delta s_{subway} + c + u \quad (7)$$

where $D(s, s_{highway}, s_{subway})$ is employment density, is the distance between a given unit and Tiananmen, $s_{highway}$ is the distance from the unit to the nearest highway, s_{subway} is the distance from the unit to the nearest subway station, c is a constant and u is a random error term.

According to the regression results, both highways and subway stations have a significant positive impact on employment density at a 99% confidence level (Table 6).

Table 6. Transport availability and employment density.

Dependent Variables	ln (Employment Density)
constant	9.361 ***
	(78.71)
s	−0.109 ***
	(−15.91)
$s_{highway}$	−0.071 ***
	(−2.61)
s_{subway}	−0.099 ***
	(−5.78)
Adjusted R^2	0.359
F-test	301.06
Sample number	1610

*** denotes 1% significance level; *t*-values are in parentheses.

3.4. Employment Density Varies by Direction

We also examined directional variation in employment density (Figure 6). We select nine directions with potential employment subcenters, as well as six directions with no specific high-employment unit in the suburbs [52] and choose grids intercepted with each direction to draw the employment density profile for each direction (Figure 7).

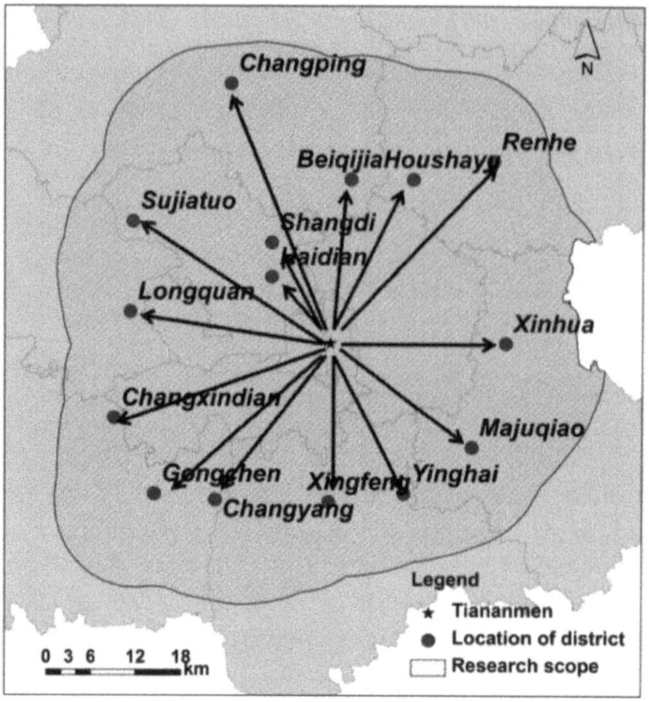

Figure 6. Directions of employment density profile.

We can make several observations based on Figure 7. First, employment density seems to decline differently along different headings. Second, despite the overall downward trend from the center to the suburbs, in most directions, the employment density first increases and then decreases. Third, the employment density increases again in the remote suburbs in several directions, suggesting the existence of employment subcenters. Fourth, in some directions, the employment density drops very dramatically beyond a certain distance to the city center, producing a cliff-shaped profile. This suggests that employment may gather within a certain area.

These observations suggest that the spatial structure of employment shows both monocentric and polycentric characteristics. Employment density decreases from about 30,000 employees/km^2 within the 3rd Ring Road to 563 employees/km^2 in the outermost ring zone, suggesting a monocentric structure. At the same time, peaks of employment density in certain suburban areas suggest a polycentric pattern. We investigate this further in the next section.

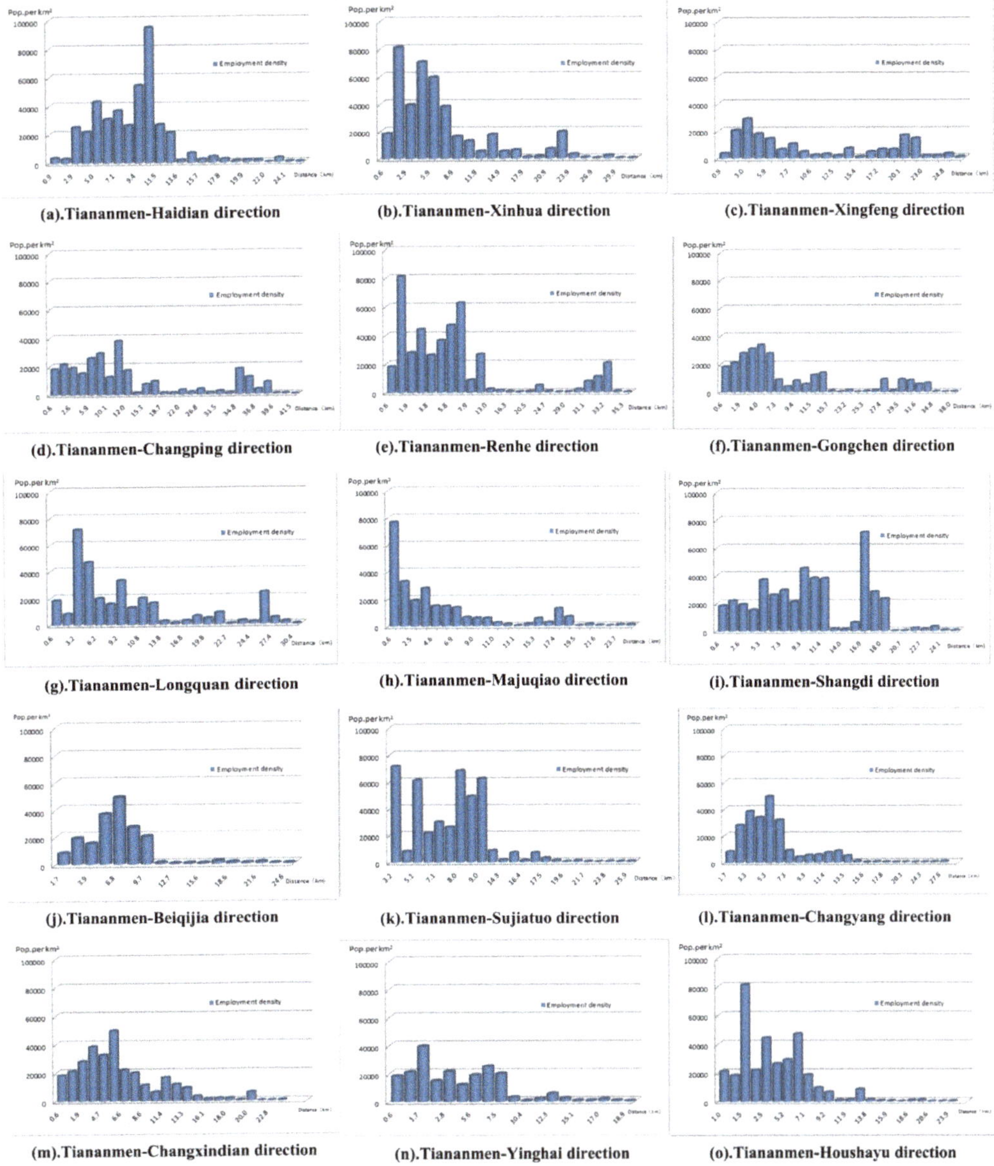

(a).Tiananmen-Haidian direction
(b).Tiananmen-Xinhua direction
(c).Tiananmen-Xingfeng direction
(d).Tiananmen-Changping direction
(e).Tiananmen-Renhe direction
(f).Tiananmen-Gongchen direction
(g).Tiananmen-Longquan direction
(h).Tiananmen-Majuqiao direction
(i).Tiananmen-Shangdi direction
(j).Tiananmen-Beiqijia direction
(k).Tiananmen-Sujiatuo direction
(l).Tiananmen-Changyang direction
(m).Tiananmen-Changxindian direction
(n).Tiananmen-Yinghai direction
(o).Tiananmen-Houshayu direction

Figure 7. Directional profiles of employment density.

4. Identification of Employment Subcenters

4.1. Results from the Monocentric Model

Table 7 reports the results of model Equation (2). The estimation adopts White's [53] method to calculate standard errors and the covariance matrix to eliminate a potential heteroscedasticity problem.

Table 7. Regression result of the monocentric model.

Dependent Variable: ln (Employment Density)	Research Area	Within the 6th Ring Road	Within the 5th Ring Road	Within the 4th Ring Road	Within the 3rd Ring Road
Constant	9.552 ***	10.391 ***	11.486 ***	10.695 ***	10.392 ***
	(83.04)	(60.05)	(61.45)	(66.87)	(49.60)
s	−0.146 ***	−0.391 ***	−0.269 ***	−0.123 ***	−0.043
	(−29.83)	(20.87)	(−14.86)	(−4.91)	(−1.02)
Adjusted R^2	0.333	0.349	0.450	0.131	0.005
F-test	803.21	610.42	220.72	22.15	1.36
Sample number	1611	811	273	141	72

*** denotes 1% significance level; *t*-values are in parentheses. Samples excluded the irregularly-shaped units on the edge of the research area and the units where Tiananmen is located.

We find that both the model and the coefficient of the distance to Tiananmen are statistically significant at the 99% confidence level (Table 7). For every kilometer that the distance to Tiananmen increases, employment density decreases by 14.6%. The adjusted R^2 is 0.333, indicating that this single distance variable explains 33.3% of the variation in employment density. In other words, the monocentric model works well [54] and the distance to Tiananmen is significant in explaining employment distribution in the research area.

While the monocentric model is useful in identifying employment subcenters, we wanted to clarify the scope at which it works. Accordingly, we varied the limits of the research area (Table 7). When limiting it to within the 6th Ring Road and the 5th Ring Road, the monocentric model is significant; when limiting it to within the 4th Ring Road, the monocentric model is still significant, but the explanatory power is greatly reduced; when limiting it to within the 3rd Ring Road, the monocentric model is no longer significant (Table 7). This is consistent with the direct observation from Figure 5 that, within the 3rd Ring Road and even within the 4th Ring Road, employment density is much higher, with no apparent decline away from the center. Furthermore, the explanatory power of the monocentric model is strongest within the 5th Ring Road and declines as the research area expands, which may be a result of employment subcenters in the suburbs.

4.2. *Subcenter Identification*

Based on the monocentric model, we conducted a residual analysis and identified 31 research units where the residual is significantly greater than zero (at the 95% confidence level) as employment subcenter candidates. Seven of them have lower employment density than the overall average and are removed from consideration [55]. For the remaining 24 units, when the units are very close to each other (within 3 km [56]), we selected the unit with the highest employment density to represent them. This gave us nine potential sub-centers located in *Shangdi* Street, *Haidian* Street, *Shahe* Town, *Changping* Urban District (including South Street and North Street), *Renhe* Street, *Gongchen* Street, *Xinhua* Street, *Zhangjiawan* Town and *Longquan* Street. We use the names of these streets and towns to name the nine potential subcenters in the following discussion.

4.2.1. Effects of Potential Employment Subcenters on Overall Employment Density

The results of model Equation (4) are shown on the left side of Table 8. After some tests, we include two dummy variables of the distance to the subway station and highway and one continuous variable of the distance to the airport into the regression function to control other factors' influences. The distances to Tiananmen, *Haidian* Street, *Changping* Urban District, *Renhe* Street and *Gongchen* Street have a significant impact on employment density. The distances to *Shangdi* Streets, *Longquan* Street and *Xinhua* Street are also significant, but the coefficient signs are positive, which is not consistent with the expectations. One possible problem of this is collinearity. For example, *Shangdi* Street, *Haidian* Street and *Shahe* Street are in the same direction, and the distance between them is short, so the distances to these three potential subcenters are highly correlated [57]. Another reason is that the effect of proximity to a subcenter declines rapidly with distance, as mentioned in Section 2.3.2.

To solve these problems, model Equation (5) has been estimated, and the result is shown in Table 8 on the right. The inverse of the distance between a grid cell and *Haidian* Street, *Xinhua* Street, *Renhe* Street, *Changping* Urban District and *Gongchen* Street is significant, and the effects are positive, as expected. However, the inverse distances to *Shangdi* Street, *Longquan* Street, *Shahe* Town and *Zhangjiawan* Street are not significant. These results suggest that *Haidian* Street, *Xinhua* Street, *Renhe* Street, *Changping* Urban District and *Gongchen* Street are more likely to be employment subcenters.

Table 8. Regression results of potential subcenter test.

Dependent Variable in (Employment Density)	Model Equation (4) (Distances to Subcenters)	Dependent Variable in (Employment Density)	Model Equation (5) (Inverse Distances to Subcenters)
Constant	9.086 ***	Constant	7.136 ***
	(7.01)		(29.01)
s	−0.172 ***	S	−0.111 ***
	(−7.45)		(−11.85)
s Haidian Street	−0.265 ***	1/s Haidian Street	3.469 ***
	(−4.61)		(4.21)
s Shangdi Street	0.302 ***	1/s Shangdi Street	0.568
	(4.32)		(0.65)
s Longquan Street	0.089 ***	1/s Longquan Street	1.722
	(4.72)		(1.56)
s Xinhua Street	0.065 **	1/s Xinhua Street	1.716 ***
	(2.15)		(2.86)
s Renhe Street	−0.101 ***	1/s Renhe Street	6.111 ***
	(−5.89)		(5.63)
s Changping Urban District	−0.073 ***	1/s Changping Urban District	6.821 ***
	(−3.24)		(7.42)
s Shahe Town	0.019	1/s Shahe Town	−0.871
	(0.35)		(−0.95)
s Zhangjiawan Street	0.019	1/s Zhangjiawan Street	1.462
	(0.78)		(1.62)
s Gongchen Street	−0.063 ***	1/s Gongchen Street	4.735 ***
	(−3.45)		(5.63)
x Airport	0.088 ***	x Airport	0.054
0–1 km from subway station	1.389 ***	0–1 km from subway station	1.783 ***
	(7.67)		(10.01)
1–3 km from subway station	0.835 ***	1–3 km from subway station	0.828 ***
	(5.62)		(5.35)
0–1 km from the highway	0.362 ***	0–1 km from the highway	0.331 ***
	(2.91)		(2.64)
1–3 km from the highway	0.134	1–3 km from the highway	0.037
	(1.08)		(0.312)
Adjusted R^2	0.441	Adjusted R^2	0.437
F-test	83.35	F-test	82.03
Sample number	1062	Sample number	1062

*** and ** denote the 1% and 5% significance level, respectively; *t*-values are in parentheses.

4.2.2. Effects of Potential Subcenters on Local Employment Density

The results demonstrate that most models are significant at the 95% confidence level, with a few exceptions, including the models within 5 km of *Renhe* Street and *Shahe* Town and models within 5 km and 10 km of *Zhangjiawan* Town (Table 9). All coefficient signs of the subcenters are as expected.

Each of the five subcenters identified in Section 4.2.1 has a strong influence on local employment density within certain distances. *Changping* Urban District and *Xinhua* Street significantly affect employment density within a radius of 5 km, 10 km and 15 km. *Haidian* Street and *Gongchen* Street are significant within a radius of 5 km and 10 km, but their influence is unclear when the radius is expanded to 15 km. *Renhe* Street is somewhat unique, given that its effects are significant at 10 km and 15 km, but not significant within a radius of 5 km. It could be that employment density does not decline significantly within 5 km of *Renhe* Street. For the other potential subcenters, *Shangdi* Street and *Longquan* Street influence employment density within a radius of 5–10 km, while *Shahe* Town and *Zhangjiawan* Street have little effect on local employment density.

According to these results, the nine potential employment subcenters can be divided into three types. The first includes *Haidian* Street, *Xinhua* Street, *Changping* Urban District, *Gongchen* Street and *Renhe* Street. These five candidates clearly influence local and overall employment distribution. The second type includes *Shangdi* Street and *Longquan* Street, which influence only local employment distribution. The third type includes *Shahe* Town and *Zhangjiawan* Street, which have high employment density, but little effect on employment distribution. According to McMillen [28], a subcenter should have a significant effect on overall employment. We therefore identify the first group of candidates as Beijing's employment subcenters (Figure 8).

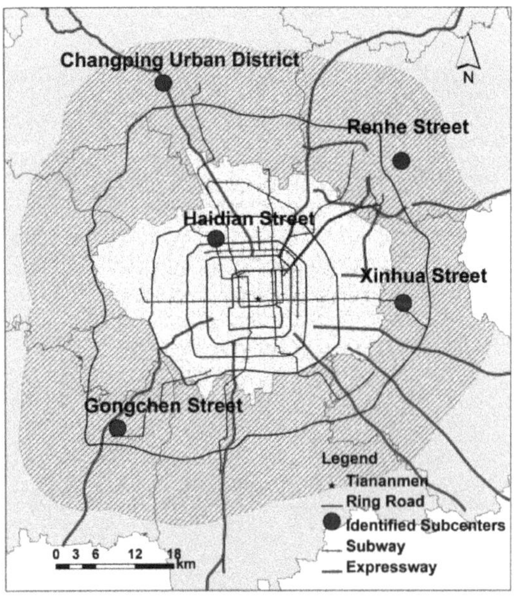

Figure 8. Beijing's confirmed employment subcenters.

Table 9. Local regression results of employment subcenters.

Dependent Variable	5 km	10 km	15 km	5 km	10 km	15 km	5 km	10 km	15 km
	Haidian Street			*Changping* Urban District			*Renhe* Street		
Constant	9.750 ***	12.911 ***	11.844 ***	5.187	9.453 ***	9.682 ***	12.946 ***	9.971 ***	10.548 ***
	(−71.49)	(30.12)	(40.83)	(1.67)	(7.72)	(10.29)	(3.52)	(9.03)	(9.39)
S	−0.480 ***	−0.289 ***	−0.244 ***	0.119	−0.062	−0.095 ***	−0.195	−0.089 **	−0.106 ***
	(−17.63)	(−12.11)	(−18.3)	(1.36)	(−1.96)	(−4.13)	(−1.83)	(−2.04)	(−3.59)
s subcenters	−0.301 **	−0.124 **	−0.038	−0.206 ***	−0.300 ***	−0.146 ***	−0.026	−0.215 ***	−0.213 ***
	(−2.1)	(−2.54)	(−1.83)	(−4.30)	(−5.06)	(−4.21)	(−0.11)	(−2.90)	(−4.86)
Adjusted R^2	0.664	0.536	0.551	0.34	0.18	0.09	0.042	0.079	0.126
F	35.63	81.19	186.55	9.83	13.28	11.36	1.70	5.04	13.68
Sample number	37	141	355	34	116	203	34	96	177
	Gongchen Street			*Xinhua* Street			*Longquan* Street		
Constant	3.581	7.558 ***	7.579 ***	10.558 ***	9.097 ***	9.527 ***	18.405 ***	11.98 ***	10.79 ***
	(1.14)	(7.11)	(10.36)	(3.97)	(10.79)	(18.27)	(4.18)	(10.25)	(12.70)
S	−0.157	−0.022	−0.072 ***	−0.051	−0.062 **	−0.127 ***	−0.335	−0.209 ***	−0.204 ***
	(−1.42)	(−0.65)	(−4.53)	(−0.46)	(−2.00)	(−7.90)	(−1.89)	(−4.52)	(−8.93)
s subcenter	−0.559 **	−0.231 ***	−0.016	−0.730 ***	−0.238 ***	−0.074 **	−1.097 ***	−0.256 ***	−0.038
	(−2.29)	(−3.18)	(−0.48)	(−2.71)	(−3.61)	(−2.38)	(−2.83)	(−2.72)	(−0.82)
Adjusted R^2	0.113	0.068	0.065	0.145	0.103	0.188	0.338	0.212	0.261
F-test	2.99	5.51	10.36	3.80	8.56	31.02	6.64	12.90	32.24
Sample number	33	125	270	35	132	260	30	99	184
	Shangdi Street			*Shahe* Town			*Zhangjiawan* Town		
Constant	12.650 ***	12.346 ***	11.490 ***	9.945 ***	8.854 ***	8.10 ***	7.240	8.862 ***	10.599 ***
	(8.28)	(21.13)	(35.16)	(3.63)	(11.49)	(18.74)	(1.78)	(3.77)	(7.71)
S	−0.230	−0.265 ***	−0.234 ***	−0.157	−0.123 ***	−0.085 ***	−0.040	−0.103	−0.155 ***
	(−1.41)	(−9.96)	(−18.73)	(−1.56)	(−4.79)	(−6.82)	(−0.32)	(−1.51)	(−3.89)
s-subcenter	−0.295 **	−0.069	−0.018	−0.044	−0.036	0.006	−0.086	−0.066	−0.102
	(−2.56)	(−1.23)	(−0.74)	(−0.02)	(−0.64)	(−0.82)	(−0.32)	(−0.66)	(−1.92)
Adjusted R^2	0.19	0.422	0.55	0.013	0.135	0.132	0.007	0.028	0.080
F-test	4.98	51.29	179.21	1.22	11.53	23.24	1.12	1.19	7.71
Sample number	37	140	293	35	136	292	29	61	156

*** and ** denote the 1% and 5% significance level, respectively; *t*-values are in parentheses.

4.3. Polycentric Model

Based on the five identified subcenters, we use model Equation (5) to explain the overall employment density distribution. The regression results are shown in Table 10. To compare the polycentric and monocentric models, a monocentric model with added control variables has also been conducted here.

The polycentric model and the coefficients of all independent variables are significant at a 99% confidence level. A partial F-test suggests that the five subcenter distance variables together are significantly more than zero, indicating that these identified subcenters influence the overall employment distribution. However, adding the distance to the five subcenters improves the entire model's explanatory power by only four percent: the adjusted R^2 of the polycentric model is 0.434, compared to the monocentric model's adjusted R^2 of 0.393.

Table 10. Comparison of the monocentric and polycentric models.

Dependent Variable (Employment Density)	Monocentric Model (Add Control Variables)	Polycentric Model
Constant	7.660 ***	7.176 ***
	(33.53)	(30.49)
S	−0.096 ***	−0.112 ***
	(−11.67)	(−12.46)
1/s *Haidian* Street		2.953 ***
		(4.17)
1/s *Changping* Urban District		5.904 ***
		(7. 28)
1/s Xinhua Street		2.002 ***
		(2.71)
1/s Renhe Street		6.601 ***
		(5.82)
1/s Gongchen street		4.693 ***
		(5.68)
x Airport	0.069	0.096
	(1.17)	(1.34)
0–1 km from subway station	2.009 ***	1.760 ***
	(11.53)	(10.07)
1–3 km from subway station	0.896 ***	0.779 ***
	(5.95)	(5.38)
0–1 km from the highway	0.560 ***	0.373 ***
	(4.04)	(3.03)
1–3 km from the highway	0.152	0.013
	(1.26)	(0.29)
Adjusted R^2	0.393	0.434
F-test	174.29	110.98
Sample number	1602	1602

*** denotes the 1% significance level; *t*-values are in parentheses.

5. Discussion and Conclusions

Although many cities in developed countries have polycentric or even dispersed employment patterns [41,58–62], the situation in Beijing appears to be different. First, the city still has very strong monocentric characteristics. The single factor of a cell's distance to Tiananmen explains 33.3% of its employment density, a figure that increases further when peripheral cells are excluded (34.9% inside the 6th Ring Road, 45.0% inside the 5th Ring Road). Second, the city has a very large employment center, which is centered at Tiananmen and generally overlaps with the 4th Ring Road. We call this center "large" because it has an area of about 300 km^2 (6.8% of the research area) and accounts for 58.7% of employment, whereas other cities' downtown areas can often be treated as points. Third, despite the identification of five sub-centers, it is difficult to conclude that the city has achieved a polycentric structure, as the polycentric model improves the explanatory power of the overall employment distribution by only 4% over the monocentric one [63,64]. Based on these findings, we conclude that the spatial structure of Beijing is still monocentric, but may be in transition to a polycentric pattern.

Beijing has tried to foster subcenters and to achieve a decentralized spatial development pattern to a certain degree; however, our study suggests that the success of these efforts has been limited. The monocentric model still explains the spatial structure of the city's employment very well, and adding the identified subcenters improves its explanatory power by only four percentage points. At the same time, though four of the five identified subcenters—*Xinhua* Street, *Changping* Urban District, *Gongchen* Street and *Renhe* Street—are located in the planned new towns, the other three new suburb towns have not achieved development as expected. We speculate that several factors contribute to Beijing's strongly monocentric character and the formation of its large employment center. First, the central city enjoys better infrastructure than other areas and hosts most of the central and city government departments, schools, hospitals, museums, public facilities and parks. Second, the ring-and-radials road structure plays a very important role in facilitating the concentration of people and economic activities in the central city. Third, and more important, the power of agglomeration remains decisive. As a result, new development still prefers locations close to the central city.

With these somewhat disappointing results in mind, several inferences may be drawn from the findings of this study. First, the monocentric characteristics of employment distribution suggest that the agglomeration economy still plays a dominate role in Beijing, and so planning interventions and heavy public investment in decentralized development may result in a loss of economic efficiency [65]. We feel that planning tools are useful in guiding and regulating development, but market needs and the efficiency of public investment also deserve careful considerations. Given the fact that the city center still accommodates most of the employment, the government should also pay attention to further improving the infrastructure and facilities of existing developed areas, thus making the city a more attractive and livable place. Second, given that the employment is still concentrated in the city center based on the findings of this paper while the population has already redistributed to the suburban area according to the literature suggests the possibility of a more imbalanced job-housing spatial distribution and longer commuting distances, which also deserves more attention when making plans and policies.

Acknowledgments: This research is supported by "the Fundamental Research Funds for the Central Universities". The authors gratefully thank the reviewers and the editor. We also thank Matt Turner, A-xing Zhu in the Department of Geography University of Wisconsin-Madison for their additional guidance, and we acknowledge the help provided by Haoran Jin, Xin Yao, Yuncheng Huang, Pingping Ma, Xin Tan and Xiaoqing Yang in the School of Geography Beijing Normal University for their constructive suggestions for this paper. Thanks also to Jacob Fleming for his language polishing and editing.

Author Contributions: Daquan Huang developed the original idea and contributed to the research design, writing and modification. Zhen Liu was responsible for data collection and processing, writing and modification. Xingshuo Zhao contributed to the research design, writing, modification and provided guidance. All authors have read and approved the final manuscript. The authors are very grateful to the anonymous referees and the editors for their helpful comments on an earlier version of this paper.

Conflicts of Interest: The authors declare no conflict of interest.

References and Notes

1. Bai, X.M.; Shi, P.J.; Liu, Y.S. Realizing China's urban dream. *Nature* **2014**, *509*, 158–160.
2. Hu, Z. Review and outlook of Population Scale in Beijing. *Urban Stud.* **2011**, *18*, 8–10. (In Chinese)
3. Hu, Z. Review and Recognition on Urban Size of Beijing. *J. Urban Reg. Plan.* **2011**, *2*, 1–18. (In Chinese)
4. Huang, D.; Jin, H.; Zhao, X.; Liu, S. Factors Influencing the Conversion of Arable Land to Urban Use and Policy Implications in Beijing, China. *Sustainability* **2015**, *7*, 180–194.
5. Sun, C.; Zheng, S.Q.; Wang, R. Restricting driving for better traffic and clearer skies: Did it work in Beijing? *Transp. Policy* **2014**, *32*, 34–41.
6. Wen, H.; Sun, J.; Zhang, X. Study on Traffic Congestion Patterns of Large City in China Taking Beijing as an Example. *Procedia-Soc. Behav. Sci.* **2014**, *138*, 482–491.
7. Cai, H.; Xie, S.D. Traffic-related air pollution modeling during the 2008 Beijing Olympic Games: The effects of an odd-even day traffic restriction scheme. *Sci. Total Environ.* **2011**, *409*, 1935–1948.
8. Hou, L.F.; Wang, S.; Dou, C.; Zhang, X.; Yu, Y.; Zheng, Y.N.; Avula, U.; Hoxha, M.; Diaz, A.; McCracken, J.; *et al.* Air pollution exposure and telomere length in highly exposed subjects in Beijing, China: A repeated-measure study. *Environ. Int.* **2012**, *48*, 71–77.

9. Tian, L.; Zhang, W.; Lin, Z.Q.; Zhang, H.S.; Xi, Z.G.; Chen, J.H.; Wang, W. Impact of Traffic Emissions on Local Air Quality and the Potential Toxicity of Traffic-related Particulates in Beijing, China. *Biomed. Environ. Sci.* **2012**, *25*, 663–671.
10. Dang, Y.; Liu, Z.; Zhang, W. Land-based interests and the spatial distribution of affordable housing development: The case of Beijing, China. *Habitat Int.* **2014**, *44*, 137–145.
11. Yu, L.; Cai, H.P. Challenges for housing rural-to-urban migrants in Beijing. *Habitat Int.* **2013**, *40*, 268–277.
12. The urban planning aims to evacuate both the employment and population from the central city. For employment, the government, on one hand, encourages industries move to the suburb; on the other hand, it sets up many development zones in the suburb. For population, many residential communities have been built with the boom of the real estate in the suburb.
13. Huang, D.Q.; Wan, W.; Dai, T.Q.; Liang, J.S. Assessment of industrial land use intensity: A case study of Beijing economic-technological development area. *Chin. Geogr. Sci.* **2011**, *21*, 222–229.
14. Dong, G. Sixty years and twenty years—Review and Prospect of the modernization development process in Beijing (2). *Beijing Plan. Rev.* **2010**, *5*, 168–171. (In Chinese)
15. Dong, G. Sixty years and twenty years—Review and Prospect of the modernization development process in Beijing (1). *Beijing Plan. Rev.* **2010**, *6*, 177–180. (In Chinese)
16. Kuang, W.; Liu, J.; Shao, Q.; Sun, C. Spatio-temporal patterns and driving forces of urban expansion in Beijing Central City since 1932. *J. Geo-Inf. Sci.* **2009**, *4*, 428–435. (In Chinese)
17. Beijing is composed of fourteen districts and two counties. The central city covers all of Xicheng District and Dongcheng District and part of Haidian District, Chaoyang District, Fengtai District, and Shijingshan District. Nine of the ten other suburban districts and counties have one new town each. Daxing district has two new towns: Huangcun new town, where Daxing district government is located, and Yizhuang new town, where Beijing Economic-Technological Development Area is located.
18. Feng, J.; Zhou, Y. The Growth and Distribution of Population in Beijing Metropolitan Area (1982–2000). *Acta Geogr. Sinica* **2003**, *58*, 903–916. (In Chinese)
19. Feng, J.; Zhou, Y. The social spatial structure of Beijing Metropolitan Area and its evolution: 1982–2000. *Geogr. Res.* **2003**, *22*, 465–483. (In Chinese)
20. Feng, J.; Zhou, Y.; Wu, F. New trends of suburbanization in Beijing since 1990: From government-led to market-oriented. *Reg. Stud.* **2008**, *42*, 83–99.
21. Feng, J.; Wang, F.; Zhou, Y. The spatial restructuring of population in metropolitan Beijing: Toward polycentricity in the post-reform era. *Urban Geogr.* **2009**, *30*, 779–802.
22. Ma, Q.; Zhang, W. Characteristics and factors analyses of suburbanization in Beijing. *Geogr. Res.* **2006**, *25*, 121–130. (In Chinese)
23. Wang, F.; Zhou, Y. Modelling urban population densities in Beijing 1982–1990: Suburbanisation and its causes. *Urban Stud.* **1999**, *36*, 271–287.
24. Alonso, W. Location and land use: Toward a general theory of land rent. *Econ. Geogr.* **1964**, *42*, 277–279.

25. Brueckner, J.K. The structure of urban equilibria: A unified treatment of the Muth-Mills model. *Handb. Reg. Urban Econ.* **1987**, *2*, 821–845.
26. Mills, E.S. An aggregative model of resource allocation in a metropolitan area. *Am. Econ. Rev.* **1967**, *57*, 197–210.
27. Muth, R. *Cities and Housing: The Spatial Patterns of Urban Residential Land Use*; University of Chicago Press: Chicago, IL, USA, 1969.
28. McMillen, D.P. Nonparametric employment subcenter identification. *J. Urban Econ.* **2001**, *50*, 448–473.
29. Tao, R.; Jin, Y. Improvement of Data Quality Assessment of Economic Census in China. *Stat. Res.* **2009**, *26*, 8–12. (In Chinese)
30. Gu, Y.; Zheng, S.; Cao, Y. The Identification of Employment Centers in Beijing. *Urban Stud.* **2009**, *9*, 118–124. (In Chinese)
31. Liu, X.; Sun, T.; Li, G. Research on the spatial structure of employment distribution in Beijing. *Geogr. Res.* **2011**, *30*, 1262–1270. (In Chinese)
32. Sun, T.; Wang, L.; Li, G. Distributions of Population and Employment and Evolution of Spatial Structures in the Beijing Metropolitan Area. *Acta Geogr. Sinica* **2012**, *67*, 829–840. (In Chinese)
33. Beijing has three levels of governments: The municipality, the district or county, and the town (township or street). The average area of towns (townships or streets) in Beijing is about 22 km^2.
34. The First Ring is actually the walls of the Forbidden City.
35. Long, Y.; Zhang, Y.; Cui, C. Identifying Commuting Pattern of Beijing Using Bus Smart Card Data. *Acta Geogr. Sin.* **2012**, *67*, 1339–1352. (In Chinese)
36. There are eleven new towns in all, but four of them are remote from the central city, averaging more than 65 km from the city center (Tiananmen)
37. In matching the enterprise locations, "Exact matching" accounts for 79.3% and "Non-exact matching" accounts for 20.7%. It can be hard or impossible to match a company with a building exactly for two reasons. First, large companies can have multiple buildings, requiring us to use their geometrical center as the company's geographic coordinate. Second, some company addresses—for example, No.19 XX Street—are not sufficiently detailed to allow us to specify an exact match. These cases lead to some errors, but these are acceptable given the size of the research unit (1.5 km $\times$ 1.5 km).
38. Beijing Municipal Statistical Bureau. *Beijing Statistical Yearbook 2011*; China Statistics Press: Beijing, China, 2011.
39. The 560 companies that were eliminated from our dataset were selected in two ways. First, some companies were identified as not limited to Beijing by their names, such as XX headquarters. Second, for companies with more than 3000 employees, we conducted internet searches on the distribution of their activities. If we found, for example, that the company has branches or factories in many locations, we eliminated it from our data.
40. McMillen, D.P.; Lester, T.W. Evolving subcenters: Employment and population densities in Chicago, 1970–2020. *J. Hous. Econ.* **2003**, *12*, 60–81.

41. McDonald, J.F.; Prather, P.J. Suburban employment centres: The case of Chicago. *Urban Stud.* **1994**, *31*, 201–218.
42. Giuliano, G.; Small, K.A. Subcenters in the Los Angeles region. *Reg. Sci. Urban Econ.* **1991**, *21*, 163–182.
43. As more than a quarter of the towns (townships, streets) and other potential employment subcenters are smaller than 4 km^2, the grid should be smaller than 4 km^2.
44. McMillen, D.P.; McDonald, J.F. Suburban subcenters and employment density in metropolitan Chicago. *J. Urban Econ.* **1998**, *43*, 157–180.
45. McDonald, J.F. The identification of urban employment subcenters. *J. Urban Econ.* **1987**, *21*, 242–258.
46. McDonald, J.F.; McMillen, D.P. Employment subcenters and land values in a polycentric urban area: The case of Chicago. *Environ. Plan. A* **1990**, *22*, 1561–1574.
47. Small, K.A.; Song, S.F. Population and employment densities—Structure and change. *J. Urban Econ.* **1994**, *36*, 292–313.
48. Employment density around Tiananmen is very low, but we still can regard it *as the center.* First, it is the city center in city planning. The Beijing government has always taken Tiananmen as the center to design the layout of the city. Second, its scope is so small that we can regard it as a point. Third, many studies have taken Tiananmen as the center to investigate the distribution of population and employment [21,31,32].
49. Muñiz, I.; Garcia-López, M.À.; Galindo, A. The effect of employment sub-centres on population density in Barcelona. *Urban Stud.* **2008**, *45*, 627–649.
50. Heikkila, E.; Gordon, P.; Kim, J.I.; Peiser, R.B.; Richardson, H.W.; Dale-Johnson, D. What happened to the CBD-distance gradient? Land values in a policentric city. *Environ. Plan. A* **1989**, *21*, 221–232.
51. For example, the Xinhua Street which is the main part of Tongzhou new town specialized in comprehensive service, the Renhe Street which is the main part of Shunyi new town specialized in aviation logistics, the Gongchen Street which is the main part of Fangshan new town specialized in new materials industry, and the Changping urban district which is the main part of Changping new town specialized in research and development. Their developing directions have also different in the planning to undertake different functions of Beijing.
52. The nine directions are Tianmen-Haidian, Tianmen-Xinhua, Tianmen-Xingfeng, Tianmen-Changping, Tianmen-Renhe, Tianmen-Gongchen, Tianmen-Longquan, Tianmen-Majuqiao and Tianmen-Shangdi.
53. White, H. A heteroskedasticity-consistent covariance matrix estimator and a direct test for heteroskedasticity. *Econometrica* **1980**, *48*, 817–838.
54. McDonald [41] adopted seven forms of monocentric models to test the employment distribution of Chicago .The adjusted R2 is between 0.258 and 0.336 while the negative exponential model is 0.296. Small *et al.* [47] used the negative exponential model to examine the employment distribution in Los Angeles region. The adjusted R^2 is 0.395 in 1970 and 0.388 in 1980.

55. All seven of these are located in remote suburban districts, including four in Fangshan District and three in Shunyi District.
56. The selection of a 3 km boundary is somewhat subjective. Related research has no strict definition of adjacency. McMillen *et al.* [40] defined it as within 1.5 miles. Giuliano *et al.* [42] defined it as at least 0.25 miles from a common border. Here we referred to the definition of McMillen *et al.* [40], and with an eye toward the size of our study unit, selected 3 km as constituting adjacency.
57. The distance between Shangdi street and Haidian street is about 7 km and the correlation coefficient is 0.91. The distance between Shangdi street and Shahe street is about 8 km and the correlation coefficient is 0.93. The distance between Haidian street and Shahe street is about 13 km and the correlation coefficient is 0.76.
58. Cervero, R.; Wu, K.L. Polycentrism, commuting, and residential location in the San Francisco Bay Area. *Environ. Plan. A* **1997**, *29*, 865–886.
59. Glaeser, E.L.; Kahn, M.E. Decentralized Employment and the Transformation of the American City. National Bureau of Economic Research, 2011. Availiable online: http://www.nber.org/papers/w8117 (accessed on 17 August 2015).
60. Garcia-López, M.-À.; Muñiz, I. Employment decentralisation: Polycentricity or scatteration? The case of Barcelona. *Urban Stud.* **2010**.
61. Hughes, H.L. Metropolitan structure and the suburban hierarchy. *Am. Sociol. Rev.* **1993**, *58*, 417–433.
62. Veneri, P. The identification of sub-centres in two Italian metropolitan areas: A functional approach. *Cities* **2013**, *31*, 177–185.
63. Compared with monocentric model, the explanatory power of the polycentric model has a significant improvement in the research of Small *et al.* [47] and Wang *et al.* [64].
64. Wang, F.; Meng, Y. Analyzing urban population change patterns in Shenyang, China 1982–1990: Density function and spatial association approaches. *Geogr. Inf. Sci.* **1999**, *5*, 121–130.
65. Ding, C.; Bethka, K. Employment Concentration and Urban Economic Growth. *Urban Plan. Overseas* **2005**, *4*, 11–18. (In Chinese)

The Role of Villages and Townships in Informal Land Development in China: An Investigation on the City Fringe of Beijing

Pengjun Zhao and Mengzhu Zhang

Abstract: The past decades have witnessed a number of informal land developments on the urban fringe in China although many strict state regulations have been made to control this. The dual urban rural land system is widely believed to be one major determinant of informal developments in the existing literature. However, the important role of local villages and townships are often neglected. This paper aims to shed light on this by looking at the gated informal housing communities in Beijing as a case study. It investigates the role of villages and townships in informal land development and the conflicts of interest that arise with state regulations in the context of political decentralization. The results of analysis show that township governments have an ambivalent attitude or even give tacit approval to informal land development in villages since these informal developments actually bring economic benefits to local villagers and themselves. The situation seems to be worse as townships have poor fiscal capacity and a growing administrative responsibility for improvement of local development in the context of decentralization. Villages are keen to capture economic benefits from informal land development with help from private developers. As a result, a local, informal coalition between townships, villages, and private developers emerged at the grass roots level. This presents a major challenge to the state regulations designed for sustainable urban growth management.

Reprinted from *Sustainability*. Cite as: Zhao, P.; Zhang, M. The Role of Villages and Townships in Informal Land Development in China: An Investigation on the City Fringe of Beijing. *Sustainability* **2016**, *8*, 255.

1. Introduction

Informal land development in China usually refers to urban development (housing or industrial development) on land without land use permission or planning approval from the state, or development on land that does not comply with land use planning and development regulations. Informal land development in China has increasingly drawn the attention of both the public and academics. Informal housing, in particular, accounted for 20% of the total land used for housing in China in 2010 [1], while informal land development in general has seriously jeopardized sustainable development, causing the loss of arable land [2] and land loss by farmers [3], as well as spurring informal housing [4,5] and urban sprawl [6]. In this context, informal development may be a spontaneous response from villages to a shortage of fiscal revenue and the need to find ways to improve local people's livelihoods. As a result, local residents have protested strongly against the government's demolition of informal housing in villages, such as occurred in Hainan, Guangdong, Beijing, Hebei, and Jiangsu [7–11]. The issues surrounding informal land development now actually go beyond land development as such, having become a political issue for China's governments with respect to social stability.

Informal land development in China has occurred in the context of political and economic transition. This transition is characterized by several simultaneous processes, including decentralization, marketization, and privatization [12,13], which have significantly changed the relationship between government and the market. The phenomenon of urban entrepreneurialism has been recognized in the context of the decentralization and transition of the Chinese market [14], whereby local and city governments have started to transform into market-friendly agents with their key goals being to establish alliances with developers and investors to promote local economic growth. The focus of city agendas now lies on economic planning, with a shift in the role of government from public "welfare providers" to "economic development managers." In this context, local governments have been struggling to attract capital and investment. At the same time, local governments (whether a city government, a district or country government, or a township government) have gained more decision-making powers from the central government to control the resources under their jurisdiction [13]. The sale of land, in particular rural land, is one vital way that local governments can accumulate capital and obtain fiscal revenue [15].

The transition process has also changed the relationship between the central and local governments, with the state giving more power to the latter. These local governments can now focus on particular policies that promote local economic growth, rather than on state development goals. In this context, the phenomena of "local-state corporatism" [16] and "growth coalitions" [17,18] have emerged and are now predominant in local development. The former refers to the fact that local governments may now cooperate with enterprise to support key industries or programs. Unfortunately, these are usually closely related to the explicit short-term goal of economic growth, with cooperation established on the basis of irrational and unsustainable policy and often at the expense of vulnerable groups [16–19]. The latter implies that an interested coalition of local politicians and stakeholders use their political power and capital to obtain huge profits or more political power. The real estate industry, for example, is a typical battlefield on which such coalitions make a grab for and accumulate wealth [18]. In this context, rural land, which is the most valuable and tractable of resources in the urban region, is inevitably dragged into the waves of the 'growth coalition' to be developed formally or informally.

There are many factors affecting informal land development. It is widely believed that the conflict between the state government and villagers over who receives the economic benefits of land development is one major reason [2,19,20]. In many cases, the state government expropriates rural land from villagers, while offering minimal financial compensation and selling the land on the market at a much higher price. This formal land development process has been criticized for causing a considerable loss of economic benefits to villagers and a decline in the quality of life of farmers who have lost their land [20]. Therefore, informal land development is often considered as one way in which villages and villagers can obtain another source of income or protect their own welfare [3].

However, this dualistic mainstream view that regards villages and the government as two opposing sides does not tell the whole story of informal land development in China. Some puzzles remain: Why does informal land development still occur despite the central government's attempts to control it through new policy? Why are the state government's regulations for land use in rural areas ineffective and impracticable? It is widely believed that certain aspects of the existing governance system could be important explanatory factors [20,21]. However, it should also be recognized that the role of township governments, which are the primary units of government administration in China's rural areas, could be an important factor in informal land development in the rural areas. Firstly, township governments are usually powerless to control informal land developments in villages. Secondly, in many cases, township governments even support informal land development for the sake of achieving local economic growth [20,21]. The precise role of township governments could thus be a key issue in relation to informal land development in China.

However, this role is often ignored in the existing theoretical explanations, which talk of urban entrepreneurialism, "local-state corporatism" and "growth coalitions." Generally, a township may play two roles. On the one hand, a township government is the state authority at the grassroots level, and it is the direct manager of various activities also the executor of the state laws and regulations specifically in the rural area. However, the enforcement powers of township governments have been highly restrained since the introduction of China's new Constitution in 1982, which established a regime of village autonomy in rural areas. Since then, there has been a slow retreat of state government power from the internal affairs of villages [22]. Consequently, the village cadres now manage internal affairs and also negotiate about external affairs on behalf of the villagers. These cadres are selected directly by villagers through democratic elections. Moreover, a township government has no power to appoint and dismiss village cadres and intervene in the internal affairs of the village, despite the fact that the village lies directly under its administrative jurisdiction. In many cases, the role of the township government in rural areas has actually shifted from a powerful governor to a development collaborator with villages [22]. This transition has significantly weakened the executive power and authority of township governments, including their capacity to manage informal land development in villages.

At the same time, township government is also undergoing a process of transformation to a service-oriented government, driven by China's marketization and decentralization [23], while under the ambitious goals of growth driven by the city government, maintaining social stability has become the priority task of township government. The improvement of local livelihoods by undertaking to reduce unemployment, increase villagers' income, and build roads and schools has become the most important mission of township governments [6,18,20]. However, the low levels of fiscal revenue received by township governments barely match these heavy responsibilities of providing public services and infrastructure in rural areas [22].

In such a context, informal land development on collective village land is one significant way that villages can increase their income and improve their quality of life. From this perspective, township government shares the same goals as the villages, rather than those of the state and municipal governments. Supported by financial resources from informal land development, villages are able to provide for their own livelihoods, compensating for the lack of welfare from the state government in the context of the dual rural-urban social security system. Consequently, township government may exhibit a unique attitude to informal land development, which could play a positive role in improving villagers' livelihoods in rural areas.

This study thus aims to explore informal land development in China, taking a perspective that is quite different from the conventional approach to informal land development, which considers it to be the result of the existing dual urban–rural land ownership system [2,4,13,19]. In contrast, the study addresses the failure of the implementation of state government land development regulations in rural areas. The paper focuses on the roles of township government, villages and private developers in the context of the process of decentralization and marketization in China, looking at the specific case of informal land development in Doudian town in the rural area of the municipality of Beijing. Officials from the township government were interviewed for the purpose of the study. It is on the basis of these interviews that the communication and interaction between various levels of government and the conflict between state regulations and the township government's responses to informal land development in villages under its jurisdiction will then be discussed.

2. Informal Land Development in China: A Theoretical Review of the Institutional Factors

Institutions "are the rules of the game in a society, or more formally, are the humanly devised constraints that shape human interaction" [24]. When it comes to institutions related to urban planning in China, there are four levels of institutions: informal institutions, which are deeply embedded values, norms, customs, and traditions; formal institutions, which are rules codified by laws, regulations, administrative orders, and administrative statutes; governance institutions; and resource allocation institutions [25]. The paper analyses institutional factors affecting informal land development. Four types of institutional factors are addressed: land use management and planning (the dual urban–rural land system), governance structure and political decentralization, resource allocation and distribution of land benefits, and the attitudes of township government and officials to informal land development.

2.1. Seeds of Informality: the Dual Urban–Rural Land System

There has been extensive research on informal land development in China as it is increasingly drawing the attention of both the public and politicians [1,4,15]. Unlike informal land development in Latin America [26] and south Asia [27], where the core issue concerns the informal development of slums due to poverty in the context of rapid urbanization, in China, land use rights are the core issue in relation to the 'informality' of informal land development [4,28]. It is widely believed that informal land development in China is largely affected by the dual urban–rural land system [1,4,15].

This system was introduced in the 1950s. According to China's Constitution, there are two kinds of land ownership in China: state-owned land and collectively owned land. In urban areas, most land is owned by the state government and managed by a municipal government on behalf of the state. However, in rural areas, land is collectively owned by villagers. The village committee, which is selected by villagers, is responsible for the management of land in a village. Accordingly, there are two different forms of land use rights in China. The state government has full property rights to urban land, and this land use right can be transferred to the market. However, villages that own rural land do not have complete land property rights. Rural land may not be sold on the market unless the state government gives permission. If villagers and villages want to sell their land for development, they must first sell it to the state government. As a result, the state government monopolizes the land market and has exclusive power to set prices when they purchase land from the village farmers. In most cases, villagers receive a small amount of compensation for the loss of their land, while the state government gains considerable benefits in the form of land-transaction fees [1,29]. More specifically, rural land use is legally divided into three types: farmland, homestead land, and collective construction land. Farmland is limited to agricultural use. Homestead land is restricted to construction of houses that are only in resided by villagers themselves. The size and location of homestead land are fixed and allocated by township government according to household size. This means that when a family has new members, it can apply to the township government for more homestead land. The construction land in a village is only to be used to build the village's infrastructure and develop enterprise activities that promote the village's collective economy. According to the state land management regulations, any use that is inconsistent with these three types is considered to be illegal or informal. In other words, villagers actually do not have the same rights related to urban use or related developments, although they are considered to have ownership of rural land.

Informal land development in villages has several key characteristics [1]. Firstly, it involves the illegal development of rural land for urban uses; in most situations this means profitable housing or real estate development. Secondly, the development process has no permission from the land, planning, or construction authorities. The whole process, including the design, construction, and sale, remain private, and beyond any government control or authorization. Thirdly, the purpose of informal land development is to make a profit by selling the housing. These prices are far beyond the villagers' budgets and offer no improvement to their quality of life.

The dual land system has caused a state monopoly on the urban land market. Villages that have farmland use rights are not allowed to sell their land directly on the market. They must sell their land to the state and transfer their collective owned land to the state. As a result, the state government and its representative governments (provinces and municipal governments) have an exclusive right to sell land to private developers according to the existing regulations. In this sense, the state government is officially entitled to gain huge profits from selling villagers' land. This could have two main effects on informal land development. Firstly, the monopoly land market increases land prices and housing prices. The demand on informal housing with a low price is increasing, which drives informal land development. Secondly, the monopoly land market reduces villages' economic benefits from land development. This could force them to gain their own interests through informal land development. More details will be discussed in Section 3.

2.2. *Decentralization: Imbalance in Central/Local Fiscal Responsibilities*

A lack of balance in the central/local fiscal relationship increases pressure on city governments to pursue alternative sources of fiscal revenue in the context of decentralization in China. The reform of the tax distribution system in 1994 is considered a milestone of decentralization. The purpose of this system was to establish separate fiscal systems for the central and local governments. This taxation reform is considered to offer incentives to marketization by giving some autonomous tax rights to local governments. However, these autonomous rights remain very limited and this central/local tax distribution system is considered to have created an imbalance in fiscal income between central and local governments, with the former taking the greater part and the high-value taxes, leaving local government with fewer taxes of lesser value. For example, added-value tax is the most important and valuable tax in China, accounting for over 60% of total tax revenue. However, according to the new tax distribution system, the central government takes 75% of added-value tax and local government the remaining 25%. This lack of balance in the central/local tax distribution has put huge fiscal pressures on city governments. For example, in Guangzhou, which is one of the most developed cities in China, the total tax revenue in 2012 was CNY 3379 billion. However, more than 75% of this was handed over to the central government [30].

In the meantime, city governments have come to be predominantly responsible for managing local public affairs, including providing public services and welfare. This trend was further strengthened by the increase in local autonomy in the context of decentralization. Currently, the central government is responsible for managing national affairs using state taxes, while local governments are responsible for providing local public services and infrastructure relying on local tax revenue. These local public services primarily include education, medical care, social housing, roads, *etc.* As the tax revenue of city governments is rather limited, they face huge pressures in providing these local public services. In particular, rapid urbanization, which has taken the form of an annual increase of nearly 10 million people in China's cities, has only worsened the situation. This lack of balance between fiscal revenues and responsibilities is thus pushing city governments to find other ways to increase local fiscal revenues, such as by selling land. A local revenue system based on "land financing" is emerging right at this moment.

"Land financing" refers to the fact that city governments heavily depend on selling land to gain fiscal revenue, and this is accepted by the central government. Land-transaction fees can be seen as compensation for insufficient local tax revenues since 1994. However, within a short span of two decades, "land financing" has gone far beyond rational levels. In many cities, the revenue from selling land accounted for 40% of total local fiscal revenue in 2012. In some cities, such as Hangzhou and Foshan, this percentage even went as high as 70%. In the context of "land financing", city governments compete for land resources. Because urban land is limited, rural land in the suburbs has become the main source of 'land financing' over the last two decades [15].

At the same time, township governments, villages, and villagers have also been attracted by the considerable economic benefits of selling land. The tax distribution reform also created fiscal pressures on township governments. In particular, the central government abolished the national agricultural tax in 2006, which was one of the main sources of fiscal revenue of township governments. Many studies have confirmed that township governments in China are now facing a financial dilemma [22,31,32]. In rural areas, a "Xiang cai xian guan" fiscal system is applied, which means that township revenues are managed by a county government rather than by a township itself. County government is an administrative unit under the city government, but is a higher level administrative unit than a township government. According to the system, township fiscal revenues should be submitted to a county government first, which then makes a fiscal budget and transfers revenues to the township government. Although this fiscal system effectively curbs corruption, it reduces the fiscal autonomy of the township government [22]. Moreover, if a county government is in financial crisis, township revenues are often misappropriated and withheld. This is a common problem among China's township governments [22].

As a result of these factors, township governments are in a disadvantaged position when it comes to receiving a fair share of fiscal revenues. Moreover, township governments are expected to take responsibility for providing public services in their territory. In the context of the current administration system, township governments are actually required to be responsible for a variety of local affairs. They are often considered to be "small but comprehensive" [22]. As a result, township governments may give permission for informal land development to villages in order to fulfill their responsibility to improve local livelihoods and gain some fiscal income from the villages at the same time.

To sum up, decentralization is a key contributor to the growth of informal land development. On one hand, political decentralization causes a withdrawal of the state command control to local governments. A township government is given more rights to make decisions about land development inside their administrative boundary. In many cases, township government could give permission for informal development. On the other hand, political decentralization means fiscal autonomy for local governments. Both municipal governments and township governments are facing an increasing pressure to supply public services to local people in the context of rapid urbanization. Revenue from selling land and tax from property trade have become important income resources for many local governments. Township governments could permit villages for informal land development since informal land development can bring economic benefits to both villagers and township governments, and thus relieve township governments' fiscal burdens.

2.3. *Unfair Distribution of Land Benefits*

The question arises: why do villages and villagers not choose a formal way to develop their land rather than an informal method? As mentioned above, city governments, on behalf of the state government, actually monopolize the land market.

Rural land cannot be used and developed for urban activities unless the land is sold to a city government first and transferred from collectively owned to state-owned urban land; in other words, villagers do not have a land use right that encompasses urban activities. Nevertheless, the process of transferring rural land to urban land produces massive added value that mostly benefits the city government. This unfair distribution of the economic benefits of land development is a key incentive for villages and villagers to develop their own land themselves, rather than sell it to a city government [1,4,15].

According to the current land laws and regulations of China, rural land development for urban use can only be conducted by a city government through a formal procedure (Figure 1). Firstly, the city government expropriates the rural land from villages and pays land expropriation compensation to the villages/villagers. Secondly, the city government transforms the expropriated rural land into urban land by changing the planning laws or regulations that apply to it. Thirdly, the city government sells the land use right to developers and obtains land-transfer fees. Afterwards, the land is developed and the properties are sold or rented on the market. Notably, the land-transfer fee obtained by the city government and the added value of the property obtained by the developers are hundreds of times the amount of the land expropriation compensation paid to the villagers. According to current land management law, the standard for land expropriation compensation to villagers is based on the land's agricultural production value but not the added value on the market, which is equivalent to about 10 to 30 times the actual agricultural production. For example, one hectare of farmland might produce 15,000 kilograms of grain, which is worth about CNY 30,000. However, when this one hectare of farmland is sold on the market, the city government could receive revenue of about CNY 90 million. More specifically, some studies have estimated that in a formal rural land development process, villages and villagers only obtain 5% to 10% of the total benefits, while the government obtains 30% and the developer 40% [1,29]. In this context, informal land development by villages can be seen as a form of resistance to the appropriating of land benefits by the state government [1,4].

Informal land development in villages primarily occurs in one of two patterns (Figure 2). The first is when a village develops rural land in the village by itself. In this scenario, the village raises funds for housing or commercial development and the land development process is completely controlled by the village committee. After the housing is developed, the village sells the housing on the market. The second pattern is more likely to occur in a situation where funds are insufficient. In this case, villages collaborate with private developers in informal land development. Such a collaboration is usually organized in the form of a village–developer cooperation alliance. In both cases, the key purpose is to capitalize on and profit from the added value of selling rural land on the market. By taking this informal approach, villages and villagers can directly control the distribution of benefits. It has been estimated that by conducting informal rural land development, villages and villagers obtain 70%–100% of the total benefits, while the developer obtains 30% and the city government reaps no benefits [1,29].

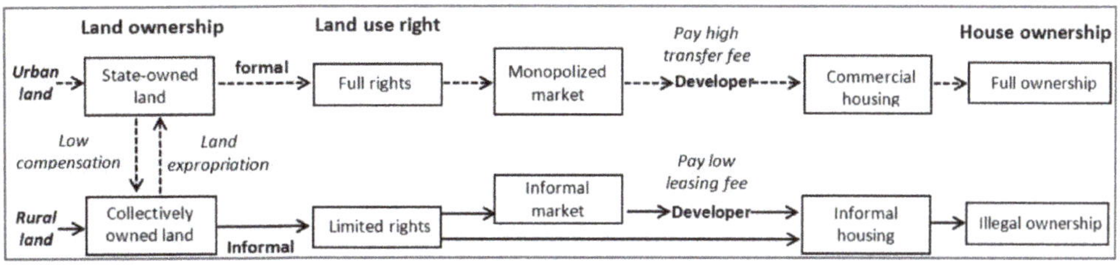

Figure 1. The process of formal and informal land development in China. Source: the authors.

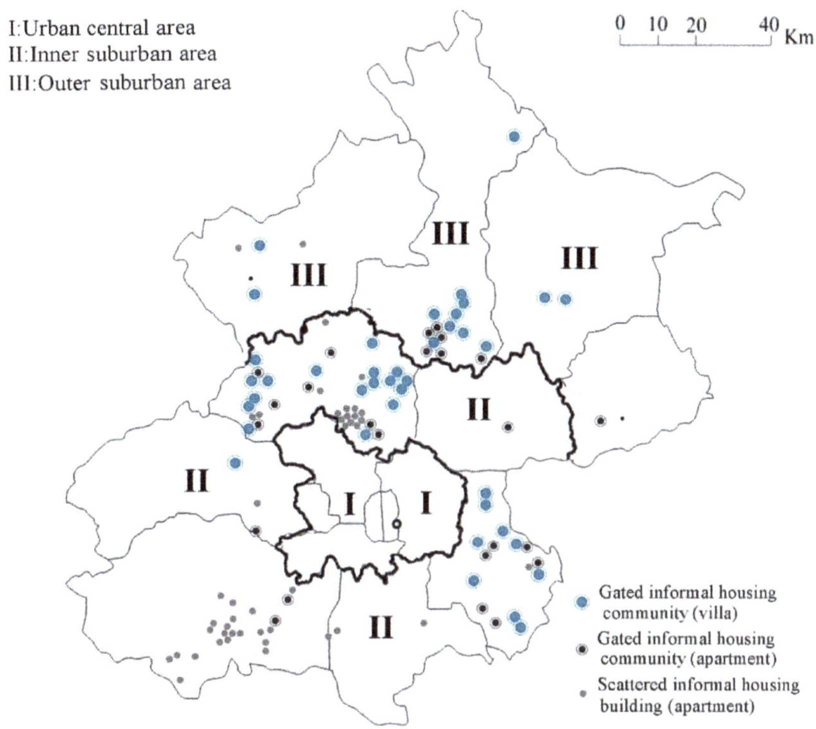

Figure 2. The location of informal housing in Beijing. Source: the authors.

2.4. *Fragmented Management across Governments*

It is noteworthy that development management has become fragmented in the process of decentralization and marketization. Different levels of government do not always share the same goals [20]. There is empirical evidence for such fragmented management of land development; for example, in Canton [33] and Hainan [20], to name just two. This can be explained from two perspectives: the complicated relationship between the central and city governments, and the changing relationship between city and township governments.

As mentioned above, economic growth is a common goal of both the central and city governments. Urban growth is in the nature of an entrepreneurial city government [34], which most Chinese city governments have now become [13]. However, the central government is also greatly concerned about enhancing its food supply and food security in the context of rapid urbanization. A rigorous goal of 90% food self-sufficiency has been highlighted by the central government. Therefore, it takes a strong position on the preservation of farmland or arable land. The Ministry of Land and Resources is a rather powerful bureaucratic force that issues strict farmland preservation goals and specific targets to city governments. Monitoring farmland preservation is carried out periodically by satellite survey. In this context, although the central and city governments share the common goal of economic growth, the bottom line of the central government on farmland protection becomes a barrier to city governments who wish to develop land to achieve growth. Under the policy of decentralization in particular, where a city government is expected to behave like a business enterprise that has to be self-financing, rural land is the most valuable resource for gaining revenue and attracting investment. Therefore, the relationship between the central and city governments is complicated when it comes to the trade-off between economic growth and farmland conservation. While city governments demand rural land for development to gain fiscal revenue, the central government aims to preserve it.

The inconsistency in the management of rural land development seems to be more problematic when looking at the relationship between township government and the higher levels of government. Township governments often have a positive attitude towards informal land development, such as in Foshan [35] and Shenzhen [21]. Township governments may not share the goal of growth with their respective city governments. As we saw above, one of the major missions for a township government is to guarantee social stability and improve the quality of life in rural areas. Selling and leasing rural land to factories and enterprises or self-developing housing for sale on the market are two important ways for township governments to create jobs and increase villagers' income. Although recognizing the illegality of informal development in villages, township governments either support or turn a blind eye to these development activities. In some cases, township government even helps villages to conceal informal land development to avoid controls from the city government and the Ministry of Land and Resources [20]. This inconsistency in the goals and needs of city and township governments in relation to informal land development occurs in many provinces, for instance in Hainan [20] and Henan [36].

Moreover, there is a difference between urban and rural land management. Urban land development is strictly regulated by a sophisticated regulation system; that is, "One approval for development location choice and two permissions for land use and building construction" (Yishuliangzheng). The whole development process, from site selection, planning, and design to construction, is strictly controlled by the city government. The Urban Planning Law enacted in 1990 ensures this management system. However, there was no law for rural land development or construction regulation until 2008 when a new Urban and Rural Planning Law came into effect. Before 2008, construction in villages was completely outside the control of state and city governments. Although this new law demands a strict management system for land development in rural areas, this system is still considered to be impractical because township governments are required to be the main management agency for rural land development.

3. Informal Land Development in Beijing

3.1. Research Method

The original investigation presented in this paper was conducted during the period from October to December 2015. The investigation is one research content of the national key project "Investigation of the roles and functions of township in China", which is funded by the Ministry of Housing and Urban & Rural Development of China. In the study, qualitative methods are applied to investigate the role of villages and townships in informal land development in China by looking at Beijing as a case study. Firstly, the overall situation of informal land development in Beijing is analyzed. Secondly, the case of Doudian town is analyzed. Several in-depth interviews with local officials were conducted to reveal the deep institutional factors involved in informal land development.

There are several reasons why Doudian town was chosen as a case study in Beijing. Doudian is a typical suburban township on the city fringe of Beijing (Figures 2 and 3). The national highway G7 and several local expressways cross Doudian. Doudian has a high level of accessibility to the Fangshan district center and the city center. There are 30 rural villages inside its administrative boundaries. Most land in Doudian is collectively owned by villages. Like other suburban townships, Doudian has experienced a rapid urbanization process since the 2000s. Thousands of migrants have moved in Doudian. The huge demand on housing brings an opportunity for informal land development. Detailed information about informal land development in Doudian is introduced in the next section.

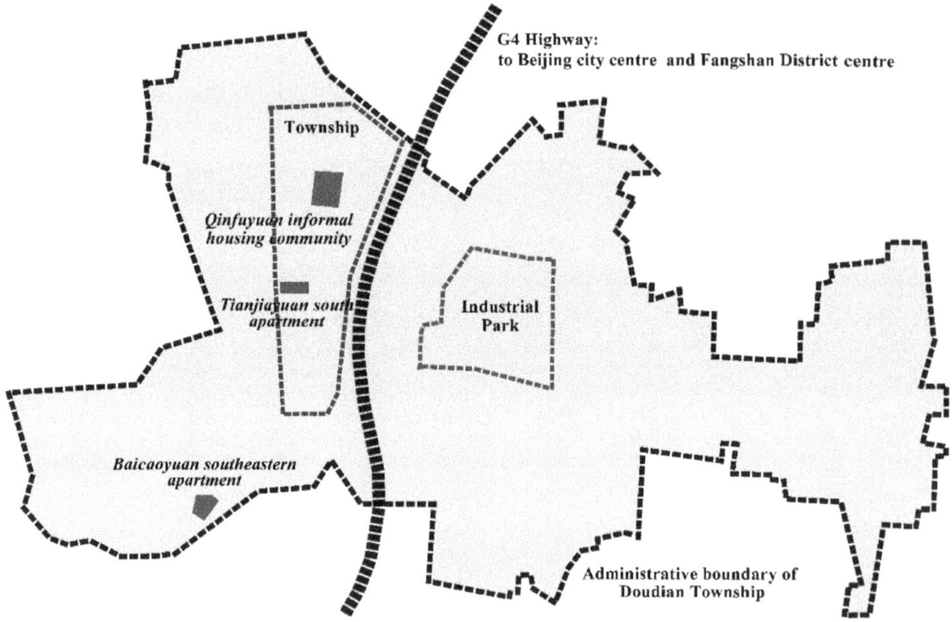

Figure 3. The layout of Doudian town.

The authors conducted face-to-face semi-structured interviews with nine officials. Four of them are in charge of urban and land use planning and management in Doudian township. Three of them are government financial officials in Doudian township. The other two interviewees are from the mayor's office and are familiar with the township land management in Doudian. These interviewees were asked how informal land development occurred in Doudian, who developed the informal gated communities, who bought houses in these informal gated communities, what the local township government's attitudes are towards the informal gated communities, why the local township cannot stop the informal land development, and other questions related to land management. Each interview lasted 30 minutes to two hours. Some questions to the interviewees were open. After the interview, two of the local officials accompanied the authors to conduct field surveys.

3.2. The Overall Situation

The city of Beijing is China's capital. It has a land area of 16,410 km^2 and had 21.5 million residents in 2014. The administrative area of Beijing consists of three areas: the central urban area, the inner suburban area, and the outer suburban area

(Figure 2). Each of these areas includes several districts or counties. In the process of rapid urban expansion, the inner suburban area has become a major new area for housing and industrial development. As a result, the municipal government of Beijing has changed the status of rural counties in the inner suburban area into urban districts, as they already have a large urban population and cater for industry. However, it should be recognized that these newly identified urban districts still have many rural towns and villages inside their administrative boundaries. Most of the land in these new urban districts is still rural land that is owned by villages.

Not surprisingly, these districts in the inner suburban area have become the major areas in which informal land development and informal housing construction have occurred, with almost 80% of informal housing in Beijing located in villages in these areas. According to the "Real estate market report" [1], the total amount of informal housing was very high, accounting for 20% of all houses sold on the real estate market in Beijing in 2012. From 2006 to 2010, a floor area of 4 million m^2 of informal housing was developed [1].

Soaring housing prices and the state government's control of real estate speculation have only boosted the informal housing market. A 'restrictive policy on housing purchase' was applied in 2009 to stop real estate speculation in Beijing. According to this policy, residents without Beijing *hukou* (urban residential registration) are not permitted to purchase housing in Beijing unless they have paid social insurance in Beijing for five years. The *hukou* system was established in the 1950s as a means of controlling population movement between rural areas and cities and between cities [37]. The *hukou* system divided the population of the whole nation into two groups: people with urban *hukou* and people with rural *hukou*. In the pre-reform era, rural residents with rural *hukou* were not allowed to migrate freely to the city and take advantage of the generally higher living standards there. Since the 1980s, reforms have been implemented, relaxing the *hukou* system and allowing migrants to work and live in the city, and this has been the main driving factor in the process of urbanization. The number of migrants from rural areas to cities increased from 70 million in 1993 to 210 million in 2008, thus exceeding 15% of the national population and accounting for 42% of all rural laborers (CSB, various years). Migrants to China's cities without local urban *hukou* are also called a "floating population". Since the 1990s, the implementation of reforms to the *hukou* system has increasingly become the role of the national government and municipal governments. The function of *hukou* as a tool of household registration has been declining. However, the *hukou*-related social welfare system and associated institutional arrangements still exist without obvious changes. Migrants without local urban *hukou* are still disadvantaged in accessing low-price housing subsidized by local governments, medical facilities, and even schools. Migrants had made up a huge part of the population in Beijing, with 7 million in 2010. Nearly 40% of these people have no

Beijing *hukou*. Most of them are low-income earners and are engaged in informal employment or temporary work with no social insurance and thus they have no capacity to buy formal housing. As a result, they have to buy informal housing, if any. Moreover, housing prices have been soaring since 2007. By 2015, the average housing price in Beijing reached CNY 30,000/m^2, meaning a two-bedroom apartment could cost as much as CNY 3 million. However, a worker's average annual wage is only CNY 77,000, with the ratio of housing price to annual household income now 20:1 on average. Housing affordability thus seriously challenges the quality of life for rural–urban migrants. With the price of informal housing usually one third of that of formal housing, there is a substantial informal housing market in Beijing.

In addition, the Ministry of Land and Resources introduced a policy limiting land supply for villas in 2006 in order to promote intensive land use. Villa real estate was thus inhibited in the formal housing market. However, there remained a huge demand for villas due to the increasing growth of the middle-class in China. In this context, informal villas developed in previously rural areas have gained popularity on the market, making up a sizable part of the total informal housing stock in Beijing (Figure 2).

The municipal city government is strongly opposed to these informal land developments, with a variety of approaches having been applied to deal with the gated informal housing communities, depending on the context (Table 1). Firstly, with respect to the gated informal housing community developments which occupy farmland, the city government claimed they must be completely demolished. An administrative form of punishment—for example, an oral warning—was delivered to township governments and villages where the informal housing had been developed. However, it is impractical to demolish this housing as it could directly cause a huge economic loss to thousands of buyers. If the informal housing in the gated informal housing communities was demolished, it would be impossible for the buyers to obtain compensation from the developers and villages because of the illegal purchase contracts. In addition, some commercial services developed for the gated informal housing community could be close as well, such as shops, private schools, medical centers, markets, *etc.* Thus, the demolition of informal housing actually encounters many difficulties and is often criticized by the public. The city government has already realized that it is very important to maintain social stability, and the demolition of informal housing may not be the best way of solving the problem. This is the reason why only seven out of 83 of the gated informal housing communities have been demolished thus far.

Table 1. The responses to the gated informal housing community from the city government of Beijing.

Situations		The Responses to Informal Housing	Case
Informal development occupying farmland	Sold out with occupancy	Strictly oppose, Punish a township government and village, Demolish	Liulinyuan flats, Daxing District
	Under construction	Strictly oppose, Punish a township government and village, Demolish	Nongjia villas, Tongzhou District
Informal development without occupying farmland	Sold out with occupancy	Oppose, Require retrofit to match state construction standards, Land use remains illegal	Fuyuan community, Changping District
	Under construction	Oppose, Confiscate and retrofit, Reuse as affordable housing or remain unused	Cunweihui community, Yanqing County

Secondly, in relation to informal developments that only occupy collective construction land in a village or the sites of the villagers' own homes, without occupying farmland, the city government has actually done little to prohibit them. In relation to the informal housing that was sold, the city government just required owners to retrofit these illegal buildings in accordance with the state construction standards. The land use of these housing sites remains illegal. In relation to the gated informal housing community that was under construction, the city government has confiscated it. In some cases, the confiscated housing of good quality has been legalized and used for the state's social housing. However, in most cases, the confiscated housing has remained empty and unused for many years.

3.3. The Case of Doudian Township

3.3.1. Development Context in Doudian

This section examines the case of Doudian town, a suburban township located in Fangshan District (Figures 2 and 3). Several informal housing projects were developed in the town, which has many advantages over other towns, such as proximity to the district center, a large population and industrial parks. The national motorway, the G4, connects the town to the city center of Beijing and other cities. Due to these advantages, Doudian town is growing fast. It had a population of 90,000 with a 64.58 km^2 land area in 2010, and it has attracted the attention of many real estate developers.

Like other suburban townships in Beijing, Doudian has witnessed the rapid growth of informal housing since the 2000s. Three informal housing sites have been confirmed by the municipal government in Doudian. They are the Qinfuyuan community in Yuzhuang village, the South Apartments in Tianjiayuan village, and the Southeast Apartments in Baicaowa village (Figure 3). Most of this informal housing was developed during the period 2002 to 2006. There are two types of informal land development: large-scale informal residential communities and scattered informal housing. The former are the gated housing communities, in which the quality of the housing is high (Figure 4). For example, Qinfuyuan is one of the largest informal gated housing communities in Doudian (Figure 3). There are 648 apartments in the community, in which nearly 3000 people live. The community is serviced by bus, shops, supermarkets, parks, *etc.*

Figure 4. Informal housing in Doudian.

3.3.2. Interviews about Informal Land Development

This study conducted in-depth interviews with local officers from the Doudian township government and the higher government of Fangshan District, investigating the roles and attitudes of the township government and the city government to these informal developments.

Theme 1: The Role and Attitudes of the Township Government

The attitude of the Doudian township government to informal housing was rather ambivalent. The officials in the government considered that it is impossible to completely eradicate informal housing and informal land development unless villagers' livelihoods are protected by alternative means. To some extent, the officials consider that the prohibition of informal land development may negatively affect villagers' livelihoods. Most villagers in Doudian are no longer active in agriculture, even though they are still peasants. Without an education or other skills, villagers are unlikely to find formal jobs in the city. At the same time, it is hard for the Doudian government to provide many employment opportunities or welfare to villagers, due to its low fiscal revenue. As a result, informal land development has become one of the most important sources of income for the villagers to ensure their livelihoods. The township government turns a blind eye, or even acquiesces to the informal land developments and informal housing in villages. The township government agrees that informal land developments have brought wealth to the villages and guaranteed villagers' livelihoods. An official revealed that one village had a CNY 100 million deposit in its account. They know the money was obtained through informal land development. However, the official stated that the township government holds a neutral position on this because the money helps the village to improve social welfare.

> *Interview 1: This village has CNY 100 million in its private account. The village may resettle villagers in multi-story buildings and lease or sell land to developers in both formal and informal ways. This is common because the village has to ensure that its villagers have a good quality of life. Although the villagers were relocated and live in multi-story buildings, they are still seen as peasants and cannot enjoy welfare benefits and insurance from the city government. This is because they are not urban residents but farmers with non-urban hukou. The old people, especially, don't have any source of income except bonuses from the village's private account. We [the township government] have not the ability to provide welfare to them either* [38].

> *Interview 2: We don't have money to provide enough jobs for the villagers. Street cleaning is the most common job that we can offer to the villagers. However, more and more villagers want this kind of job. But we don't have enough money to increase the number of such jobs, even though we need more people to clean the streets. Thus, a job position may be shared by several villagers. They are employed part-time and work at different periods of time during the day* [38].

Informal development of rural land has thus become a vital way for villagers to obtain an income and maintain their livelihood. Most villages in Doudian, especially those in close proximity to the township, are no longer involved in agricultural activities. Many villagers have leased their farmland to professional farming companies. However, these villagers are excluded from the urban social security and welfare system because they are considered to be peasants and are not granted urban *hukou*. These rural villages thus struggle to provide some form of welfare for themselves.

A collective economy lies at the root of all villages and there are many ways for a village to enhance this collective economy. One new method has arisen as a result of the introduction of a land quota policy in Beijing, designed to control the loss of land. Due to this policy, land for further development in the existing built-up area is now scarce, and the rural land in peripheral villages is becoming attractive to private developers, who approach the villages and encourage them to sell their rural land. While these private developers pay a land use right fee to the villages, it should be recognized that this transaction and thus the income from this sale of rural land is illegal and completely out of sight of the state and the municipal governments. As a result, there is also a degree of corruption among the village cadres, who appropriate land revenue for themselves. This has become a new social issue in rural areas.

The villages in Doudian town play an active and aggressive role in informal rural land development, collaborating well with private developers. However, it is interesting to note that the Doudian township government is rarely involved in any collaboration with developers. This contrasts with the conventional behavior of township governments, who generally do collaborate closely with private developers in informal land development. As mentioned above, the wealthy village that has saved CNY 100 million has a well-considered scheme to develop its land. Three years ago, the village cadres even intentionally moved the villagers from the village to new multi-story buildings, then cleared and prepared the land before a developer was chosen to develop the land. A small developer first negotiated with the village but failed. One year later, one of the best and powerful developers in China, the Vanke Real Estate Corporation, arrived on the scene and claimed they had an approach that would legalize the informal land for formal development. Subsequently, the village agreed to sell the land to the developer.

The Doudian township government had no say during the entire process. After the village and the developer had reached a deal, the township government even assisted the developer to make the illegal transaction legal. The township government actually gained a benefit in the form of a management fee that was submitted by the village. Nevertheless, the amount that the township government received was small, with most of the land-transfer fees given to the villagers by the developer in advance. Despite this, the township government did not object to the informal deal but, on the contrary, regarded this as a good way for villages to increase their income and improve their livelihoods. In addition, the township government welcomed other influential developers wishing to undertake urban development in the villages as they believed this could enhance the town's living environment, change the poor image of the town, and thus attract more investment.

Most of the buyers and residents now living in the informal housing are not locals. However, the township government has a very positive attitude to the immigrants and the buyers because they are relatively wealthy and contribute to local economic growth. Many villagers now run small businesses in the town, such as supermarkets, restaurants, and grocery stores, and immigrants are their main customers.

Interview 3: Half of the residents are immigrants. Most of them work in the city center, Changyang town center, and Liangxiang town (the government center of Fangshan district). They are important. The villagers have set up some small businesses to cater for these wealthy immigrants. Most groceries, restaurants, and other shops are operated by villagers and these immigrants are their main clients [38].

The township government even has sympathy for the informal housing owners. The infrastructure, such as electricity, gas, water, and drainage, in informal housing communities is usually provided by the developers and villages. Such infrastructure is separate from the municipal services because of the illegality of the informal housing. Most of these services are poor in quality and cause environmental pollution. The township government, indeed, has no duty to provide infrastructure and services to the residents living in the informal housing communities. However, with a strong concern for both the environment and social stability, the township government is very willing to provide infrastructure to the residents living in the informal housing communities rather than merely demolish them, as the city government wishes.

Interview 4: It is impossible to stop their (residents in informal housing) heating and water supply. They have to live there. The winter is so cold. But the pollution caused by burning coal is very serious. The drainage is so poor that sewage flows into farmland and pollutes underground water. If we had enough money, we would really like to provide good infrastructure for them [38].

Theme 2: The Weaknesses of the Township Government

As for the implementation of the regulations set by the city government, the township officials said that there is no doubt that they should implement the regulations, but that they have no enforcement powers to prevent informal housing development. The only thing that a township government can do is to warn the villagers and request them to stop informal land development before the officials from the city government come to enforce demolitions and impose fines. In other words, the land use and development management rights remain in the hands of the city government of Beijing and the land resource bureau and planning bureau of the city government remain the authorities in charge of land use and development management. However, these municipal administration departments have almost no direct relationship or communication with the Doudian township government. They communicate with the Fangshan district government, which has jurisdiction over Doudian township. This means the direct superior of the Doudian township government in relation to informal housing development is the Fangshan district government, rather than the powerful departments of the city government. This inefficient administrative structure has reduced the ability of Doudian township government to manage informal land development in its rural area.

> *Interview 5: We don't have enforcement powers. We can warn them (villagers) but it is useless. And it is really hard for us to monitor informal land development in villages. It (leasing or selling land) is their internal affair, we can't step in. And the villages are very complicated* [38].

The officials also highlighted a financial dilemma that the township government had been struggling with. The annual fiscal revenue of Doudian was approximately CNY 7.6 billion. This is relatively small compared to its expenditure needs for many local affairs. Since 2010, the township government has faced a challenge: maintaining fiscal revenue from land development as the central government introduced a national policy of "shrinking new land development and promoting land efficiency." The township is thus short of funds to support rural development. The officials stated that they must allow some villages to develop the local economy by themselves. For example, to develop a tourist village, one official emphasized, a primary task is to find funds to improve local infrastructure. Since both the township and the villages are short of funds, informal land development could be a realistic and practical way of securing start-up funds. In this situation, the township government is unlikely to oppose informal land development.

> *Interview 6: We don't have money and a land quota. It is impossible for us to develop the town economy without funds and land. If a village can transform into a tourism village, the villagers will have solid incomes. We support their ideas and their development activities. But without funds and land, how can we support them? The only thing we can do is to let them do what they want to do* [38].

The growing power of private developers is another major factor influencing informal land development in rural areas. The villages usually sell their land to developers in the form of a lease for a given period, such as 70 years. Subsequently, the developers build housing on the land and sell the houses to private buyers. This development is typically illegal since the land transfer was not legal. However, some powerful developers such as Vanke can ensure the land use is legal. They are wealthy and have the power to influence the local property market. They also have an influence on the local policymaking process at both the Doudian township level and Fangshan district level. In most cases, the developers not only pay tax to the township government but are also important contributors to local public facilities, such as schools, roads, and parks. The villages clearly realize the important role of these large developers and usually choose to collaborate with them, giving them permission for informal land development. These powerful developers, indeed, have sufficient political resources and alternative ways to make their development collaboration activities with the villages legal. When a coalition between a large developer and a village is formed, the township government has a very limited role when it comes to intervening in their development activities.

Theme 3: Irrational Land Use Policies

The officials considered that the existing land use policies designed to control land development in Doudian are unreasonable and unfeasible. Doudian town is located in the urbanized area and is relatively close to the center of Beijing. There is no need for the town to preserve a large amount of farmland. However, the existing land policy of the Ministry of Land and Resources still requires the town to preserve farmland. According to the land plan, some farmland designated for preservation is even located in the town's central area. These plans have impeded the improvement of the urban living environment and land use efficiency in Doudian. In addition, there are conflicts between the land use plan of the land bureau and the urban plan of the urban planning bureau. The land use plan is focused on the preservation of agricultural land, while the urban plan is focused on the provision of land for urban development. In some cases, these two types of plans do not match with each other. Land use plans often ignore the town's rapid urbanization. Many industrial and housing development plans cannot be put into practice because of a shortage of land.

However, it is difficult for the township government to change a land plan according to the existing land use policy. Moreover, China's land use and development is strictly managed by a land construction quota system. The central government gives a certain land development quota to provincial governments, which distribute this land quota to cities. Afterwards, the city government distributes the land quota to districts and counties, which then make plans for land use and urban development according to the land quota. Thus, a township government has no right to make a decision on land quotas. In practice, it rarely receives a sufficient quota to allow development, as most of the land quota is taken by the city government for development in the urban areas.

> *Interview 7: I don't know why the higher government is still asking us to preserve so much farmland. The villagers are no longer coming back to do farming. Some preserved farmland is even in the town center. New roads have to change direction to avoid it (preserved farmland). It is a waste and a bad thing for the town's landscape. The developers, investors, and buyers won't like such weird surroundings* [38].

Industrial land use policy is another barrier to development in Doudian. While there are industrial parks and manufacturers in Doudian town, the officials from the township government do not think the township actually benefits much from these huge industrial and development projects, which are led by the city government. Most of the taxes paid by the industrial enterprises are collected by the city government. Thus, the township government receives a small amount of the tax but contributes with a large amount of land. For example, Changan Automobile Group, one of China's largest car manufacturers, built a large-scale R&D industrial park in the town. This project was actually led and supported by the city government. Doudian township government, however, has a negative attitude to the project. The reasons are complicated. Firstly, this R&D park has an independent management committee that is directly led by the city government. The township government has no rights to intervene in the development of the industrial park. Secondly, this industrial park uses the land quota that should have been allocated to Doudian. However, the township government did not receive any land-transfer fees because the city government waived them in order to attract the manufacturer. Thirdly, this industrial park is only used for R&D, with the main production lines, sales agencies, and headquarters located in Chongqing City. This means that the industrial park only brings in a small amount of revenue to the township government.

> *Interview 8: It (the industrial park) occupies our construction land quota. The city government provides much support and the land use is legal, but we receive little tax. The manufacturer registered its office in Chongqing and there is no production here. It has its own committee. We can't manage it. Most of the benefits go to the higher (city) government; we receive little* [38].

Nevertheless, the township government acknowledged the potential benefits of the industrial park, which offers some low-end and temporary jobs to local villagers. The industrial park even provides training for villagers in various work skills. The township government agrees this could assist in reducing unemployment.

> *Interview 9: It [the industrial park] welcomes the villagers to work. These jobs are very easy and flexible. The location is close to the villages. Most villagers have to look after their elderly parents in the villages. Therefore, these jobs are very popular among villagers. The company even trains some young villagers and sends them to other factories in other cities. It's very helpful to improve their livelihoods* [38].

To sum up, the above interviews and discussions unfold the role of township governments and villages in the development of gated informal housing communities. Doudian township government has an ambivalent altitude to informal land development, although the township government is facing strong requirements from the state government to control informal land development. In most cases, township governments may give tacit consent to villages' illegal land development activities. Apparently, there are growing conflicts between the state control and local land management. One of the major reasons for this is that informal land development could bring economic income to the villages and the township. Another reason is that the township actually has limited power to control informal land development in the context of marketization. The market-oriented reforms make land management more fragmentized. Multi-scale forms of governance with multi-stakeholder arrangements (MSAs) emerged in urban development in China [39]. In this multi-stakeholder arrangement system, the conflicting interests are allowed and the negotiation between the government, village and other sectors replaces the mere command control from government. As a result, the increasing market activities in local economic and land development and multi-stakeholder arrangement weaken the management power of a township government. The situation seems to become worse when the disadvantages of the existing land use policies are taken into account. In particular, the irrational land use policies from the state bring challenges to the implementation of the land development management for township government.

4. Discussion and Policy Implications

This paper analyzed the processes behind the new trend in informal land development and related factors in China. The study took a different perspective from the conventional understanding of informal land development as the result of the existing dual urban–rural land ownership system [2,4,19]. Instead, looking at the case of Beijing and the township of Doudian in particular, the paper addressed the role of villages and townships in informal land development and the conflicts with state regulations. In particular, several different types of institutional barriers to land development management are addressed. They are: land use management and planning, governance structure and political decentralization, resource allocation and distribution of land benefits, and the attitudes of township government and officials to informal land development. Based on the analysis above, several important points in relation to institutional capacity building are discussed below.

Firstly, the conflict between state regulations and the need for local development could be the most important reason for informal land development on the city fringes. The previous literature has found that, according to the existing land development management system, the distribution of land revenues arising from the process of rural land development between villages and the state is highly imbalanced [1,4]. Most of the revenues are collected by the state and city governments, with the villages and their inhabitants only receiving a small proportion of the revenues that are produced from their land. Thus, informal land development could be considered a form of resistance by villages to the loss of economic benefits accruing from the added value of the land [16]. Grassroots governments, for example township governments, could have more direct and important effects on informal land development.

The analysis above demonstrates that township governments, indeed, have a rather ambivalent attitude towards informal land development. Rapid urbanization has meant that most villagers no longer engage in agricultural activities. However, the state offers little welfare and social security assistance to these unskilled villagers. Township governments will not prevent informal land development and even believe these informal developments could play a positive role in improving the quality of life in the rural areas. In reality, township governments have very limited power to control informal land development when a coalition between villages and powerful developers is formed. Growing marketization is giving more power to developers. In this context, the role of township governments in managing rural land development is only declining more.

Township governments do not always stand in line with the city government. As mentioned above, unlike the city government, which focuses on economic growth, township governments pay more attention to social stability and the quality of life in rural areas. To a large extent, township governments tend to share the same goals as rural villages, rather than the state and the city governments. Informal rural land development is seen as one way to increase villagers' income and obtain revenues to provide welfare and public services in rural villages. Especially in the context of decentralization, township governments are faced with a shortage in finances but an increase in their responsibilities in the provision of public services. In this context, informal land development may relieve the township government's burden of providing services in rural areas.

In this light, it is apparent that "local-state corporatism" and urban entrepreneurialism do exist in China, but may merely occur at the level of the city government. Both villagers and township governments may actually become victims of such "local-state corporatism" if they only obtain small economic benefits from land development projects led by the city government. In some instances, the city government actually expropriates scarce and valuable construction land from township governments to support city-level development schemes. This finding is inconsistent with previous studies that suggested that township governments play an active role in such "growth coalitions" [17,18].

Moreover, township governments re facing an institutional dilemma in managing land development in rural areas. Township governments are expected to be responsible for both farmland preservation and construction management in villages. However, township governments have a very limited capacity to implement municipal regulations. The process of decentralization in China has had various effects: more state powers have been transferred to provincial and city governments, while village autonomy has also increased. However, enforcement powers remain more centralized, lying not with district and township governments but at the city government level. As a result, township government suffers from a variety of weaknesses in managing informal land development in villages.

In addition, the fragmented planning system in rural areas is another factor influencing informal land development. In China, urban planning provides the legal basis of urban land use and development management. Any development and construction permissions must be issued under the statutory planning authority. However, urban planning regulations are often ignored in rural areas. One of the reasons for this is that the land is collectively owned in these rural areas. There are also conflicts between urban planning and land use planning. While the latter is focused on farmland preservation, the former is focused on fostering urban development. A township that is keen to develop its local economy is often faced with a complex process of negotiation between land use planning and urban planning authorities.

According to these findings, policy interventions that tighten regulations without institutional reform may be insufficient and impractical to curb informal land development. It is irrational to arbitrarily prohibit informal land development, because it is one important way of improving villagers' livelihoods. Policy interventions for the purpose of curbing informal land development should instead aim to: (1) meet villagers' needs for a sustainable livelihood; (2) balance fiscal revenues and fiscal responsibilities between the central and local governments; and (3) give greater administrative power and autonomy to township governments. These goals could be achieved through the following institutional changes.

Firstly, villagers' livelihoods should be further guaranteed in a sustainable manner. It is irrational to expropriate rural land by giving a one-off compensation to villagers. Many studies have indicated that peasants who have lost their land have no skills that would allow them to find a job, and they do not have an appropriate plan to use the compensation to further their livelihoods [40,41]. Once they have spent all of their compensation, they have nothing left to live on. Land transfer should thus be linked with social welfare in rural areas. In addition, city governments often fail to grant farmers urban *hukou* after their lands are expropriated. As a result, farmers who lose their land cannot enjoy basic urban welfare and social insurance, even though their lands have become part of the city. Therefore, it is necessary to adopt policies that ensure full and reasonable economic benefits to villagers when their land is expropriated.

Secondly, township government, a direct administrative unit in the rural area, should be given sufficient rights and powers to manage land development. In practice, the central government has paid a great deal of attention to the dilemmas faced by township governments. Administrative reform aiming to give more powers and liberty to township governments has already begun. At the same time, however, it is necessary to allocate more taxes to township governments in order to enhance their capacity to improve the quality of life in rural areas. Service-oriented grassroots government should be established.

Thirdly, it is not feasible to apply universal land policy to all towns across the whole country. Some towns, especially those located close to metropolitan regions, are faced with great pressure to develop and urbanize. These towns may already be in a rapid process of transformation from a rural to an urban area. The villagers in these towns are no longer engaged in farming. This means that strict farmland preservation in these towns is possibly impracticable. Therefore, different land policies should be designed and applied according to local conditions. In other words, for towns in urbanized areas, land use policy should be more flexible.

Finally, urban growth need not occur at the expense of village and township development opportunities. Land revenues from rural land development should be fairly distributed between the state government, villages, and township governments, with the latter given access to reasonable land revenue in order to build infrastructure, supply public services, and provide better welfare in the rural areas.

5. Conclusions

This paper discussed the phenomenon of informal land development, which has taken place since the 2000s, and considered the implementation of state land development regulations at the grassroots level, such as by village and township governments. In particular, it investigates the role of township government and villages in the development of gated informal housing community. The conflict between state regulations and local development needs is one of the most important factors affecting informal land development on the city fringe. Decentralization and marketization have significantly changed the roles of different levels of government. While "local-state corporatism" and urban entrepreneurialism do exist in China, this mainly occurs at the level of city government, with both villagers and township governments likely to be casualties of this "local-state corporatism" in the current land development management system.

Decentralization and marketization have significantly changed the roles of different levels of government. Township governments do not always share the goal of growth with their respective city governments. Unlike the latter, which are keen to stimulate economic growth, township governments are more concerned about the quality of life and social stability in rural areas. Consequently, township governments have an ambivalent attitude towards informal land development, which actually brings economic benefits to local villagers and themselves. The situation seems to be worse as townships have poor fiscal capacity and a growing administration responsibility for improvement of local development in the context of decentralization. Villages are keen to capture economic benefits from informal land development with help from private developers. As a result, a locally informal coalition between townships, villages, and private developers emerged at the grass-roots level.

These findings reveal the complexity underlying informal land development in China. In relation to future policy, informal land development will probably continue unless the issue of villagers' livelihoods is addressed, and the lack of balance between different levels of government in relation to fiscal revenues and responsibilities has been resolved. Policies designed to curb informal land development should integrate social welfare and long-term sustainable goals. In particular, inclusive social policies should be addressed in future rural land development policymaking.

Acknowledgments: This research is funded by the NSFC project No. 41571147.

Author Contributions: Pengjun Zhao wrote the main framework, the introduction, the discussion and the conclusion section of the paper. Mengzhu Zhang wrote the literature review and the analysis section of the paper, and contributed to the date collection.

Conflicts of Interest: The authors declare no conflict of interest.

References

1. Ren, Z.Q. Limited Property Right Problem Research in China: The Present Situation and the Way Out. Available online: http://blog.sina.com.cn/s/blog_4679d3510102dtt0.html (accessed on 1 March 2012). (In Chinese)
2. Lin, G.C.S.; Ho, S.P.S. The state, land system, and land development processes in contemporary China. *Ann. Assoc. Am. Geogr.* **2005**, *95*, 411–436.
3. Zhou, Q. *The Urban and Rural Areas of China*; China Citic Press: Beijing, China, 2013. (In Chinese)
4. Wu, F.; Zhang, F.; Webster, C. Informality and the Development and Demolition of Urban Villages in the Chinese Peri-urban Area. *Urban Stud.* **2013**, *50*, 1919–1934.
5. Zhang, L.; Zhao, S.X.B.; Tian, J.P. Self-help in Housing and Chengzhongcun in China's Urbanization. *Int. J. Urban. Reg. Res.* **2003**, *27*, 912–938.
6. Zhao, P. Managing urban growth in a transforming China: Evidence from Beijing. *Land Use Pol.* **2011**, *28*, 96–109.
7. Nandao Evening News. Sanya City Yesterday Demolished 12000 Square Meters of Limited Property Right Houses. Available online: http://ndwb.hinews.cn/html/2018/14/content_7_1.htm (accessed on 14 August 2014). (In Chinese)
8. Xinkuaibao News. Give all to Purchase Limited Property Right House, 40 Workers Lose Everything. Available online: http://epaper.xkb.com.cn/view/752641 (accessed on 31 December 2011). (In Chinese)
9. Hangzhou Daily. Beijing Started to Demolish Limited Property Right Houses. Available online: http://hzdaily.hangzhou.com.cn/dskb/html/2016/03/content_880826.htm (accessed on 3 June 2010). (In Chinese)

10. Beijing Daily. Hebei Publicly for the First Time Demolished Limited Property Right Houses. Available online: http://finance.chinanews.com/house/2013/16/5621008.shtml (accessed on 6 December 2013). (In Chinese)
11. Jiangsu News. Compensation for Demolition of Limited Property Right Houses in Tiexinqiao raised questions. Available online: http://jsnews.jschina.com.cn/system/2014/12/24/023065072.shtml (accessed on 23 December 2014). (In Chinese)
12. Chow, G.C. *China's Economic Transformation*, 2nd ed.; Blackwell Publishing Ltd.: London, UK, 2007.
13. Wei, Y.D. Decentralization, marketization and globalization: The triple process underlying regional development in China. *Asian Geogr.* **2001**, *20*, 7–23.
14. Chien, S.; Wu, F. The Transformation of China's Urban Entrepreneurialism: The Case Study of the City of Kunshan. *Cross Current. East Asian Hist. Cult. Rev.* **2011**, *1*, 1–28.
15. Hua, S. *Transformation of Urbanization and Land Trap*; Oriental Press: Beijing, China, 2013. (In Chinese)
16. Zhu, J. Urban physical development in transition to market. The case of China as a transitional economy. *Urban Aff. Rev.* **2000**, *36*, 178–196.
17. Zhang, T. Urban Development and a Socialist Pro-growth coalition in Shanghai. *Urban Aff. Rev.* **2002**, *37*, 475–499.
18. Zhu, J. Local growth coalition: the context and implications of China's gradualist urban land reforms. *Int. J. Urban Reg. Resear.* **1999**, *25*, 534–548.
19. Guo, X. Land expropriation and rural conflicts in China. *China Q.* **2001**, *166*, 422–439.
20. Wang, Y.; Scott, S. Illegal Farmland Conversion in China's Urban Periphery: Local Regime and National Transitions. *Urban Geogr.* **2008**, *29*, 327–347.
21. Li, L.; Li, X. Redevelopment of urban villages in Shenzhen, China–An analysis of power relations and urban coalitions. *Habitat Int.* **2011**, *35*, 426–434.
22. Zhou, Z. The Integration of Urban and Rural Development in the Process of Development of Township Government Governance. Ph.D.'s Thesis, Soochow University, Suzhou, China, 2014. (In Chinese)
23. Xu, Y.; Zhu, T. Service-oriented township government: reasons and construction. *China Publ. Admin.* **2009**, *294*, 86–89. (In Chinese).
24. North, D.C. *Institutions, institutional change, and economic performance*; Cambridge University Press: Cambridge, UK, 1990.
25. Zhao, P. The evolution of the urban planning system in contemporary China: An institutional approach. *Int. Dev. Plan. Rev.* **2015**, *37*, 269–287.
26. Gilbert, A.; Ward, P. *Housing, the State and the Poor*; Cambridge University Press: New York, NY, USA, 1985.
27. Nijman, J. A study of space in Mumbai's slums. *Tijdschrift voor Economische en Sociale Geografie* **2010**, *101*, 4–17.
28. Zhu, J. From land use right to land development right: Institutional change in China's urban development. *Urban Stud.* **2004**, *41*, 1249–1267.

29. Liu, R.; Wong, T.; Liu, S. The informal housing market in Beijing's rural areas: Its formation and operating mechanism amidst the process of urbanization. *Geogr. Resear.* **2010**, *29*, 1355–1368. (In Chinese)
30. China's National Bureau of Statistics (CNBS). *China's Fiscal Statistics Yearbook*, 2012; China's National Bureau of Statistics: Beijing, China, 2013. (In Chinese)
31. Zhou, F.; Zhao, Y. Analyses of the Local Government Finance: A Case Study in Central and Western China. *China Rural Surv.* **2003**, *4*, 25–37. (In Chinese)
32. Tan, Q. Fiscal Decentralization and Functions of the Township Government. *China Rural Surv.* **2002**, *2*, 2–20. (In Chinese)
33. Tan, X.; Yuan, Q.; Lu, B. Analysis of the Participation Mechanism of Villagers in Urban Village Reconstruction: A Case Study of Liede Village in Guangzhou City. *Trop. Geogr.* **2012**, *32*, 618–625. (In Chinese)
34. Boyle, M.; Hughes, G. The politics of urban entrepreneurialism in Glasgow. *Geoforum* **1994**, *25*, 453–470.
35. Yang, L.; Yuan, Q. A Model of Rural Urbanization Based on Development of Collectively owned Land at the Village Level: The Case of Nanhai, Foshan. *Urban Plann. Forum.* **2012**, *204*, 34–41. (In Chinese)
36. Zynews. Zhandian Township Government Involves in Informal Housing Development. Available online: http://news.zzynews.com/2014-09/10/content_9701162.htm (accessed on 10 September 2009). (In Chinese)
37. Zhao, P.; Howden-Chapman, P. Social inequalities in mobility: the impact of the hukou system on migrants' job accessibility and commuting costs in Beijing. *Int. Dev. Plann. Rev.* **2011**, *32*, 363–384.
38. Zhao, P.; (Peking University, Beijing, China); Zhang, M.Z.; (Peking University, Beijing, China). Personal Communication, 2015.
39. Zhao, P.; Lu, B.; Woltjer, J. Conflicts in urban fringe in the transformation era: An examination of performance of the metropolitan growth management in Beijing. *Habitat Int.* **2009**, *33*, 347–356.
40. Chen, H.; Chen, X.; Xie, Y. Occupational Differentiation and Influencing Factors of Land-lost Farmers during Urbanization. *China Popul. Resour. Environ.* **2013**, *23*, 72–79. (In Chinese)
41. Zhao, H.; Tong, X. Self-identity of the Passive-urbanized Group in the Process of Obtaining Urban Adaptability and Modernity: An empirical study on 561 land-displaced peasants in Nanjing. *Sociol. Stud.* **2006**, *23*, 86–106. (In Chinese)

Urban Residential Land Use Reconstruction under Dual-Track Mechanism of Market Socialism in China: A Case Study of Chengdu

Yongchun Yang, Deli Zhang, Qingmin Meng and Corrin McCarn

Abstract: We study urban residential land use changes by analyzing the massive migration and relocation of two typical social classes: employees in government departments (EGD) and urban demolition displaced households (UDDH). After the reform and opening-up of China that has taken place during the last 30 years, the residential land use of both the EGD and UDDH groups has been notable in terms of the changing urban landscape in China's cities. A considerable number of studies highlight the large scale relocations of weaker groups usually through sample surveys at a microscopic scale, which sheds light on the relationships between market forces and government intervention and power. However, employees hired by state government departments and related branches (*Shiye Danwei*) have been neglected. Bridging the empirical research gap and using Chengdu as a case study, we compare residential relocations of EGD and UDDH groups in Chengdu. Our analysis based on field surveys conducted from 2009–2013 indicates that the relocations of EGD and UDDH are spatially agglomerated due to China's unique dual-track mechanism driven by market forces and government power since the late 1970s. The study shows that most of the UDDH are migrated from urban centers to fringes, while a large number of EGD still agglomerate close to urban centers. Government interventions differentiate residential relocations of EGD and UDDH, and market mechanisms reinforce these relocations. Potential problems caused by the dual-track mechanism are finally discussed and summarized.

Reprinted from *Sustainability*. Cite as: Yang, Y.; Zhang, D.; Meng, Q.; McCarn, C. Urban Residential Land Use Reconstruction under Dual-Track Mechanism of Market Socialism in China: A Case Study of Chengdu. *Sustainability* **2015**, *7*, 16849–16865.

1. Introduction

Since 1978, one of the most significant features of China's transition from a planned to a market economy is the emergence of an enormous increase of migrants [1], which has been a hot topic attracting considerable attention in the literature from the early 1980s. Fast economic growth has brought not only large scale rural-urban migrations within metropolitan areas but also city wide massive migrations demonstrating new differentiation in the process of suburbanization or urban residential space restructuring (URSR) [2–6]. Instead of relying on the state or state-owned enterprises (*Shiye Danwei*) for welfare housing, the commercialization of urban housing establishes a free urban housing market [7–10] resulting in the government having allowed urban residents to buy and sell their houses since late 1980s [11]. Most families must turn to the urban commodity housing market to satisfy their accommodation needs [7,8,12–15], and a city wide influx of migration and mobility has appeared since the reform and opening up of China . Migration and mobility in tandem with massive urban development and urban fabric restructuring have created the phenomenon of the redistribution of different social strata.

Although suburbanization is common in China's large cities [2], the intensive gentrification has led to a new trend of increasing numbers of the upper and middle class living in the inner city. This trend also results in the lower class and displaced individuals moving away from city centers in China. Illegal land expropriation and forced relocation have become one of the most severe social conflicts in China [16], and unfair compensation is currently one of the most criticized issues [13,17]. The regeneration of old city areas has resulted in an increasing extent of relocation to suburban areas, which has also pushed the process of suburbanization. China has experienced three periods of rapid urban expansion, respectively in the early 1980s, early 1990s, and beginning of the 2000s, with construction of roads and industrial sites. This new construction has encroached on massive suburban and rural areas and resulted in many farmers becoming landless. The three waves of urban expansion, also called the three Chinese "enclosure" movements, have caused a total amount of 7.65 million hectares from 1987–2005 and 1.72 million hectares from 2005–2011 of agricultural land to be converted into construction land [18,19], and urban sprawl has resulted in a total of more than 40 million farmers becoming landless with about 2 million individuals being displaced each year [17].

The structural changes in urban residential space are caused by gradual and complex institutional reform, which is closely related to political, economic, and social development. Current research indicates three types of study scenarios: the market force hypothesis, the government power hypothesis, and the synthesizing hypothesis. Firstly, some studies consider that it is market forces, namely social income polarization and the privatization reform of urban housing, after initial reformation and opening up, that has led to the large scale relocation and relational changes of urban residential space [2]. To some extent, market forces are a decisive factor which limit the availability of residential space [15].

Secondly, some scholars consider that the social factors driving urban space changes, such as the individual's preference for residential space, lead the coalition in some cities. Examples of this influence would be land use for the illegal construction of urban villages in Xi'an and Guangzhou [20]. Given the gradual nature of China's institutional reform, the state-led marketization and transition periods of decentralization and globalization will not only result in the formation of emerging social classes but also influence the choice of residence of different social classes [21,22], which would redefine the structure of urban residential spaces in China. For instance, the availability of residential space is a state-led process in which local governments want to maximize their income from the real estate boom [17] and local officials tend to collect more land sale revenues from the urban land market [15].

Thirdly, many scholars believe that urban space changes are the result of the interaction and integration of government forces, market forces, and social forces [5,6,23]. Since the reformation and opening up, China has entered the transition from a planned economy to a market economy with gradual institutional reform. This transition has resulted in a dual-track mechanism—both government interventions and market forces come into play in the transformation of the social-economic system. The balance between government forces and market forces will continue to be present, and the changes of residential locations and relocation of different social classes reflect the roles and interactions of government, market, and related social forces. The relocation of urban demolition displaced households (UDDH) and the floating population living in "urban villages" is closely related to the current market mechanisms and government interventions and power in China [15].

A comprehensive analysis of residential relocation of UDDH and employees in government departments (EGD) at a city scale to better understand residential location and relocation is needed in order to explore the roles of government and market forces in both migratory processes and urban spatial restructuring. Existing empirical case studies are typically based on a limited number of districts or communities and focus on the relocation of individuals in either the upper or lower economic class. In this study, we do a comparative analysis of residential relocations of EGD and UDDH groups in order to develop an understanding of urban residential spatial restructuring by using the city of Chengdu as our study area.

Chengdu is one of the National Famous Historical and Cultural cities in China. Chengdu is also one of the third economic growth poles in western China, and in 2014 Chengdu had about 14 million inhabitants in its metropolitan area; 252 multinational corporations out of the world's top 500 have invested in Chengdu, which makes it rank third in China for attracting the top 500 companies. From the period of reform and opening up, Chengdu has experienced significant changes and rapid phases of urban expansion and social, spatial and structural changes, making it of high interest for urban research and administration study. Additionally, the residential relocation of UDDH and EGD among different cities in China since the 1990s has experienced a common progressive approach, although UDDH's relocation in Beijing and Shanghai could be more significant to urban spatial restructuring due to large scale relocations there. Our study is based on the data integration of face-to-face interviews (e.g., location, construction time, built-up area, building levels, overall housing units, land area for UDDH and EGD, and interviewee's attitudes to relocation and comments on related policy) and census data, which additionally is spatially examined with remote sensing data and GIS analysis; then, we conducted our empirical study, using and combining the path dependence, market socialism, and market transformation factors.

This paper is organized into sections below: we present a theoretical approach of the dual-track mechanism of government and market and its roles for the differentiation of residential relocations between EGD and UDDH in urban spatial restructuring in Section 2. In Section 3, we introduce our field survey including face-to-face interviews taking place in Chengdu. We summarize and analyze the surveyed data in Section 4. In Section 5, we discuss the driving force mechanism of the differentiation of residential relocations between EGD and UDDH, and explore the possible effects on urban residential spatial restructuring and potential problems related to marketization and government roles by comparing and contrasting with some international studies. The paper concludes with a concise summary in Section 6, and dictates that additional studies are needed to explore impacts on social, economic, and environmental development.

2. An Urban Migration/Relocation Model of Two Social Classes under China's Imperfect Market

In a free market economy model, the market is the only dominant force in resource allocation and price determination with the government simply regulating the market in order to ensure legal and fluid operation of market mechanisms. Existing market economic systems of the United States and other Western countries are chiefly in line with this theoretical model. This requires the various economic elements to be free flowing and an entire price marketization of products and services. The essence of the free market economy model is the liberalization and fairness of a market economy with the exception of market monopolies and illegal government interventions.

However, the gradual mode of China's institutional transformation has resulted in a distinctive market socialism, which is an incomplete market economy model [23–26]. This incompleteness of China's market socialism not only reflects various restrictions of several key resources or supplies by the government, but also the incomplete price marketization of key products and services. This has also led to the market monopoly formation of state-owned enterprises and strong government intervention; in fact, the urban real estate market is a typical case in China. Urban land leasing systems were established in the 1980s, but local authorities have been responsible for about 70% of fiscal expenditures after the 1994 tax and fiscal reform [27], which brings about the formation of an incomplete land supply market. Since the 1980s, it has become the local officials' responsibility to develop local economies, especially GDP growth, which is used to assess the performance of local governments. By taking advantage of increasing real estate prices and the skyrocketed land bidding prices, local governments attempt to boost revenue [28]. A fruitful source of revenue for local governments is to seek additional revenue from selling more land areas through public auction.

Additionally, local governments have adopted compulsory price and administrative levies to requisite rural land and develop it into urban land [29,30]. However, the conversion of rural land rights from the collective to the state is still deemed to be an administrative allocation rather than a market transaction, owing to China's ambiguous land rights [31] and the land leasing system only being applicable to state owned land. In other words, "the formal, legal framework of China's rural collective ownership remains weak and ambiguous" [32]. The compensation offered to the collective and landless farmers is far below the "real" market value of the land after conversion and unreasonably fails to provide a means for them to live in the cities [17]. Landless farmers lose their social networks, steady agricultural income, and the landowner's sentimental attachment to the land [33]. The compensation as part of a new act for the assembly of urban land released by China's state council in 2011 to replace the old one announced in 2001 is regulated as "no lower than the market price of the similar property in the vicinity", which indicates a low limit for the compensation schedule [34]. Land transactions have been in the form of bidding auctions since the 1990s, but the lowest price and yearly total supply of auction land depends on the local government [35,36]. Local governments need to get enough capital from the land market to ease financial pressure, support the construction of infrastructure, and compensate economic losses in order to provide favorable conditions to attract projects and investments, e.g., low industrial land premium and tax relief, *etc.* [37]. Thus, a sophisticated dual-track mechanism of market forces and government intervention and power formed in an incomplete market economy has largely affected China's real estate market and the process of residential relocations of different social classes such as EGD and UDDH against a backdrop of urban space expansion and reconstruction [38,39].

According to Alonso's theoretical model of urban growth and housing supply [40], urban spaces were roughly divided into the city center, inner residential area, industrial zone, and suburban residential zone through the interplay between the housing market and urban expansion. Urban centers in China have more comprehensive and advanced infrastructure with convenient services such as commercial, education, emergency care, health and sanitation services, *etc.* Under the context of dual-track urbanization in China [41], the rich often like to live as close as possible to the city center and/or the suburbs or rural areas or new urban areas with better environmental and service facilities [42], rather than move to the relatively poor suburbs or rural areas with much lower levels of infrastructure. Usually, the real estate prices farther away from the city center are much cheaper and more affordable for people with a lower economic standing. Therefore, a trend of urban residential space restructuring has emerged in China: the lower economic class gradually moves to the periphery, and the upper and middle classes increasingly agglomerate in the city center and move to the periphery or new urban areas at the same time. In this way, Chinese cities have simultaneously exhibited trends of rapid suburbanization and growing prosperity of city centers since the 1980s [39].

Under the context of dual-track urbanization in China [41], EGD residential units are seldom replaced but almost always renovated at their original sites, new areas adjacent to their work place, or even old residential areas. Moreover, the prices of this housing sold to EGD are only about 1/2 to 1/3 of market values because government subsidies adopt the standards of affordable housing and do not count land leasing prices in the interest of marketizing these parcels. In contrast, the parcels of residential units of UDDH are always marketed after demolition, and UDDH have to pay the expensive market prices of new housing. It is impossible for governments to pay out a reasonable price to the relocating individuals, but they are settled in 'affordable housing' or resettlement housing in the low-cost areas, which are typically lower economic urban districts away from urban centers. Therefore, EGD are becoming spatially centralized or move to new urban areas, but the UDDH group is gradually relocating away from the urban center and becoming increasingly suburbanized.

3. Study Area and Field Survey

Centered in the Chengdu Plain, Chengdu is a very large city in West China. The main city of Chengdu is characterized as one single center with three circle structures, a typical "Ring + Radiation"of transportation network and land use. Most cities in China share this common spatialstructure (Figure 1).

According to the 1981, 2001, and 2013 Chengdu Statistical Yearbook (http://tongji.cnki.net/ overseas/engnavi/NaviDefault.aspx), non-agricultural population of Chengdu increased from 1.92 million in 1980 to 3.46 million in 2000 and to 7.17 million in 2012, and its average rate of growth is 16.6% per year; simultaneously, the Chengdu built-up area was expanded from 60 km^2 in 1980 to 231 km^2 in 2000 and to 515.5 km^2 in 2012, and the average rate of growth is 28.5% per year. Urban regeneration and suburbanization have not only promoted spatial expansion and regeneration, but also have profoundly changed the social space structure. More importantly, the changes to residential areas of EGD and UDDH clearly reflect the city's political-economic and socio-cultural mechanisms. The resettled individuals in Chengdu are generally divided into urban demolition inhabitants and landless peasants on the urban fringes.

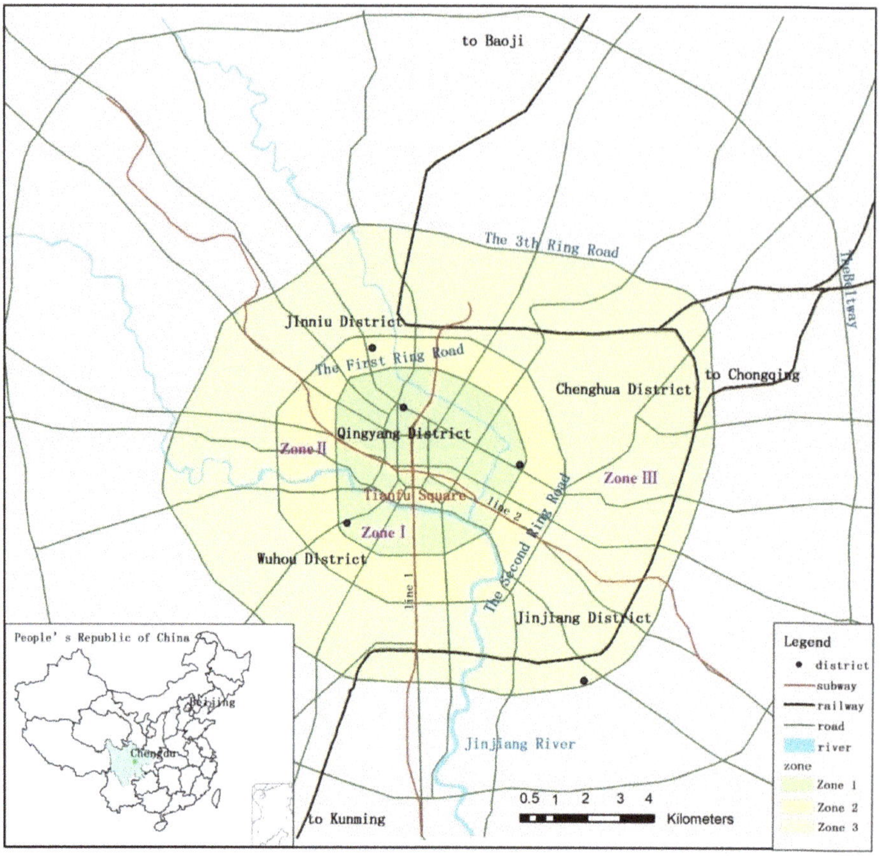

Figure 1. Location and urban spatial structure of Chengdu.

Our survey includes two parts. We first investigated about 300 interviewees, including 17 government officials, 15 enterprise managers, and the related residents by sampling typical residential areas for EGD and UDDH (Table 1) via face-to-face interviewing, which was conducted from May to October during 2010–2011 and in June and July 2013.We mainly chose interviewees about 40 years of age and older who experienced urban renewal and construction or urban demolition, *etc.* Female interviewees constituted 51% of the total, and interviewees with high school education and below accounted for about 40%. Questions concerning government policies and tools for residential relocation and migration, reasons for relocation, attitudes about housing construction enterprises, price decisions of affordable housing and resettlement housing, government employees and resettlement residents' attitudes to resettlement compensation, and opinions on current and optimal relocation sites were included in the interview.

The second survey is for detailed field data collection of residential parcels, including the spatial data and attribute data of separated living quarters and courtyards (e.g., sometime called Chinese quadrangles, a basic and traditional pattern used for residences and housing complexes) with house numbers in the survey area, and we conducted this survey from May to October in 2012. The surveyed attributes included construction time, housing type, residential parcel scale, land use type before the conversion to residential land, and the related details of Chengdu's residential courtyards or units. The scope of the sites and spaces of all residential areas were labeled and then geovisualized by using digital aerial photos and Landsat TM images.

Table 1. The second survey of a sample of residents for UDDH.

Num.	Name	Chinese name	Building time	Acreage (m^2)
1	Xiti Road community	西体路街区	Before 1980	93,795.9
2	No.6 community of 3rd section of Ring	居民房	19r 81–1985	9088.66
3	East 1st Road of Qingyang Community	青羊东一路小区	1986–1990	6235.18
4	Shiren district	石人小区	1991–1995	129,516
5	Mudian Road community	牧电路小区	1991–1995	75,626.8
6	Xiaojiahe community	肖家河	1991–1995	63,124.5
7	Dongguang district	东光小区	1991–1995	375,415
8	Changshouyuan community	长寿苑社区	1996–2000	140,613
9	No.202 community of Xinyi West Street	新义西街社区	1996–2000	126,089
10	Ruixiangyuan community	瑞康苑	1996–2000	43,426.2
11	NO.3 community of Jingtianlu	经天路社区	2001–2005	530,979
12	Jingzhu district	荆竹小区	2001–2005	302,120
13	NO.9 community of Jinsha	金沙路9号	2001–2005	127,728
14	Shengdeng Village Apartments	圣灯村公寓	2001–2005	189,273
15	Haitong community	海桐	2006–2010	189,930
16	Shufeng Garden community	蜀风花园城	2006–2010	125,890
17	River Phoenix community	沙河凤凰城	2006–2010	159,114
18	Orient New city community	东方新城	2006–2010	96,993.8
19	Orient Benefit city	东方惠城	After 2011	49,650.7
20	Jincui Garden community	锦翠花园	After 2011	37,721.4

Our study area is within the 3rd ring road of Chengdu (Figure 1). The urban fringes of Chengdu have been changing as part of the city's rapid urbanization process, and the city proper of Chengdu before the reform refers to the urban area by the edge of Jinjiang River and Fu River. Since 1980, both the rapid growth of urban population and the built-up area have promoted the formation of a new city proper which surrounds the single center of Tianfu Square and forms a "Ring and Radiation" spatial expansion pattern. The construction of ring roads in Chengdu reflects the urban sprawl process. The First Ring Road (FRR) was open to traffic in 1986, the Second Ring Road (SRR) in 1993, and the Third Ring Road (TRR) in 2002. In this study, we call the region within FRR Zone I, the area between FRR and SRR Zone II, the area between SRR and TRR Zone III. Surveying the residential households, we find that city dwellers usually distinguish their residential locations by relative distance to Tianfu Square and the nearest ring road. Currently, the citizens of Chengdu accept the region within SRR as the main part of the city, the southern and western regions between SRR and TRR as the urban built-up areas, and the northern and eastern regions between SRR and TRR as the urban fringe. The regions beyond TRR are the suburbs or marginal areas, which recently have continued to be developed from farm lands and villages to sparse built-up areas with a few resettlements.

4. Data Analysis and Results

Using ArcGIS 10.1, Landsat TM, digital aerial photos images, and our surveyed location and population information, we digitize the surveyed residential parcels, build topology, add the attribute data, spatially merge the adjacent lands with the same attributes, and generate a residential land database. A total of 2911 residential parcels have been surveyed, and we create a residential parcel and attribute geodatabase for spatial mapping and analysis.

The results of our spatial analysis show that the fastest increasing period of residential land expansion in Chengdu is in line with the expansion of resettlement areas (Figure 2). Residential parcels for resettlement are roughly distributed in different districts, but areas within Zone I only account for 6.76% of the total resettlement area, 25.75% in Zone II, and 67.49% in Zone III (Table 2). The construction sequence indicates the resettlement housing for UDDH gradually advances towards the urban fringes and the majority of new resettlement areas are closer in proximity to TRR (Table 2). The resettlement land proportion in Zone III is 20.87% in 1990, 46.20% in 1995, 79.74% in 2000, 95.48% in 2005, 99.02% in 2010, 100% in 2013; but the corresponding proportions in Zone II rapidly decline from about 54% in 1981–1995 to 17.77% in 2000, 4.46% in 2005, 0.98% in 2010, and the corresponding proportions in Zone I further rapidly decline from 63.06% in 1980 to 26.95% in 1985, 22.99% in 1990, 5.55% in 1995, and 0.06% in 2005.

The reform and opening up of China have lasted more than 30 years, and the provincial, municipal, and district residential units, in accordance with the principle of adjacency to workplaces, decrease significantly from the city center to the outskirts of the city. The construction times indicate that the government units and their residential courtyards in Chengdu were first established in the 1950s, and most existing residential parcels were built in 1980–1998 (Table 3). As such, 96.3% of existing EGD residential areas are within SRR (*i.e.*, Zone I and II) and, among them, 69.45% within FRR (*i.e.*, Zone I).

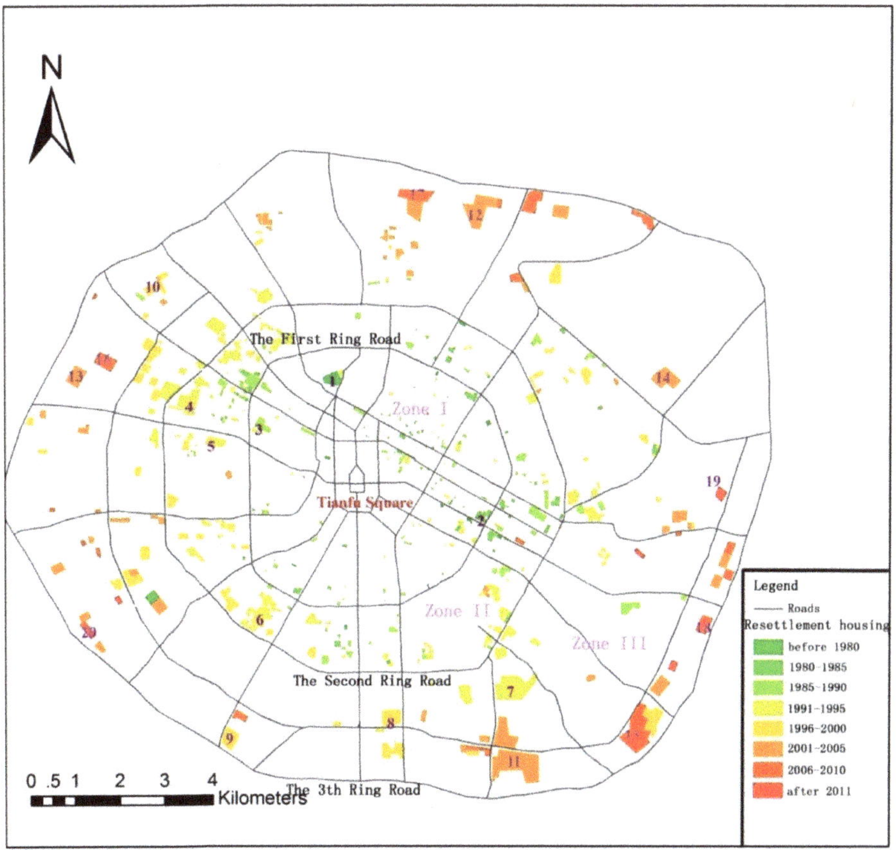

Figure 2. A spatial-temporal map of the resettlement housing land use for UDDH.

Table 2. The spatial-temporal changes of resettlement housing land use.

Construction time	Zone I (%)	Zone II (%)	Zone III (%)
Before 1980	63.06	36.94	
1981–1985	26.95	58.51	14.54
1986–1990	22.99	56.14	20.87
1991–1995	5.55	48.25	46.20
1996–2000	2.49	17.77	79.74
2001–2005	0.06	4.46	95.48
2006–2010		0.98	99.02
2011–2013			100.00
	6.76	25.75	67.49

We divide Chengdu EGD residential parcels into three types of provincial, municipal, and district units and analyze them respectively (Table 3, Figure 3). The government units typically occupy desirable locations, mainly within traditional urban centers, and have optimal natural and social environments. The construction peak of EGD residential units was during 1981–1995. Among the total EGD residential units, the provincial authorities occupied 62.27% of these in 1985, 58.62% in 1990, and 52.06% in 1995; while the municipal authorities occupied only 35.84% in 1985, 33.33% in 1990, and 36.99% in 1995.

Our surveyed data show that the higher the administrative level, the faster the construction and/or regeneration the residential land is conducted. In other words, the provincial EGD have the first priority for the construction and occupation of parcels of high quality, the second is the municipal units, and the third is the district units. The number of provincial EGD residential units in Zone I reaches 48.46% of the total EGD residential area, but the number of municipal authorities amounts to only 18.52% (Table 3). District-level authority units only account for 7.72% of the total EGD residential units within the total surveyed area.

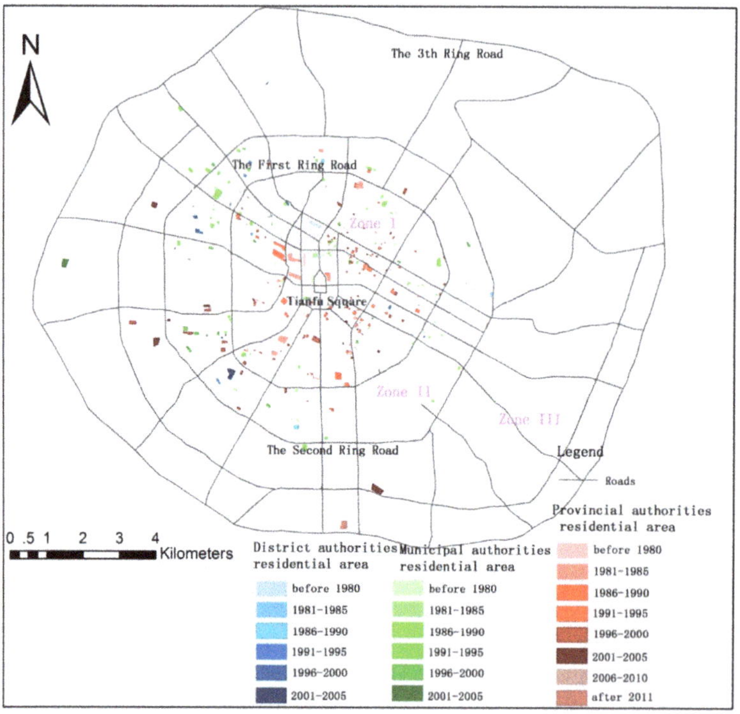

Figure 3. The residential distribution of employees in government departments (EGD).

Table 3. Spatial-temporal changes of residential land use for the three levels of government.

	Zone I (%)			Zone II (%)			Zone III (%)		
	Provincial	municipal	district	Provincial	municipal	district	Provincial	municipal	district
Before 1980	39.13	30.43	8.70	4.35	13.04	4.35	0	0	0
1981–1985	60.38	20.75	0	1.89	15.09	1.89	0	0	0
1986–1990	56.32	20.69	2.30	2.30	12.64	3.45	0	2.30	0
1991–1995	46.58	16.44	2.74	5.48	20.55	5.48	0	2.74	0
1996–2000	39.19	16.22	2.70	14.86	16.22	8.11	1.35	1.35	0
2001–2005	30.00	0	0	20.00	0	10.00	30.00	0	10.00
2006–2010	25.00	0	0	25.00	0	0	25.00	25.00	0
After 2011	48.46	18.52	2.47	6.79	15.12	4.94	1.54	1.85	0.31

Notes: The number of residential parcels is used for statistical analysis.

EGD residential units are chiefly concentrated in the city center, although relocations also show a trend of moving out, such as Tianfu New Area and development zones and some suburbs. In Zone I, the proportion of the provincial residential area to the total constructed residential area in the same period decreased from 60.38% in 1985 to 25% in 2010, the percentage of municipal residential area decreased from 20.75% in 1985 to 16.22% in 2005 and 0% in 2010, and the percentages of the district residential area were all less than 3% in different periods (Table 3). In 2006–2010, the provincial residential units in total accounted for 75%, 25% each located in Zone I, Zone II and Zone III; only 25% of the municipal residential area was located in Zone III. In fact, because some government departments were moved outside of Zone I, correspondingly the related EGD residential units were relocated outside of Zone I too. For example, the Chengdu government planned and constructed the Tianfu New Area in Zone III and the Chengdu municipal administrative offices moved to the Tianfu New Area in 2008, which resulted in the construction of some EGD residential units close to the Tianfu New Area in Zone III.

Currently, the provincial and municipal office units mainly agglomerate in the city center around Tianfu Square, and respectively form two centralized areas: one area around the Sichuan Provincial Communist Party Committee Departments adjacent to the commercial street and another area around the Sichuan Provincial Government Departments in Duyuan Street. EGD units at Yangshi Street, Commercial Street, West Yuhe Street, and Duozi Lane are located near the Sichuan Provincial Party Committee, Chengdu Municipal Party Committee and Municipal Government (e.g., the former location of Municipal Government and Party Committee was moved to the Tianfu New Area in 1998, but the old residential units remain). Centering on the provincial government, residential units are distributed in other areas along Duyuan Street for Provincial Science and Technology Department,

Provincial Education Department, Provincial Food Authority, Provincial Finance Department, and other departments. Many EGD residential units are adjacent to beautiful scenic spots in the city center, such as Du Fu Thatched Cottage and Huanhuaxi Spot.

EGD residential buildings generally are three to four stories high built in the 1960–1980s. After 1990, most of these buildings were gradually rebuilt at their original sites except for Kuanzhai Lane. Following renovations, these parcels of EGD residential units have not been marketed, and thus most of the parcels have not been relocated. In other words, EGD residential parcels are still government residential units after China's public housing privatization started and the welfare housing supply was discontinued in 1998.In addition, our survey results show that EGD residential areas still take the traditional residential unit role, whether they are old country yards or new ones bought from the market, and these parcels are managed by the same government unit.

Our survey did not determine the number of EGD and UDDH. Fortunately, we found that the housing land use area of EGD and UDDH is 1,981,217 m^2 and 10,999,290 m^2 which is respectively 3.71% and 20.59% of the total housing land use area (53,420,544 m^2) of the city in 2012. This indicates among all urban housing, about one fifth or more resulted from UDDH that occurred in the last three decades causing major urban spatial reconstruction and significantly changing the urban landscape and social and cultural spatial structures.

We compared the differences between EGD and UDDH at each zone in the last three decades (Figure 4). Before 1980, there is very little difference between the residential areas of EGD and UDDH. However, significant differences occurred during the three decades in five-year increments from 1981–1985 to 2006–2010. Within Zone 1, EGD development nearly doubled that of EGD; while in Zone 2 and Zone 3, UDDH development was much higher than that of EGD. It is apparent that in the last three decades, EGD residential land was aggregated in Zone I, but UDDH was concentrated in Zone 3.

We then used a simple t-test to statistically assess the continual five-year changes of developed residential area and summarized the results in Table 4. The p-values of the t tests were less than 0.05 which indicated the differences in the averages of developed residential areas between UDDH and EGU are significant. In Zone 1, the average development of the residential area in a five-year period of EGU is 1.65 of the area of UDDH; in Zone 2, the average developed area of UDDH is 3.74 times the area of EGU; and in Zone 3, the average developed UDDH area in a five-year period is 28.17 times the area of EGU. The Zone 1 centered EGU and Zone 3 aggregated UDDH pattern showed a significant urban residential segmentation in the last 30 years since the opening up (Figure 4 and Table 4).

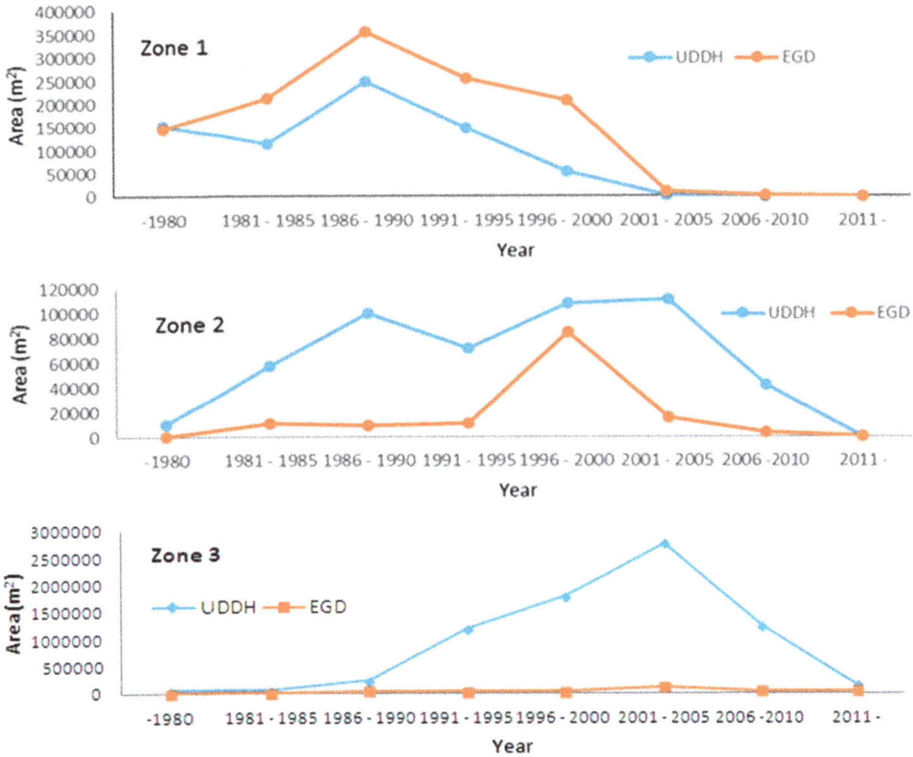

Figure 4. Residential land development of UDDH and EGD across the three zones in Chengdu.

Table 4. A two-sample t test to analyze residential areas of UDDH and EGU from before 1980 to after 2010 in five-year increments.

	Zone 1		Zone 2		Zone 3	
	UDDH	EGU	UDDH	EGU	UDDH	EGU
Mean (m^2)	89,578	148,118	62,139	16,636	938,582	33,316
Variance	8,278,095,788	17,625,808,150	1,838,364,540	773,392,826	972,745,807,356	1,069,300,829
p-value	0.018		0.004		0.016	

5. Discussion

According to the path dependence theory proposed by New Institution School as the representative of the North [43], China's gradual institutional reform model has resulted in the formation of a typical dual social structure [23,42]: (1) the incomplete or semi-marketization group who still enjoys certain welfare benefits under the old regime, such as the staff of government apartments and other related units in education and medical agencies, *etc.* (*Shiye Danwei*) and (2) the complete marketization group outside the old regime, such as private businesses, workers, farmers, and other social classes [41,44,45]. The two groups rely on different social welfare support, such as health care, education, and employment [46]. Because two mechanisms of marketization and administration in resource allocation still exist in China, the semi-marketization group with close relationships to government are likely to get access to development resources and opportunities by taking advantage of administrative tools and adjusting market mechanisms. In the housing market, government departments on behalf of the semi-marketization group use their political dominance to take advantage of the renewal of old residential areas and the construction of new residential units in the vicinity of their workplaces in order to access better social and natural environments close to the city center or in more desirable suburbs with good infrastructure. In addition, some government departments recently have occupied affordable housing. However, the complete marketization crowd composed of most workers, rural migrant workers, and rural collectives, *etc.* are gradually weakening and moving to less desirable locations, especially the suburbs located near areas of demolition and complete renovation. The complete marketization crowds are reluctant but must move out of their original residential areas due to the impact of social income polarization, the shortage of resources and information, and essentially the lack of political power.

On behalf of the State in assigning resources, the three levels of government make full use of their power to choose desirable settlements for their employees, which shows the clear path dependence that they have inherited from the socialist period [23]. Residential units for EGD are generally close to their offices and the regions possessing beautiful natural environments, and historical and cultural heritage. EGD residential units are most often adjacent to the urban center and/or scenic spots in suburbs and/or in new urban areas. Thus, these residential units are usually retained in places with convenient transportation, beautiful environments, and heritage. The provincial departments follow historical traditions, making use of the existing old government office sites, and also constructing new offices in or close to the city center in order to facilitate the management of the whole province. Municipal departments are generally in the vicinity of the provincial government units in the city center or in new urban areas in order to easily manage the whole city. District authorities are located in the district center or at sites with beautiful environments within their jurisdictions, which are convenient for jurisdictional service and management.

Prior to 1990, EGD residential units were concentrated in better locations in the city center, while the resettlement of the marketization crowd was scattered across the city area. Since 1990, the land-leasing policy has been implemented, the city planning efforts have been strengthened to promote local economic development, and therefore the resettlement areas have gradually advanced into urban fringe areas in an agglomerated format with increasing amounts of farmland being occupied. It has greatly affected the living standards of the resettled residents due to the lack of development of infrastructure. In the process of urban space expansion, EGD has still occupied the better residential locations, but the remaining population, especially the lower economic and weaker classes, have gradually moved out to the suburbs in the past 30 years. This somehow confirms the point of the Neo-Marxist theory: the essence of urban spatial configuration is the physical manifestation of the social status of each class in a city, e.g., the differentiation of residential areas is the physical manifestation of changing relationships among various social groups.

Unlike the EGD residential units, residential location choices of the vast majority of the complete marketization crowd, such as rural-urban migrants and urban fringe farmers under the market mechanism continue to become narrow and limited, and their residences are pushed to the city edge. In other words, in the process of urban demolition and renovation, most ordinary citizens in old settlements are increasingly migrating further away from the city center to lower income areas, where there is limited social, economic, and cultural infrastructure.

According to our survey, over 80% of relocated residents such as people in Dongguang District and Jingzhu District are not satisfied with the amount of monetary compensation they receive for demolition, and think that monetary compensation provided by government or developers needs to be increased roughly at least 30%–50% in order to meet the market value of their old housing and maintain their living expenses, not considering the loss of social relations, *etc.* Most of the farmers, such as those in Shengdeng Village Apartments, believe the government should not expropriate their entire farmlands and they are more reluctant to give up their former housing plots (*Zhaijidi*). Similarly, more than 90% of relocating farmers consider land prices for expropriation should be increased by at least 30%–50%, and that the government should guarantee the protection of their livelihood (e.g., job assistance) and strengthen the management and supervision of the arable land expropriation compensation retained by village-level governments.

Most of the interviewees of UDDH agreed that they expect to be relocated to a new urban area. They believe that most of the buildings in their current residential area are very old and were built before 1980. Families typically live in tiny spaces with old utilities, limited parking lots, and poor security monitoring, internet, and other infrastructure. They hope that they can be relocated to a community close to central business centers or new urban centers to improve their life environment and quality. By relocation, they also expect to increase the built-up areas of their new houses, retain housing values, and they even hope that the new houses can be beneficial to their next generations' development. The UDDH care more about the socioeconomic benefits. However, there are still some persons who, are typically older than 70 years old, would not like to be relocated due to the loss of social connections. Our survey indicated that 68% of UDDH agreed that relocation typically helps improve their lives' quality except for in those rural areas which are too far away from urban centers. However, we also found in the interview that about 3% UDDH think the government exploited them, and they did not agree that the State has the right to possess, use, and dispose land.

In opposition, all government officials surveyed in this study think that Chengdu's existing compensation standards containing the municipal government premium and subsidies have surpassed the national and provincial standard of compensation, because all land belongs to the country and the demolished houses are old and in poor condition. Most officials also privately acknowledge that land leasing revenue (actually including rent gap) is indeed one of the main financial resources which allow governments to improve infrastructure, public spending, and attract investment, and stated that the government cannot "make bricks without straw". A few officials confessed that "the government has been also trying to guarantee the lives of landless peasants, but because of peasants' low levels of education (mostly middle school or even primary school level), the employment problem is still serious". Some officials said that "Chengdu municipal government is doing well and the relevant demolition contradictions or conflicts are much less than other cities in China". Local government's and/or developer's compensation for demolition focuses on the existing old housing in accordance with the construction area of residential houses and household demographic situation, but ignores employment, transportation, education, and other social and cultural factors.

It comes to be understood that the majority of resettled populations have increasingly been moved away from the city center, and they are migrated and resettled by government and/or developers. It is a compromise made by stakeholders since the low prices of land parcels and housing on the outskirts are seen as acceptable by developers and governments, as the majority of individuals relocating have low incomes. Also, rural collective land is easily occupied because collectively-owned land has been assigned low compensation levels under expropriation, and rural landless farmers with low political power have no choice but to accept them. On the other hand, low monetary compensation according to total building area of old and small demolished houses often results in the original occupants, usually low-income and even laid-off workers, having great difficulty remaining in their original locations closer to the city center where commercial housing has high prices. In addition, through administrative means, the government has relocated the poor and weak groups to the agglomerated resettlement areas. Therefore, the market mechanism seemingly forces the majority of relocating individuals to urban fringe areas, which is primarily driven by an incomplete socialist market regime and the state-owned land institutional setting.

EGD has had a much better fate than UDDH in the process of residential space restructuring. EGD residential units are almost always renovated or constructed at nearby sites through administrative assignment. For example, residential units owned by the Sichuan Provincial Finance Department *etc.* still surround Tianfu Square or are close to Chunxi Road where land leasing prices are the highest in Chengdu. This leads to a low possibility for large-scale EGD residential relocations. For example, new EGD residential units between South SRR and TRR have been constructed since 2008, because Chengdu municipal administrative departments moved to the Tianfu New Area in 2008, which had been planned and built since 2005. These new residential units have been sold to EGD in the form of commercial houses with affordable prices, and their establishment has driven the development of so-called commercial houses in that special neighborhood. However, some officials or EGD have admitted the prices of these so-called commercial houses are 30%–50% lower than the market values, because land prices are not included due to their construction at their original locations.

We can summarize two modes of housing supply in the current urban residential market. (1) EGD under the incomplete market of China have access to housing supplied in a new residential area by renovating old residential units, reconstructing new residential units in government office areas, subsidizing housing prices, and reserving the typical features on the socialist stage by means of administrative allocation. (2) Most UDDH can own housing by means of two ways: one way is to purchase cheaper commercial houses in suburbs in the real estate market, since UDDH only receive low monetary compensation in the real estate market; another way is to buy resettlement housing far from the city center.

In this study, we identified two main classes of citizens, the EGD and the UDDH, and spatially examined their housing patterns in the process of urban spatial reconstruction under a paradigm of China's dual-track mechanism. We understand that it is difficult to absolutely separate market forces from government power and interventions in China, and likewise it is not easy to absolutely separate the effects of individual behavior, including government forces from market roles, in a typical market economic system. In China, government plays a dominant role and individuals such as the UDDH group have to conform due to lack of financial and social support as we discussed above; but in a market economic system, individuals' behavior often plays leading roles, with the government and market then following suit. For example, in the USA, the middle class prefers suburban environments and increasing proportions of the middle class are moving to suburban areas, and the market elements and government policy are in line with this type of migration against the backdrop of an open market. The related social and cultural factors further strengthen these types of changes which have constitute the suburbanization process in the USA.

Our study is a complex Chinese version of "The City as a Growth Machine" [47]. However, the general city is the areal merchandise representing the interests of land-based owners, who always try to profit through increases in the land use of the area. While this occurs, the central and local governments play different critical roles at the expense of competing localities to achieve economic growth. Local governments or municipalities are basically growth machines that produce wealth for those in power by exploring real estate at the taxpayers' expense. Now urban growth, local economic development, and those who lead and control these two processes, are still central to the politics of cities [48]. However, the spatial process in Chengdu is not as complex as the residential segmentation in Europe, where social and ethnic factors are critical for neighborhood selections [49,50], which determined that the spatial patterns after urban restructuring due to relocations and segregation in those cities (e.g., Dutch cities) were not as those in the USA and the zoning patterns that we analyzed in this study. The segregation of UDDH away from the center of Chengdu with higher land prices seems close to that of the income-based residential patterns occurring in Israeli cities, which were modeled using an agent-based model according to the Schelling model of segregation [46,51] that showed the wealthy would not reside near poorer neighbors. However, in our field survey, we did not find any direct conflicts between EGD and UDDH, and no one in EGD mentioned s/he would not like to reside close to UDDH. We also found from our interview that EGD or UDDH did not think that they were segmented in the process of relocation, but they agreed that it may be a trend for the wealthy and the poor to be spatially separated in the future. We understand that in the opening up process, EGD would not like to be suburbanized due to lack of sufficient social and cultural infrastructure, and that UDDH have to be suburbanized due to lack of sufficient money and the sky-rocketing land prices within or close to central business districts.

6. Conclusions

Urban spatial organization and remodeling essentially reflect political and socio-economic relationships and their changes [52–54]. The evolution of the social production model is a changeable spatial process [54]. China's gradual reform, under the multi-actor governance structure that involves complex relations among different stakeholders, has created an incomplete state-oriented socialist market which has led to the existence of an intermingled dual mechanism of market forces and administrative means. In the flawed real estate market of China, governments can treat differently the residential relocations of employees in government departments (EGD) who are the part of the semi-marketization class, and the urban demolition displaced (UDDH) who are part of the marketization class, which has undoubtedly reduced the impartiality of urban space use in the urban residential spatial restructuring process.

Market monopoly and land leasing under certain administrative allocations of land resources have caused a path dependence of urban social spatial restructure. EGD in China's incomplete market have gained access to housing by renovating original, old residential units or constructing new residential units in government office areas, which is reminiscent of the features of urban landscapes on the socialist stage. EGD purchase their housing at low prices but in much better locations, usually close to the city center with access to high quality services and social, economic, and cultural infrastructure. However, UDDH with low monetary compensation either have to purchase commercial houses in the open real estate market or buy resettlement housing provided by government or developers. Both commercial housing and resettlement housing of low prices are typically far away from the city center in suburban areas with poor social, economic, and cultural infrastructure. The UDDH, mainly including most workers and farmers, have become the lower income and politically weaker class in urban areas and are easily forced to move to the suburban areas due to lack of sufficient political support.

China's central government has recognized the issues of segregation and marginalization due to this dual-track mechanism. A recent report issued by the Development Research Center of the State Council makes the case for market-based allocation of land, people, and capital [55]. In this report, a new model of urbanization for China pointed out two closely related priority areas 'reforming land management and institutions', which emphasizes more efficient use of land that requires stronger property rights for farmers, much higher compensation for land requisition, new and updated mechanisms for converting rural construction land to urban uses, and market-driven pricing for urban land allocation. The process of 'reforming urban planning and design', which promotes better use of existing urban land through flexible zoning, with smaller and mixed land use plots, would result in denser and more efficient urban development.

Under the transition from limited access societies to open access societies [15], central governments and local governments would pay more attention to the fairness of the institutional base, since the path dependence of residential relocations of different classes such as EGD and UDDH driven by the intermingled government intervention and market forces has caused a series of problems. Most of UDDH who belong to the lower income and politically weaker classes face further segregation and high unemployment risk in the process of dual-track urbanization [17,56,57]. EGD are increasingly aggregated in the urban centers or certain suburbs and new urban areas with a high quality natural environment, and social and cultural infrastructure. It seems further institutional reforms, especially of and ownership and land leasing systems, need to be conducted in order to decrease the social inequity and better distribute the benefits from urban demolition and land development among different social strata.

Acknowledgments: The authors would like to thank the National Science Foundation of China for the project support (#41171143 and 41571155). The authors also appreciate the work of the other field survey members including graduate students Lu Hong, Wang Hongguang, and Yang Pei; we offer thanks to the Chengdu Municipal Construction Bureau, Urban Planning Bureau of Chengdu, Chengdu Municipal Development and Reform Commission, and the Bureau of Transportation for the help that they could provide. The authors also appreciated Dr. Youqin Huang's review of an early version of this paper. We also appreciate the three anonymous reviewers' useful comments that help enhance the quality of this paper.

Author Contributions: Yongchun Yang and Qingmin Meng designed the research, Yongchun Yang wrote an initial manuscript, Qingmin Meng and Yongchun Yang revised the paper, Qingmin Meng finalized the paper; Qingmin Meng and Deli Zhang analyzed the data, and Deli Zhang was responsible for the field survey, interview, and data collection; Corrin McCarn edited the English.

Conflicts of Interest: The authors declare no conflict of interest.

References

1. Li, B.Q.; Shin, H.B. Intergenerational Housing Support between Retired Old Parents and their Children in Urban China. *Urban Stud.* **2013**, *50*, 3225–3242.
2. Feng, J.; Zhou, Y.X. Restructuring of socio-spatial differentiation in Beijing in the transition period. *Acta Geogr. Sin.* **2008**, *63*, 829–844.
3. Ma, L.J.C. Urban transformation in China, 1949–2000: A review and research agenda. *Environ. Plan. A* **2002**, *34*, 1545–1569.
4. Ma, L.J.C. Economic reforms, urban spatial restructuring, and planning in China. *Prog. Plan.* **2004**, *61*, 237–260.
5. Ma, L.J.C. Urban administrative restructuring, changing scale relations and local economic development in China. *Political Geogr.* **2005**, *24*, 477–497.
6. Wu, F. China's changing urban governance in the transition towards a more market-oriented economy. *Urban Stud.* **2002**, *39*, 1071–1093.
7. Chen, J.H.; Fei, G.; Ying, W. One decade of urban housing reform in China: Urban housing price dynamics and the role of migration and urbanization, 1995–2005. *Habitat Int.* **2011**, *35*, 1–8.
8. Lee, J. From welfare housing to home ownership: the dilemma of China's housing reform. *Hous. Stud.* **2000**, *15*, 61–76.
9. Wang, Y.P. Urban housing reform and finance in China: A case study of Beijing. *Urban Aff. Rev.* **2001**, *36*, 620–645.
10. Wang, Y.P.; Murie, A. The process of commercialization of urban housing in China. *Urban Stud.* **1996**, *33*, 971–989.
11. Lim, G.C.; Lee, M.H. Political ideology and housing policy in modern China. *Environ. Plan. C* **1990**, *8*, 477–487.
12. Zhang, X.Q. Urban land reform in China. *Land Use Policy* **1997**, *14*, 187–199.
13. Zhang, X.Q. *Privatization: A Study of Housing Policy in Urban China*; Novo Science Publishers: New York, NY, USA, 1998.

14. Zhang, X.Q. Institutional transformation and marketisation: the changing patterns of housing investment in urban China. *Habitat Int.* **2006**, *30*, 327–341.
15. Li, J. Land sale venue and economic growth path: Evidence from China's urban land market. *Habitat Int.* **2014**, *41*, 307–313.
16. Jeong, J.H. From illegal migrant settlements to central business and residential districts: Restructuring of urban space in Beijing's migrant enclaves. *Habitat Int.* **2011**, *35*, 508–513.
17. He, S.; Liu, Y.T.; Chris, W.; Wu, F. Property rights Redistribution, Entitlement Failure and the Impoverishment of Landless Farmers in China. *Urban Stud.* **2009**, *46*, 1925–1949.
18. National Bureau of Statistics of China (NBSC). Chinese Statistical Year Book 2009. Available online: http://www.stats.gov.cn/tjsj/ndsj/2009/indexeh.htm (accessed on 1 December 2015).
19. National Bureau of Statistics of China (NBSC). Chinese Statistical Year Book 2013. Available online: http://www.stats.gov.cn/tjsj/ndsj/2013/indexeh.htm (accessed on 1 December 2015).
20. Wang, X.Z. The relationship between land use and society power factors of inland key cities in China: A case study of Xi'an. *Acta Geogr. Sin.* **1998**, *53*, 175–184. (In Chinese)
21. Wei, Y.H.D. Regional development in China. *Eurasian Geogr. Econ.* **2007**, *48*, 16–36.
22. Wei, Y.H.D. Restructuring for growth in urban China: Transitional institutions, urban development, and spatial transformation. *Habitat Int.* **2012**, *36*, 396–405.
23. Yang, Y.C. Chinese cities in transition: Mixed spatial structure produced by a hybrid transitional model. *Geogr. Res.* **2015**, *34*, 2021–2034. (In Chinese)
24. Golubchikov, O.; Badyina, A.; Makhrova, A. The hybrid spatialities of transition: Capitalism, legacy and uneven urban economic restructuring. *Urban Stud.* **2014**, *51*, 617–633.
25. Golubchikov, O. World-city entrepreneurialism: Globalist imaginaries, neoliberal geographies, and the production of new St Petersburg. *Environ. Plan. A* **2010**, *42*, 626–643.
26. Wu, J.L.; Yu, K.P.; Fogel, R. *China in the Next 30 Years*; Central Compilation and Translation Press: Beijing, China, 2012. (In Chinese)
27. Tao, R.; Yang, D.L. The revenue imperative and the role of local government in China's transition and growth. 2008. Available online: http://web.ntpu.edu.tw/~guan/courses/growth/ran.pdf (accessed on 17 December 2015).
28. Peng, R.; Wheaton, W.C. Effects of restrictive land supply on housing in Hong Kong: An econometric analysis. *J. Hous. Res.* **1994**, *5*, 263–291.
29. Lu, G. Achievements and challenges: 30 years of housing reforms in the People's Republic of China. 2010. Available online: http://papers.ssrn.com/sol3/papers.cfm?abstract_id=1619161 (accessed on 17 December 2015).
30. Xu, Y.; Tang, B.S.; Chan, E.H.W. State-led land requisition and transformation process of rural villages in transitional China: A case study of Shanghai. *Habitat Int.* **2011**, *35*, 57–65.
31. Ho, P. *Institutions in Transition: Land Ownership, Property Rights and Social Conflict in China*; Oxford University Press: Oxford, UK, 2005.

32. Po, L. Property Rights Reforms and Changing Grassroots Governance in China's Urban–Rural Peripheries: The Case of Changping District in Beijing. *Urban Stud.* **2011**, *48*, 509–528.
33. Heller, M.; Hills, R. Land assembly districts. 2008. Available online: http://www.harvardlawreview.org/wp-content/uploads/pdfs/heller_hills.pdf (accessed on 17 December 2015).
34. He, Z.Y.; Asami, Y. How Do Landowners Price their Lands during Land Expropriation and the Motives Behind It: An Explanation from a WTA/WTP Experiment in Central Beijing. *Urban Stud.* **2014**, *51*, 412–427.
35. Leung, C.K.Y.; Chen, N.K. Intrinsic cycles of land price: a simple model. *J. Real Estate Res.* **2006**, *28*, 293–320.
36. Li, J.; Chiang, Y.H. What pushes up China's real estate price? *Int. J. Hous. Mark. Anal.* **2012**, *5*, 161–176.
37. Tao, R.; Su, F.B.; Liu, M.X.; Cao, G.Z. Land leasing and local public finance in China's regional development: Evidence from prefecture-level cities. *Urban Stud.* **2010**, *47*, 2217–2236.
38. Li, J.; Chiang, Y.H.; Choy, H.T.L. Central-local conflicts and property cycle: A Chinese style. *Habitat Int.* **2011**, *35*, 126–132.
39. Yang, Y.C. *Gradual Institution Change and Geographical Space Involution*; Lanzhou University Press: Lanzhou, China, 2013. (In Chinese)
40. Alonso, W. *Location and Land Use*; Harvard University Press: Cambridge, USA, 1964.
41. Liu, Y.G.; Yin, G.; Ma, L.J.C. Local state and administrative urbanization in post-reform China: A case study of Hebi City, Henan Province. *Cities* **2012**, *29*, 107–117.
42. Ding, C.R. Policy and praxis of land acquisition in China. *Land Use Policy* **2007**, *24*, 1–13.
43. North, D.C. *Institutions, Institutional Change and Economic Performance*; Sdx Joint Publishing Company: Shanghai, China, 1994. (In Chinese)
44. Wu, F. Residential relocation under market-oriented redevelopment: The process and outcomes in urban China. *Geoforum* **2004**, *35*, 453–470.
45. Bian, Y.; Logan, J.R.; Lu, H.L.; Pan, Y.K.; Guan, Y. "Work Units"and the commodification of housing: Observations on the transition to a market economy with Chinese characteristics. *Soc. Sci. China Engl. Ed.* **1997**, *18*, 28–35.
46. Clark, W.A.V.; Rivers, N. Community choice in large cities: Selectivity and ethnic sorting across neighborhoods. In *Understanding Neighborhood Dynamics: New Insights for Neighbourhood Effects Research*; Van Ham, M., Manley, D., Bailey, N., Simpson, L., Maclennan, D., Eds.; Springer: New York, NY, USA, 2012.
47. Molotch, H. The city as a growth machine: Toward a political economy of place. *Am. J. Sociol.* **1976**, *82*, 309–332.
48. Jonas, A.E.G.; Wilson, D. *The Urban Growth Machine: Critical Perspectives Two Decades Later*; State University of New York Press: Albany, NY, USA, 1999.
49. Bolt, G.; Van Kempen, R.; Van Weesep, J. After urban restructuring: relocations and segregation in Dutch cities. *J. Econ. Soc. Geogr.* **2009**, *100*, 502–518.

50. Bold, G.; van Kempen, R. Dispersal patterns of households who are forced to move: desegregation by demolition: A case study of Dutch Cities. *Hous. Stud.* **2010**, *25*, 159–180.
51. Hatna, E.; Benenson, I. Geosimulation of income-based urban residential patterns. In *Advanced Geosimulation Models*; Marceau, D.J., Benenson, I., Eds.; 2011; Available online: http://www.casa.ucl.ac.uk/ Advanced%20Geosimulation%20Models.pdf (accessed on 1 December, 2015).
52. Harvey, D. *The Condition of Post Modernity*; Blackwell: Oxford, UK, 1989.
53. Rex, J.; Moore, R. *Race, Community and Conflict: A study of Sparkbrook*; Oxford University Press: London, UK, 1967.
54. Lefebrve, H. *The Production of Space*; Blackwell: Oxford, UK, 1991.
55. The World Bank and the Development Research Center of China's State Council. Urban China: Toward Efficient, Inclusive and Sustainable Urbanization. 2014. Available online: http://www.worldbank.org/en/country/china/publication/urban-china-toward-efficient-inclusive-sustainable-urbanization (accessed on 1 December 2015).
56. Liu, Y.T.; Wu, F.L. Urban poverty neighbourhoods: Typology and spatial concentration under China's market transition, a case study of Nanjing. *Geoforum* **2006**, *37*, 610–626.
57. Schelling, T.C. Dynamic models of segregation. *J. Math. Sociol.* **1971**, *1*, 143–186.

Section III: Effects of Urban Development and Land Use Change

Walkability, Land Use and Physical Activity

Yehua Dennis Wei, Weiye Xiao, Ming Wen and Ran Wei

Abstract: Physical activity (PA) promotes healthy life and contributes to sustainable development. In this paper, we rely on the Utah Household Travel Survey data and analyze the determinants of PA in terms of neighborhood land use, accessibility to transportation, and socio-demographic status in Salt Lake County, Utah, United States using four-component walkability indices at various geographic scales. We find that PA is associated with neighborhood land use and social demographic status, including the compact design of the neighborhood. The results also indicate that land use mix is insignificant, and that current neighborhood design only supports people's 20-min walk. The spatial lag model reveals significant spatial autocorrelation of PA but the barely improved R^2 validates the dominant effect of neighborhood land use and social demographic status on people's walking behavior.

Reprinted from *Sustainability*. Cite as: Wei, Y.D.; Xiao, W.; Wen, M.; Wei, R. Walkability, Land Use and Physical Activity. *Sustainability* **2016**, *8*, 65.

1. Introduction

The United States is troubled by a high obesity rate, although the rate at which it is increasing appears to have slowed [1]. The same trend is evident in Salt Lake County in which approximately 25.4% of adults are obese, and childhood obesity rates are also on the rise [2]. High obesity prevalence has clearly become a legitimate public health concern, motivating researchers to examine the determinants of obesity problems. Given that physical inactivity is widely considered as a major risk factor for overweight and obesity [3,4], increasing physical activity (PA) would help to promote citizens' health and contribute to sustainable development.

A large body of literature has emerged to address the effects of social environment and urban form on PA. Evidence shows that physical inactivity results from the synthesis of urban amenity, socio-demographic status, personal preference, and genetic factors [5]. Density, diversity and accessibility are driving factors of PA, and residents living in neighborhoods characterized by higher residential density and land-use diversity tend to walk more [6]. Mixed land use, higher network connectivity and denser residential space can help to improve neighborhood walkability [7]. PA is also associated with neighborhood social and environmental factors, such as socio-demographic status, transportation, land-use pattern, and neighborhood design [8–13].

This study first maps the spatial distribution of walking behavior and land-use factors and then explores the spatial autocorrelation and clustering of walking behavior. Second, we test the validity of four land use measures to predict walking behavior at different geographical scales. Last, we examine the relationships between walking behavior and neighborhood social and physical characteristics, including transportation and socio-demographic status.

2. Literature Review

2.1. Physical Activity: Benefits and Measurements

PA is defined as any bodily movement produced by skeletal muscles that results in energy expenditure [14]. Regular PA is regarded as an important component of a healthy lifestyle [15]. The benefits of PA are well documented, and it is widely recognized that increasing PA can help to prevent excessive weight gain [12,16–19], improve disease control [15,20,21], reduce the symptoms of depression [22], and more generally, enhance health. The American College of Sports Medicine and the Centers for Disease Control and Prevention recommended 30-min of moderate or vigorous PA (e.g., walking and jogging) per day for five days a week for 18–65 year-old-healthy adults [23].

Developing a reliable tool to quantify PA is a priority. A wide range of methods in the form of subjective or objective tools are available to capture PA. The most commonly used tools are the subjective tools like surveys and questionnaires. Baecke *et al.* [24] classify habitual PA into 29 items, which concern occupation, movement, and PA during leisure time, including sports and sleeping habits. By evaluating over 10 questionnaires about PA, Jacobs *et al.* [25] call for a more comprehensive questionnaire about different kinds of PA. The "last 7 day recall", which includes questions about people's PA participation in the last seven days, proves to be suitable for PA research [26]. There are many resources for survey-based and self-reported measurement of PA, some of which are published and accessible online [27]. Compared with subjective measures, objective measures are more precise and reliable [28]. However, the high cost of the techniques entailed in collecting objective PA data impedes its wide application. Accelerometer is one of the widely used objective tools in PA research and performs well [29–32]. Using both objective and subjective PA measurement tools is encouraged in future PA work [31].

2.2. Determinants of PA

PA participation in the US is not adequate, and one in three adults and four in five adolescents do not meet levels of PA recommended by public health guidelines [14,32–34]. The etiology of individuals' participation in PA, including socio-demographic status and individual characteristics, has drawn much attention. Age is one of the most examined demographic factors. Teenagers' participation in PA declines drastically as they grow up. PA participation tends to be stable in young and middle-aged adulthood, then increases a little after middle-age adulthood and declines in old age [14,35–37]. Gender is another important PA correlate, although the observed pattern has not been always consistent [34,38–41]. PA is also linked to family structure at least among children [42]. Finally, PA prevalence is higher among Caucasians than African-American and Hispanics [4,43].

The relationship between urban form and travel behavior is a key concern of urban planning [44]. Among various urban forms, mixed land use and public transit are found to facilitate residents' travel-based PA [45]. The sprawl index, a linear combination of 22 land-use variables, seems to be associated with PA, as people living in sprawling metropolitan areas tend to walk less and weigh more [46]. High levels of PA are detected in the neighborhoods with highly mixed land use, high density and great accessibility to work opportunity [47–49]. Street design features such as street connectivity, street safety, sidewalk and crosswalk in the neighborhoods are also examined with results showing that people walk more in a well-connected and safe street [47–49]. Furthermore, neighborhood amenities like access to parks are also linked to PA, with park amenity characteristics appearing to be better determinants of PA than park accessibility [50,51].

2.3. Walkability and Land-Use

Walking is recommended as one of the best forms of moderate-vigorous physical activities. Compared to vigorous PA, walking is suitable for people at all ages [17,23]. In order to promote walkability, the "5D" theory, which includes population density, pedestrian-friendly design, diversity of destinations, destination accessibility, and distance to transit, is proposed as a measure of urban form, and walkability increases with the increase in the values of the "5D" variables [9,47,52–54]. The original walkability index is proposed to include three variables: street connectivity, net residential density, and land use mix, which respectively describes the pedestrian-friendly design, population density and diversity of neighborhood land use [11]. The retail floor area ratio is then added to achieve a more favorable result [12]. This four-component walkability index is used in many studies concerning walkability [55–61]. However, the application of the four-component walkability index is still limited. A context-specific walkability index is proposed by Mayne *et al.* [58] and Christian *et al.* [10]. Mayne *et al.* [58] test the validity of the context-specific walkability index and find that the context-specific walkability index can maintain about 90% of the full walkability index. Likewise, a three-component walkability index developed based on a PA study in Sweden stresses that future policies concerning the built environment should be based on context-specific factors [62].

Among these four components of walkability index, land-use mix is the most emphasized. For instance, by testing the effect of multi-category land-use mix on obesity, Brown *et al.* [7] argue that walkable land-use relates to healthy weight.

As a component of walkability index, varying combination of land-use mix is also suggested to affect the strength of the relationship with different types of walking [10]. Geographical scale is also a key element in calculating land-use mix [63,64]. However, land-use mix is not always a significant indicator of walkability [10], and the performance of land-use mix depends on the types of trips and study areas [10,58].

In summary, two limitations in the current literature are noteworthy. First, little work has been carried out to take spatial autocorrelation into consideration and to employ spatial regression models. Both these methods are important because walking is a spatial behavior with spatial clustering and variation. Second, the effect of context-specific walkability indices constructed on various geographical scales is less studied and remains largely unknown.

3. Data and Methodology

3.1. Study Area and Data Source

The validity of land use and walkability indices is evaluated in the Salt Lake County, which covers over 2000 km^2 and has a population of about 1 million in 2010. This research primarily relies on the Utah Household Travel Survey (UHS), which was a travel diary-based survey conducted in 2012 by multiple state government agencies and organizations, including Wasatch Front Regional Council, Utah Department of Transportation, and Utah Transit Authorities. This survey took three days to record 101,404 biking, driving and walking trips taken by 27,064 individuals living in 9155 households across the State of Utah. About 2800 households in Salt Lake County participated in this survey [65]. However, people in some households did not report their walking trips or their walking behavior. After checking thes data, 158 households in Salt Lake County with complete information were selected as samples in this research. The spatial distribution of these 158 households is presented in Figure 1. The spatial data, including locations of public facilities, road network, and bus and light rail stations, come from the Utah Automatic Geographic Reference Center. Additionally, the land use data at the parcel level is from Salt Lake County 2012 tax assessor's Computer assisted mass appraised (CAMA) data.

3.2. Measures

According to previous research on PA [8,10,11], the average of an individual's walking minutes is a proper dependent variable to measure walking behavior. Since it is a trip-level data and the neighborhood environment is measured at household level, the data are aggregated into household level.

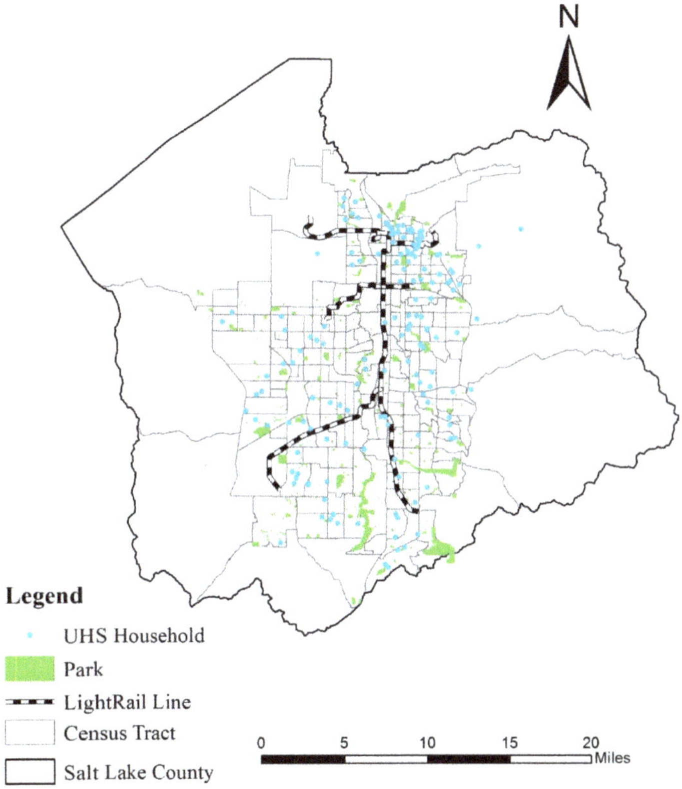

Figure 1. Study area.

With the records of people's walking trips, the frequency of walking every week and the household size, the average value of people's walking minutes for each household can be estimated and used as the dependent variable. Due to the fact that the dependent variable is skewed, the natural log transformation is used to achieve a continuous dependent variable with approximate normal distribution. We consider the independent variables as the accessibility to transportation, amenities, socio-demographic conditions (household level), land use, and walkability index (Table 1). Transportation, amenities, land use and walkability index are calculated using ArcGIS, the most popular commercial GIS software, and Python, an open-source programming language.

Table 1. Definition of variables.

Dependent Variables	Description
PA	Ln (Average time of people's walking trips)
Predictor Variables	Description
Accessibility (AS)	Ln (Distance to closest bus station) Ln (Distance to closest light rail station)
Amenity (A)	Ln (Distance to closest park) Ln (Distance to closest worship)
Socio-demographic Conditions (SC)	Average Age
	Average Education level Vehicles per person Percentage of adult workers Average Household Income (4-caterogry data)
Land use and Walkability index	Dwelling Density Street Connectivity Land-Use Mix Retail Floor Area Ratio

3.3. Statistical Methods

Ordinary least squares (OLS) modeling is performed to examine the relationship between average time of walking trips and walkability index, transportation, amenities, and socio-demographic status. However, since previous studies have shown that people's walking minutes largely depend on neighborhood environment [11,47], walking minutes are likely to be spatially autocorrelated. Spatial autocorrelation test is also conducted, which confirm the existence of such spatial autocorrelation. The Moran's I is 0.26 with statistical significance. Given that ignoring the spatial autocorrelation could cause biased and inconsistent estimates for the regression coefficients [66], spatial lag model is employed to incorporate the spatial autocorrelation of walking time by introducing a spatially lagged dependent variable in the regression equation as follows:

$$y = \rho \times W \times y + \beta \times x + \varepsilon \quad (1)$$

y is the walking time to be regressed, ρ is the spatial autoregressive coefficient, and W is the spatial weight matrix defining neighbors for each location. In this case, the point based method, Delaunay triangulation is employed to define the spatial weight matrix W. According to the method, the circumcircle of any triangle contains no point in its interior [67]. If a leg of a Delaunay triangle connects observation i and j, it sets the elements in the unweighted spatial weight matrix D (i, j) = 1 [68]. This will naturally result in a symmetric matrix. With a row-standardized spatial weight matrix, $W \times y$ is the average of the neighbors' walk minutes, which is usually referred to a spatially lagged dependent variable. x and β are the independent variables and their corresponding coefficients. ρ is the coefficient for the spatially lagged dependent variable. ε is a vector of the error term. In this model, the spatial autocorrelation is measured as an independent variable in the regression model and the spatial regression model can get better model performance and predictive accuracy [69].

4. Walkability Index and PA

Walkability is initially operationalized as a composite of four environmental attributes within the neighborhood area, which is defined by the road network buffer with specified bandwidth:

Dwelling density: Residential parcels within the buffer area are selected and then the total area of the selected parcels and the total number of dwelling units are summarized. Dwelling density is finally calculated by dividing the number of dwelling units by the total area of residential parcels within the neighborhood. The high dwelling density can not only improve the accessibility of complementary activities, but also compress the living space [56].

Street connectivity: Connectivity is measured by the number of intersections per square kilometer within the network buffer. Only the intersections with three or more unique intersecting streets are included in the street connectivity calculation.

Land-use mix: Land-use mix is used to evaluate the evenness in the distribution of land use. If the land use is evenly distributed in the network buffer, the value of land use mix would be 1. If there is only one type of land use in the network buffer, the value would be 0. The method to calculate land use mix and the relationship between land use mix and walkability is well discussed by Christian *et al.* [10]. The land use mix described in this paper is calculated with four land use types: residential, commercial, industrial, and recreational.

Retail floor area ratio: the parcels in commercial use are used to calculate the retail floor area ratio. Salt Lake County 2012 tax assessor's computer-assisted mass appraisal (CAMA) database provides the ground floor area for each parcel. Retail floor area ratio is the total ground floor area divided by the total area in household network buffer. High retail floor area ratio means more optional destinations for shopping and more local employed opportunities within the walking distance [56]. The parking space would be compressed as well and walking is more encouraged.

While selecting a proper geographic scale is discussed in the previous studies, little thought is given regarding how or why these geographic scales should be selected to define neighborhood areas [70]. The importance of geographical scale and multilevel approaches are also emphasized in PA research [6,71,72]. Yamada *et al.* [63] have employed three different geographical scales, 1-km network buffer, block group, and census track, to demonstrate that different geographical scales could lead to different walkability measures. Moudon *et al.* have employed four scales of neighborhood to identify walkable neighborhoods [73]. Schlossberg and Brown have also studied the relationship between walkability indicators and transit-oriented development at two geographical scales (0.25 mile and 0.5 mile) [74]. Since this research aims to create a context-specific walkability index, multi-scale tests on the performance of these four components are conducted to explain PA. A variety of bandwidths is used to buffer the road network to define neighborhood area, including 400 m (5-min walk), 600 m, 800 m (10-min walk), 1000 m, 1200 m (15-min walk), 1500 m, 1600 m (20-min walk), and 1700 m. The OLS model is used to examine the relationship between walkability index and walking minutes with different geographical scales. The results are shown in Figures 2 and 3.

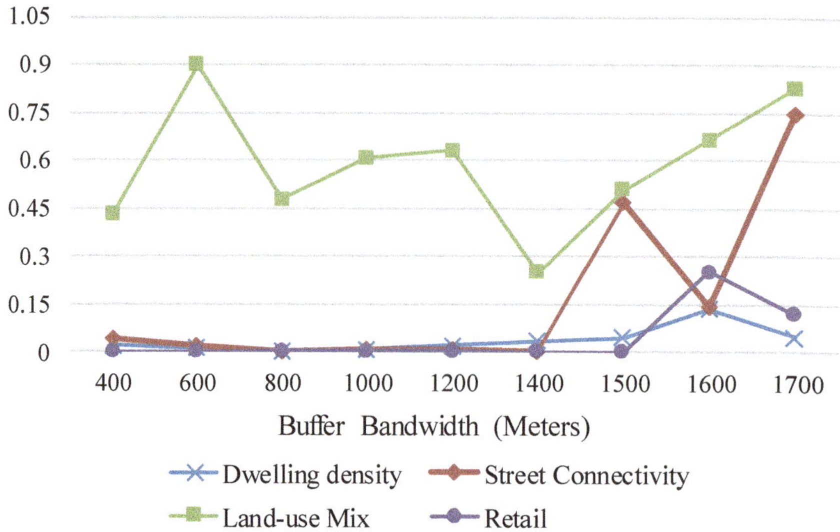

Figure 2. The result of multi-scale test (*p*-value).

Figure 2 shows the significance of each component. The significance of dwelling density, street connectivity and retail floor area ratio decrease dramatically when the bandwidth of the network buffer reaches about 1600 m. The R^2 also experiences a rapid decline around 1600 m, implying that the neighborhood environments, including street connectivity, the commercial land use, and population density, do not affect people's walking trips beyond 1600 m. In other words, the neighborhood environments only affect walk trips within 20-min walking distance in Salt Lake County, and longer trips usually depend on some other factors, such as self-selection. In fact, according to Root's research on geographic scale, in many U.S. urban areas, people routinely travel 5 to 10 miles to carry out many daily activities, such as shopping, taking children to school, or exercising [75]. According to the trip records, the average walking time is about half an hour for each person per day. Comparatively, Salt Lake County is now at a low walkability level and the future development of neighborhood environment is promising to improve walkability.

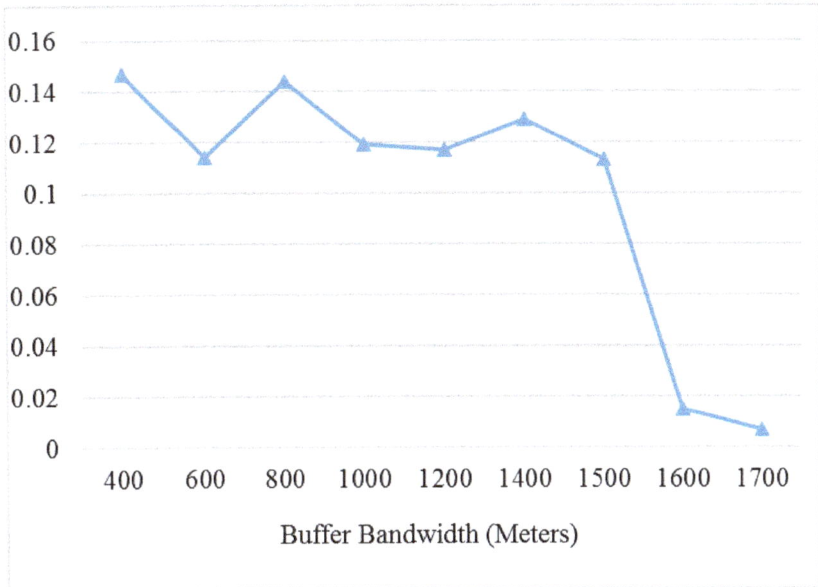

Figure 3. The result of multi-scale test (R^2).

Another interesting point here is that the land use mix is not significant at any geographical level. Figure 4 shows the spatial clusters of physical activities in Salt Lake County using the local indicator of spatial autocorrelation (LISA). The spatial clusters of land use mix are displayed in Figure 5. Clearly, the clusters of high physical activities are not consistent with those of large land use mix, suggesting that land use mix is not a good indicator of PA or the relationship between land use mix and PA is not linear. Research in Salt Lake County and Australia reveal that the even mixture of land use does not necessarily lead to better walkability. The performance of land use mix is largely affected by the land use category, the type of walking, and the study area [8,10].

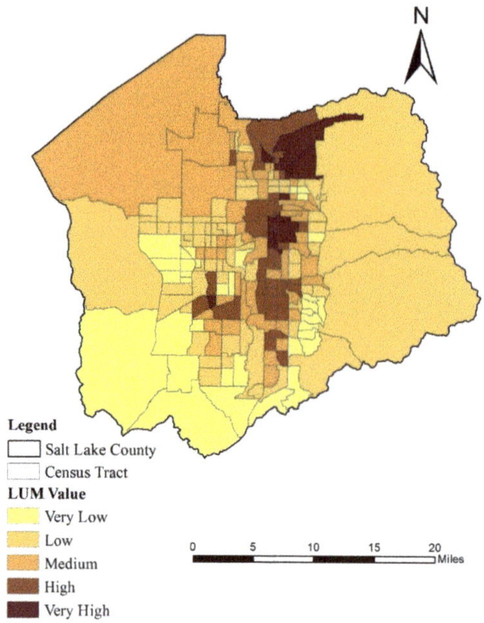

Figure 4. Spatial distribution of land use mix (LUM) in Salt Lake County.

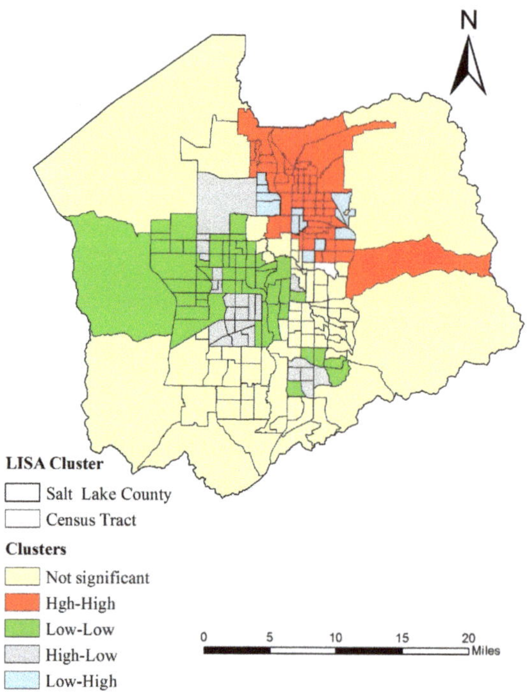

Figure 5. Spatial clusters of PA in Salt Lake County.

5. Result of Statistical and Spatial Analyses

Based on the results of the performance of walkability index under different bandwidths, the neighborhood is defined as the 800-m road network buffer. The OLS linear regression model is used to examine the determinants of the PA in Salt Lake County and the results are shown in Table 2. The results of variance inflation factor (VIF) test for all the variables are all below 5, indicating that there is no alarming multi-collinearity issue. The adjusted R^2 is 0.453 while the p-value for F-statistics is significant, suggesting the great fitness of this regression model.

Table 2. Ordinary least squares (OLS) regression results.

Variable	Coefficient	*p*-Value
	Accessibility	
Distance to bus stop	−0.037	0.670
Distance to light rail stop	−0.061	0.685
	Amenity	
Distance to park	−0.152	0.191
Distance to worship	0.232	0.04
	Socio-demographic Conditions	
Average AGE	0.005	0.067
Average education level	0.735	0.0003
Vehicles per person	−0.146	0.181
Percentage of adult worker	0.232	0.152
Household income	−0.037	0.116
	Land use	
Dwelling density	1381	0.021
Street connectivity	−3198	0.187
Land use mix	−0.127	0.316
Retail floor area	12.98	0.027
	Assessment	
R^2	0.4452	

According to Figure 3, land use factors can explain about 15% of the variance of the dependent variables, so the land use indicators play a vital role in explaining people's walking behavior. Among these four factors, dwelling density and retail floor area ratio show great significance and both contribute positively to PA. Higher population density indicates more walkable communities. The commercial land use is also a key element in promoting walkability and thus the compact design of the commercial land use could help to improve neighborhood walkability.

Figure 6 shows the spatial distribution of the dwelling density and retail floor area ratio. There are many similarities of the spatial distributions of these two variables in Salt Lake County. High dwelling density regions concentrate around the downtown and university areas while these regions also possess high retail floor area ratio.

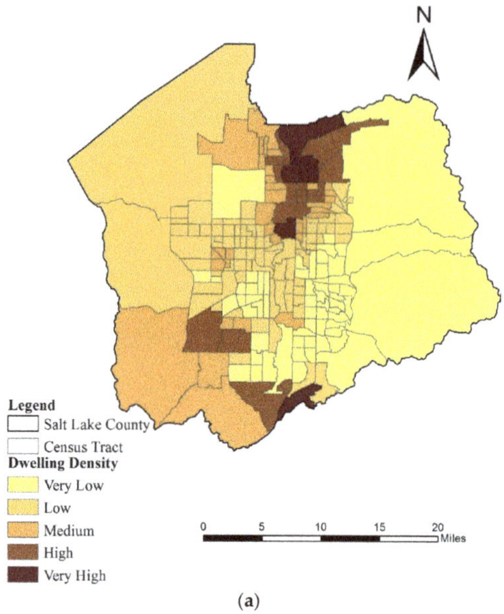

Figure 6. (**a**) Spatial distribution of dwelling density in Salt Lake County. (**b**) Spatial distribution of retail floor area ratio in Salt Lake County.

In addition, the socio-demographic factors are significant in the OLS regression model. Age and education level are important determinants of walkability. Age has a positive effect on walkability, and the young families tend to walk less than comparatively older families. Education level is the most significant determinant in this model, as well-educated family walk more. This confirms the importance of education in promoting walkability [75].

To account for the spatial autocorrelation of PA in Salt Lake County, the spatial lag model is implemented and the results are presented in Table 3. Comparing these two models, the R^2 is improved from 0.45 to 0.46, which is not a significant improvement. However, the spatial lag variable is significant.

Table 3. Spatial lag regression result.

Variable	**Coefficient**	
	OLS Linear Regression Model	**Regression Model with Spatial Lag**
	Accessibility	
Distance to bus stop	−0.037	−0.013
Distance to light rail stop	−0.061	−0.011
	Amenity	
Distance to park	−0.152	−0.129
Distance to worship	0.232 *	0.221 *
	Social economic status	
Average AGE	0.005 *	0.006 *
Average education level	0.735 ***	0.717 ***
Vehicles per person	−0.146	0.143
Percentage of adult worker	0.232	0.174
Household income	−0.037	−0.036
	Land use	
Dwelling density	1381 *	1134 *
Street connectivity	−3196	−2346
Land use mix	−0.127	−0.093
Retail floor area	12.98 *	12.3 *
Rho	/	0.0022 *
	Assessment	
R^2	0.4452	0.4593

Note: *** Significant at 0.01 level; ** Significant at 0.05 level; * Significant at 0.1 level.

6. Discussion

As we have discussed in the literature review, little work has been done to examine the impacts of geographical scales when testing the PA-walkability link. The 800-m road network buffer is usually considered to be the best choice to define the neighborhood [11,57]. However, the underlying reasons for the selection of this bandwidth is less discussed. This paper tests the performance of the four land use indicators, confirms the validity of the 800-m bandwidth, and finds the threshold for defining neighborhood (1600-m network buffer). On average, it is estimated that the 800-m walk takes 10 min and the 1600-m walk takes 20 min. However, 30-min moderate-vigorous PA on at least five days each week is recommended for adults [23,34] and 60 min for the youth [76]. Hence, most of the current neighborhood designs are unlikely to be sufficiently facilitate neighborhood residents to meet the PA recommendation.

This study employs the four-component walkability index as the land use indicators to explore people's walking minutes in Salt Lake County. With the 800-m network buffer defined as the neighborhood, the walkability index can explain about 15% variance of PA without any other control variables. The walkability index's contribution to the variance of PA is larger than what is reported in the research conducted in Atlanta and King County, Washington, wherein it respectively explains 10.7% and 8.3% variance of PA [11,12]. Hence, people's walking behavior relies on neighborhood land use more significantly in Salt Lake County. Since the neighborhood land use cannot meet the general requirement of at least 30 min moderate-vigorous PA, improving the density of the neighborhood land use will play an important role in promoting walkability in the future.

As the urban sprawl phenomenon is looming large in the United States, increasing population density and land use diversity is recommended to policy makers because the compact city with medium-high density housing can facilitate the sustainability of the city [77]. Numerous studies have documented that increasing the density can decrease vehicle mile of travel (VMT) and create short trips [12,47,52]. The created short trips will make people walk more in the neighborhoods. In the northeastern area of Salt Lake County, the density is relatively high. But in other areas, the density is low. The government and policy makers should pay more attention to increasing the density in these areas.

Among the four traditional variables in walkability index, the land use mix is not significant at various geographical scales. One reason is that the relationship between land use mix and PA depends on the study area and trip type. After checking the destinations of the 490 trips taken by the 158 households, we find that over 200 trips are for exercising or walking with the dog. These trips are less likely to be affected by the mix-used land and are more likely to be influenced by neighborhood land use, personal preference and social demographic status. Another reason might be that Salt Lake County is at low mix land use level. The 2800 households covering the entire Salt Lake County are considered to calculate the four-category land use mix. The average value of land use mix in Salt Lake County is 0.22, while the land use mix values are 0.31 in King County and 0.38 in Atlanta [11,12]. Compared to these two well-developed regions, the mix use is relatively low in Salt Lake County.

Previous studies on the relationship between walkability and neighborhood environment rarely consider spatial autocorrelation. However, based on the result of Moran's I test, spatial autocorrelation should not be ignored. If we ignore the spatially lagged term, the coefficients in the regression model would be biased [78]. In this research, both OLS and spatial lag regression models are implemented and the regression results affirm that spatial lag regression model is a good method in this research. However, spatial lag regression model only considers the spatial autocorrelation in dependent variables (lagged dependent variable). Other spatial regression models should be employed in further studies.

7. Conclusions

As the obesity epidemic becomes a serious problem in the United States, physical inactivity, a risk factor of obesity, has drawn much attention. Promoting PA is good for sustainable development and people's health. Walking is a sustainable transport mode that is available to people without any impending costs, and can significantly increase the physical fitness of people. In recent years, several studies are being undertaken to examine the determinants of people's PA.

In this paper, PA is predicted by land use and walkability index, accessibility to transportation, and socio-demographic status. The walkability indices at various geographic scales are constructed and tested, suggesting that the spatial threshold for defining neighborhood for the purpose of examining PA is 1600 m. In addition, the dwelling density, street connectivity, and retail floor area ratio are significant determinants for PA while the land use mix is not. We also find that the best bandwidth for defining neighborhood is 800 m. An OLS linear regression model and a spatial lag model are employed to examine the relationship between neighborhood land-use, socio-demographic status, and PA. The results show that land use and socio-demographic status are the dominating determinants of PA. Density and compact design for land use are key elements in promoting PA and therefore efforts are still needed to improve density and compact design.

These findings should help researchers, policy makers and practitioners gain a better understanding of the determinants of PA so that effective interventions and programs can be implemented to achieve the sustainable development of cities. In a contrast to sprawling urban pattern, compact urban form and design can significantly improve neighborhood walkability and promote people's preference for walking.

In future work, more urban evolution factors should be examined such as urban land expansion, which also influences PA and sustainability [79].

Acknowledgments: We would like to acknowledge the funding of the CSBS Interdisciplinary Research Pilot Program at the University of Utah and the National Natural Science Foundation of China (41329001). This research was also partly supported by the National Institute of General Medical Sciences of the National Institutes of Health under award number R01CA140319-01A1 issued to Ming Wen.

Author Contributions: Yehua Dennis Wei and Weiye Xiao contributed to research design and data analysis. Yehua Dennis Wei, Weiye Xiao, Ming Wen and Ran Wei wrote the paper. All authors have read and approved the final manuscript.

Conflicts of Interest: The authors declare no conflict of interest.

References

1. Ogden, C.L. Prevalence of obesity in the United States, (2009–2010). *NCHS Data Brief* **2012**, *82*, 1–8.
2. Salt Lake County Health Department. Obesity Prevention. Available online: http://slco.org/slcohealth/programs/beActiveAndHealthy/obesity.html (accessed on 6 September 2015).
3. Katzmarzyk, P.T.; Gledhill, N.; Shephard, R.J. The economic burden of physical inactivity in Canada. *Can. Med. Assoc. J.* **2000**, *163*, 1435–1440.
4. Crespo, C.J.; Smit, E.; Andersen, R.E.; Carter-Pokras, O.; Ainsworth, B.E. Race/ethnicity, social class and their relation to physical inactivity during leisure time: Results from the Third National Health and Nutrition Examination Survey, 1988–1994. *Am. J. Prev. Med.* **2000**, *18*, 46–53.
5. Bauman, A.E.; Reis, R.S.; Sallis, J.F.; Wells, J.C.; Loos, R.J.; Martin, B.W.; Lancet Physical Activity Series Working Group. Correlates of physical activity: Why are some people physically active and others not? *Lancet* **2012**, *380*, 258–271.
6. Saelens, B.E.; Sallis, J.F.; Black, J.B.; Chen, D. Neighborhood-based differences in physical activity: An environment scale evaluation. *Am. J. Public Health* **2003**, *93*, 1552–1558.
7. Frank, L.D.; Engelke, P.O. The built environment and human activity patterns: Exploring the impacts of urban form on public health. *J. Plan. Lit.* **2001**, *16*, 202–218.
8. Brown, B.B.; Yamada, I.; Smith, K.R.; Zick, C.D.; Kowaleski-Jones, L.; Fan, J.X. Mixed land use and walkability: Variations in land use measures and relationships with BMI, overweight, and obesity. *Health Place* **2009**, *15*, 1130–1141.
9. Brown, B.B.; Smith, K.R.; Hanson, H.; Fan, J.X.; Kowaleski-Jones, L.; Zick, C.D. Neighborhood design for walking and biking: Physical activity and body mass index. *Am. J. Prev. Med.* **2013**, *44*, 231–238.
10. Christian, H.E.; Fiona, C.B.; Nicholas, J.M.; Matthew, W.K.; Mark, L.D.; Paula, H.; Anura, A.; Giles-Corti, B. How important is the land use mix measure in understanding walking behavior? Results from the RESIDE study. *Int. J. Behav. Nutr. Phys. Act.* **2011**.

11. Frank, L.D.; Schmid, T.L.; Sallis, J.F.; Chapman, J.; Saelens, B.E. Linking objectively measured physical activity with objectively measured urban form. *Am. J. Prev. Med.* **2005**, *28*, 117–125.
12. Frank, L.D.; Sallis, J.F.; Conway, T.L.; Chapman, J.E.; Saelens, B.E.; Bachman, W. Many pathways from land use to health: Associations between neighborhood walkability and active transportation, body mass index, and air quality. *J. Am. Plan. Assoc.* **2006**, *72*, 75–87.
13. Lee, R.E.; Booth, K.M.; Reese-Smith, J.Y.; Regan, G.; Howard, H.H. The Physical Activity Resource Assessment (PARA) instrument: Evaluating features, amenities and incivilities of physical activity resources in urban neighborhoods. *Int. J. Behav. Nutr. Phys. Act.* **2005**.
14. Caspersen, C.J.; Powell, K.E.; Christenson, G.M. Physical activity, exercise, and physical fitness: Definitions and distinctions for health-related research. *Public Health Rep.* **1985**, *100*, 126–130.
15. Pate, R.R.; Pratt, M.; Blair, S.N.; Haskell, W.L.; Macera, C.A.; Bouchard, C.; Buchner, D.; Ettinger, W.; Heath, G.W.; King, A.C. Physical activity and public health: A recommendation from the Centers for Disease Control and Prevention and the American College of Sports Medicine. *J. Am. Med. Assoc.* **1995**, *273*, 402–407.
16. Andersen, R.E.; Crespo, C.J.; Bartlett, S.J.; Cheskin, L.J.; Pratt, M. Relationship of physical activity and television watching with body weight and level of fatness among children: Results from the Third National Health and Nutrition Examination Survey. *J. Med. Assoc.* **1998**, *279*, 938–942.
17. Fan, J.X.; Brown, B.B.; Hanson, H.; Kowaleski-Jones, L.; Smith, K.R.; Zick, C.D. Moderate to vigorous physical activity and weight outcomes: Does every minute count? *Am. J. Health Promot.* **2013**, *28*, 41–49.
18. Smith, K.R.; Brown, B.B.; Yamada, I.; Kowaleski-Jones, L.; Zick, C.D.; Fan, J.X. Walkability and body mass index: Density, design, and new diversity measures. *Am. J. Prev. Med.* **2008**, *35*, 237–244.
19. Warburton, D.E.; Nicol, C.W.; Bredin, S.S. Health benefits of physical activity: The evidence. *Can. Med. Assoc. J.* **2006**, *174*, 801–809.
20. Gordon-Larsen, P.; Nelson, M.C.; Page, P.; Popkin, B.M. Inequality in the built environment underlies key health disparities in physical activity and obesity. *Pediatrics* **2006**, *117*, 417–424.
21. Powell, K.E.; Thompson, P.D.; Caspersen, C.J.; Kendrick, J.S. Physical activity and the incidence of coronary heart disease. *Annu. Rev. Public Health* **1987**, *8*, 253–287.
22. Dunn, A.L.; Trivedi, M.H.; O'Neal, H.A. Physical activity dose-response effects on outcomes of depression and anxiety. *Med. Sci. Sports Exerc.* **2001**, *33* (Suppl. 6), S587–S597.
23. Haskell, W.L.; Lee, I.M.; Pate, R.R.; Powell, K.E.; Blair, S.N.; Franklin, B.A.; Macera, C.A.; Heath, G.W.; Thompson, P.D.; Bauman, A. Physical activity and public health: Updated recommendation for adults from the American College of Sports Medicine and the American Heart Association. *Circulation* **2007**, *116*, 1081–1093.

24. Baecke, J.A.; Burema, J.; Frijters, J.E. A short questionnaire for the measurement of habitual physical activity in epidemiological studies. *Am. J. Clin. Nutr.* **1982**, *36*, 936–942.
25. Jacobs, D.R., Jr.; Ainsworth, B.E.; Hartman, T.J.; Leon, A.S. A simultaneous evaluation of 10 commonly used physical activity questionnaires. *Med. Sci. Sports Exerc.* **1993**, *25*, 81–91.
26. Craig, C.L.; Marshall, A.L.; Sjöström, M.; Bauman, A.E.; Booth, M.L.; Ainsworth, B.E.; Pratt, M.; Ekelund, U.; Yngve, A.; Sallis, J.F.; *et al.* International physical activity questionnaire: 12-country reliability and validity. *Med. Sci. Sports Exerc.* **2003**, *35*, 1381–1395.
27. Borowski, L.A.; Bowles, H.R. Resources for locating and selecting self-report measures of physical activity. *J. Phys. Act. Health* **2012**, *9*, S91–S92.
28. Ellery, C.V.L.; Weiler, H.A.; Hazell, T.J. Physical activity assessment tools for use in overweight and obese children. *Int. J. Obes.* **2014**, *38*, 1–10.
29. Brown, B.B.; Wilson, L.; Tribby, C.P.; Werner, C.M.; Wolf, J.; Miller, H.J.; Smith, K.R. Adding maps (GPS) to accelerometry data to improve study participants' recall of physical activity: A methodological advance in physical activity research. *Br. J. Sports Med.* **2014**, *48*, 1054–1058.
30. Hendelman, D.; Miller, K.; Baggett, C.; Debold, E.; Freedson, P. Validity of accelerometry for the assessment of moderate intensity physical activity in the field. *Med. Sci. Sports Exerc.* **2000**, *32*, S442–S449.
31. Swartz, A.M.; Strath, S.J.; Bassett, D.R.; O'Brien, W.L.; King, G.A.; Ainsworth, B.E. Estimation of energy expenditure using CSA accelerometers at hip and wrist sites. *Med. Sci. Sports Exerc.* **2000**, *32*, S450–S456.
32. Troiano, R.P.; Berrigan, D.; Dodd, K.W.; Masse, L.C.; Tilert, T.; McDowell, M. Physical activity in the United States measured by accelerometer. *Med. Sci. Sports Exerc.* **2008**, *40*, 181–188.
33. Fan, J.X.; Kowaleski-Jones, L.; Wen, M. Walking or dancing patterns of physical activity by cross-sectional age among US women. *J. Aging Health* **2013**, *25*, 1182–1203.
34. Hallal, P.C.; Andersen, L.B.; Bull, F.C.; Guthold, R.; Haskell, W.; Ekelund, U.; Lancet Physical Activity Series Working Group. Global physical activity levels: Surveillance progress, pitfalls, and prospects. *Lancet* **2012**, *380*, 247–257.
35. Bradley, R.H.; McRitchie, S.; Houts, R.M.; Nader, P.; O'Brien, M. Parenting and the decline of physical activity from age 9 to 15. *Int. J. Behav. Nutr. Phys. Act.* **2011**.
36. Zick, C.D.; Smith, K.R.; Brown, B.B.; Fan, J.X.; Kowaleski-Jones, L. Physical activity during the transition from adolescence to adulthood. *J. Phys. Act. Health* **2007**, *4*, 125–137.
37. Wen, M.; Su, D. Correlates of Leisure-Time Physical Activity Participation Among Latino Children and Adolescents with Acanthosis Nigricans. *J. Immigr. Minor. Health* **2015**, *17*, 1330–1336.
38. Bellocco, R.; Jia, C.; Ye, W.; Lagerros, Y.T. Effects of physical activity, body mass index, waist-to-hip ratio and waist circumference on total mortality risk in the Swedish National March Cohort. *Eur. J. Epidemiol.* **2010**, *25*, 777–788.

39. Lee, P.H.; Macfarlane, D.J.; Lam, T.; Stewart, S.M. Validity of the international physical activity questionnaire short form (IPAQ-SF): A systematic. *Int. J. Behav. Nutr. Phys. Act.* **2011**.
40. Löllgen, H.; Böckenhoff, A.; Knapp, G. Physical activity and all-cause mortality: An updated meta-analysis with different intensity categories. *Int. J. Sports Med.* **2009**, *30*, 213–224.
41. Trost, S.G.; Pate, R.R.; Sallis, J.F.; Freedson, P.S.; Taylor, W.C.; Dowda, M.; Sirard, J. Age and gender differences in objectively measured physical activity in youth. *Med. Sci. Sports Exerc.* **2002**, *34*, 350–355.
42. Quarmby, T.; Dagkas, S. Children's engagement in leisure time physical activity: Exploring family structure as a determinant. *Leis. Stud.* **2010**, *29*, 53–66.
43. Marshall, S.J.; Jones, D.A.; Ainsworth, B.E.; Reis, J.P.; Levy, S.S.; Macera, C.A. Race/ethnicity, social class, and leisure-time physical inactivity. *Med. Sci. Sports Exerc.* **2007**, *39*, 44–51.
44. Boarnet, M.; Crane, R. The influence of land use on travel behavior: Specification and estimation strategies. *Transp. Res. Part A Policy Pract.* **2001**, *35*, 823–845.
45. Barton, H. Land use planning and health and well-being. *Land Use Policy* **2009**, *26*, S115–S123.
46. Ewing, R.; Schmid, T.; Killingsworth, R.; Zlot, A.; Raudenbush, S. Relationship between urban sprawl and physical activity, obesity, and morbidity. *Am. J. Health Promot.* **2003**, *18*, 47–57.
47. Ewing, R.; Cervero, R. Travel and the built environment: A synthesis. *Transp. Res. Rec. J. Transp. Res. Board* **2001**, *1780*, 87–114.
48. Heath, G.W.; Brownson, R.C.; Kruger, J.; Miles, R.; Powell, K.E.; Ramsey, L.T.; Task Force on Community Preventive Services. The effectiveness of urban design and land use and transport policies and practices to increase physical activity: A systematic review. *J. Phys. Act. Health* **2006**, *3*, S55–S76.
49. Salvo, D.; Reis, R.S.; Stein, A.D.; Rivera, J.; Martorell, R.; Pratt, M. Peer Reviewed: Characteristics of the Built Environment in Relation to Objectively Measured Physical Activity Among Mexican Adults, 2011. *Prev. Chronic Disease* **2014**, *11*, 1–16.
50. Kaczynski, A.T.; Potwarka, L.R.; Saelens, B.E. Association of park size, distance, and features with physical activity in neighborhood parks. *Am. J. Public Health* **2008**, *98*, 1451–1456.
51. McCormack, G.R.; Rock, M.; Toohey, A.M.; Hignell, D. Characteristics of urban parks associated with park use and physical activity: A review of qualitative research. *Health Place* **2010**, *16*, 712–726.
52. Cervero, R.; Kockelman, K. Travel demand and the 3Ds: Density, diversity, and design. *Transp. Res. Part D Transp. Environ.* **1997**, *2*, 199–219.
53. Ewing, R.; Cervero, R. Travel and the built environment: A meta-analysis. *J. Am. Plan. Assoc.* **2010**, *76*, 265–294.
54. Humpel, N.; Owen, N.; Leslie, E. Environmental factors associated with adults' participation in physical activity: A review. *Am. J. Prev. Med.* **2002**, *22*, 188–199.

55. Frank, L.D.; Sallis, J.F.; Saelens, B.E.; Leary, L.; Cain, K.; Conway, T.L.; Hess, P.M. The development of a walkability index: Application to the Neighborhood Quality of Life Study. *Br. J. Sports Med.* **2010**, *44*, 924–933.
56. Leslie, E.; Coffee, N.; Frank, L.; Owen, N.; Bauman, A.; Hugo, G. Walkability of local communities: Using geographic information systems to objectively assess relevant environmental attributes. *Health Place* **2007**, *13*, 111–122.
57. Manaugh, K.; El-Geneidy, A. Validating walkability indices: How do different households respond to the walkability of their neighborhood? *Transp. Res. Part D Transp. Environ.* **2011**, *16*, 309–315.
58. Mayne, D.J.; Morgan, G.G.; Willmore, A.; Rose, N.; Jalaludin, B.; Bambrick, H.; Bauman, A. An objective index of walkability for research and planning in the Sydney Metropolitan Region of New South Wales, Australia: An ecological study. *Int. J. Health Geogr.* **2013**.
59. Van Dyck, D.; Deforche, B.; Cardon, G.; de Bourdeaudhuij, I. Neighbourhood walkability and its particular importance for adults with a preference for passive transport. *Health Place* **2009**, *15*, 496–504.
60. Van Dyck, D.; Cardon, G.; Deforche, B.; Sallis, J.F.; Owen, N.; de Bourdeaudhuij, I. Neighborhood SES and walkability are related to physical activity behavior in Belgian adults. *Prev. Med.* **2010**, *50*, S74–S79.
61. De Meester, F.; van Dyck, D.; de Bourdeaudhuij, I.; Deforche, B.; Sallis, J.F.; Cardon, G. Active living neighborhoods: Is neighborhood walkability a key element for Belgian adolescents? *BMC Public Health* **2012**.
62. Sundquist, K.; Eriksson, U.; Kawakami, N.; Skog, L.; Ohlsson, H.; Arvidsson, D. Neighborhood walkability, physical activity, and walking behavior: The Swedish Neighborhood and Physical Activity (SNAP) study. *Soc. Sci. Med.* **2011**, *72*, 1266–1273.
63. Yamada, I.; Brown, B.B.; Smith, K.R.; Zick, C.D.; Kowaleski-Jones, L.; Fan, J.X. Mixed land use and obesity: An empirical comparison of alternative land use measures and geographic scales. *Prof. Geogr.* **2012**, *64*, 157–177.
64. Duncan, M.J.; Winkler, E.; Sugiyama, T.; Cerin, E.; Leslie, E.; Owen, N. Relationships of land use mix with walking for transport: Do land uses and geographical scale matter? *J. Urban Health* **2010**, *87*, 782–795.
65. Tian, G.; Ewing, R.; Greene, W. Desire for Smart Growth: A Survey of Residential Preferences in the Salt Lake Region of Utah. *Hous. Policy Debate* **2015**, *25*, 446–462.
66. Anselin, L. Spatial regression. In *The SAGE Handbook of Spatial Analysis*; Fotheringham, S., Rogerson, P., Eds.; SAGE: Los Angeles, CA, USA, 2009; pp. 255–276.
67. Lee, D.T.; Schachter, B.J. Two algorithms for constructing a Delaunay triangulation. *Int. J. Comput. Inf. Sci.* **1980**, *9*, 219–242.
68. Pace, R.K.; LeSage, J.P. Chebyshev approximation of log-determinants of spatial weight matrices. *Comput. Stat. Data Anal.* **2004**, *45*, 179–196.
69. Yu, D.; Wei, Y.H.D.; Wu, C. Modeling spatial dimensions of housing prices in Milwaukee, WI. *Environ. Plan. B Plan. Des.* **2007**, *34*, 1085–1102.

70. Root, E.D. Moving neighborhoods and health research forward: Using geographic methods to examine the role of spatial scale in neighborhood effects on health. *Ann. Assoc. Am. Geogr.* **2012**, *102*, 986–995.
71. Sallis, J.; Owen, N. *Physical Activity and Behavioral Medicine*; Sage Publications: Thousand Oaks, CA, USA, 1999.
72. Spence, J.C.; Lee, R.E. Toward a comprehensive model of physical activity. *Psychol. Sport Exerc.* **2003**, *4*, 7–24.
73. Moudon, A.V.; Lee, C.; Cheadle, A.D.; Garvin, C.; Johnson, D.; Schmid, T.L.; Weathers, R.D.; Lin, L. Operational definitions of walkable neighborhood: Theoretical and empirical insights. *J. Phys. Act. Health* **2006**, *3*, S99–S117.
74. Schlossberg, M.; Brown, N. Comparing transit-oriented development sites by walkability indicators. *Transp. Res. Rec. J. Transp. Res. Board* **2004**, *1887*, 34–42.
75. Heath, G.W.; Parra, D.C.; Sarmiento, O.L.; Andersen, L.B.; Owen, N.; Goenka, S.; Montes, F.; Brownson, R.C.; Lancet Physical Activity Series Working Group. Evidence-based intervention in physical activity: Lessons from around the world. *Lancet* **2012**, *380*, 272–281.
76. Strong, W.B.; Malina, R.M.; Blimkie, C.J.; Daniels, S.R.; Dishman, R.K.; Gutin, B.; Hergenroeder, A.C.; Must, A.; Nixon, P.A.; Pivarnik, J.M. Evidence based physical activity for school-age youth. *J. Pediatr.* **2005**, *146*, 732–737.
77. Ancell, S.; Thompson-Fawcett, M. The social sustainability of medium density housing: A conceptual model and Christchurch case study. *Hous. Stud.* **2008**, *23*, 423–442.
78. Anselin, L. *Spatial Econometrics: Methods and Models*; Kluwer: Dordrecht, The Netherlands, 1988.
79. Gao, J.; Wei, Y.H.D.; Chen, W.; Yenneti, K. Urban Land Expansion and Structural Change in the Yangtze River Delta, China. *Sustainability* **2015**, *7*, 10281–10307.

Risky Business: Sustainability and Industrial Land Use across Seattle's Gentrifying Riskscape

Troy D. Abel, Jonah White and Stacy Clauson

Abstract: This paper examines the spatial and temporal trajectories of Seattle's industrial land use restructuring and the shifting riskscape in Seattle, WA, a commonly recognized urban model of sustainability. Drawing on the perspective of sustainability as a conflicted process, this research explored the intersections of urban industrial and nonindustrial land use planning, gentrification, and environmental injustice. In the first part of our research, we combine geographic cluster analysis and longitudinal air toxic emission comparisons to quantitatively investigate socioeconomic changes in Seattle Census block-groups between 1990, 2000, and 2009 coupled with measures of pollution volume and its relative potential risk. Second, we qualitatively examine Seattle's historical land use policies and planning and the growing tension between industrial and nonindustrial land use. The gentrification, green cities, and growth management conflicts embedded within sustainability/livability lead to pollution exposure risk and socioeconomic vulnerability converging in the same areas and reveal one of Seattle's significant environmental challenges. Our mixed-method approach can guide future urban sustainability studies to more effectively examine the connections between land use planning, industrial displacement, and environmental injustice. Our results also help sustainable development practitioners recognize that a more just sustainability in Seattle and beyond will require more planning and policy attention to mitigate obscured industrial land use conflicts.

Reprinted from *Sustainability*. Cite as: Abel, T.D.; White, J.; Clauson, S. Risky Business: Sustainability and Industrial Land Use across Seattle's Gentrifying Riskscape. *Sustainability* **2015**, *7*, 15718–15753.

1. Seattle's Contradictory Geographies

Two decades ago, Seattle Washington developed a comprehensive plan with the ambitious title "*Toward A Sustainable Seattle: A Plan for Managing Growth 1994–2014*" [1]. It was the first comprehensive plan in the US that expressly aimed for sustainability [2]. In its most recognized depiction, sustainable development involves the simultaneous achievement of economic growth, environmental protection, and equitable development. First popularized in 1987 at the international level, sustainability was defined as: "development that meets the needs of the present without compromising the ability of future generations to meet their own needs" [3]. Inspired by this special issue's tenet that sustainable economic development and smart growth involves equitable urban land development and the effective mitigation of its conflicts, we use a sociohistorical analysis of Seattle, WA to examine the restructuring of urban land use patterns, gentrification trends, and the city's shifting environmental riskscape. Recognized as one of the world's leaders in planning for urban sustainability with an "Emerald City" nickname, Seattle provides an excellent case study to examine the dilemmas of planning for and mitigating sustainability's land use conflicts in urban industrial geographies.

We draw on a growing literature that focuses on the limitations of sustainability's popular conceptions. Dobson [4,5] argued that social justice and environmental sustainability were irreconcilable. Likewise, Agyeman [6] described the dissonance between the discourses of "Just Sustainability" and an "Environmentalist-Stewardship Sustainability". Others focus on the contradictions embedded within the sustainability concept itself [7–9]. For instance, Campbell [10] portrayed sustainability as a "planner's triangle" where economic growth, environmental protection, and social equity are corner points divide by three dynamic conflicts. A resource conflict pits economic growth and efficiency against environmental protection on one side. On the second side, a development conflict separates social justice and environmental protection. Third, a property conflict divides economic growth and social justice. Godschalk [11] would extend the planner's triangle with a fourth point to form what he called the Sustainability/Livability Prism that illuminates three additional land use planning challenges: (1) a growth management conflict; (2) a green cities conflict; and (3) a gentrification conflict.

First, a growth management conflict involves the tension between livability and economic growth and those pushing for managing development *versus* those committed to unfettered market processes. The green cities conflict is a second tension between the development of natural and built environments. The stewardship of natural resources like water quality can overshadow the quality of the built environment and its influence on public health for example. Third, a gentrification conflict entails the dissonance between those committed to urban revitalization against advocates of protecting poorer neighbourhoods from displacement. By many accounts, Seattle seems to have avoided all of these conflicts and this Emerald City has been touted for its sustainability planning and outcomes because it launched the first indicators program [12,13], ranked first in a national index [14] of sustainable policy development, earned third place in another ranking of cities [15], and was described by Krueger and Ageyman [16] as a model of "actually existing" sustainability. However, a growing number of studies question Seattle's environmental reputation. After examining comprehensive plans from 30 cities, Berke and Conroy [17] ranked Seattle in the middle of the pack. Sanders [18] observed "The downside of (Seattle's) greener urbanism may have been its tendency to reinforce a trend toward a more fractured landscape in a city that would become increasingly out of reach to working and middle-class families". More recent studies found that the city's "...pollution riskscape and urban development burdens were skewed toward the city's most socially vulnerable residents" [19] and Seattle was producing a "gentrified sustainability" [20].

Likewise, Dierwechter [21] observed that the city's planning should be called "smart segregation" instead of smart growth while Dooling [22] used Seattle's history of public green space development to introduce a research agenda on ecological gentrification that illuminates losers in the green cities conflict. In fact, uneven ecological degradation and urban inequity have always been a centerpiece of Seattle's environmental history according to Klingle [23]. He described how the city's ecologically motivated cleanup of Lake Washington lead to more pollution for Seattle's other major watershed, the Duwamish River. Klingle's account of this South Seattle tributary and its nearby communities illuminated how the city has been unable to avoid a major feature of American urban geography—environmental injustice. Such contradictory accounts led us to examine the evolving structures of Seattle's land use, development, and the city's environmental riskscape and address some of the questions posed in this special issue. What are the underlying processes driving Seattle's industrial land use restructuring? How can both quantitative GIS and qualitative methods be combined in a case study design to study the interactions of land use planning and sustainability? How can mixed-methods be combined to better inform urban sustainability policies that are more equitable?

Environmental Inequities and Gentrification

Environmental inequities for racial and economic groups has been the focus of four decades of environmental justice scholarship. In this field's watershed study, sociologist Robert Bullard [24] introduced the distributional scrutiny of race and pollution to social science in a case study of Houston, TX. He described how the city's landfills were located in predominately African-American neighbourhoods and concluded that institutionalized housing market discrimination, a lack of zoning, and government permitting decisions led to the city's "...black neighbourhoods becoming the 'dumping ground' for the area's solid waste". Bullard's analysis quickly received attention by national politicians and policy practitioners leading to a study [25] of landfill siting in eight Southern states by the US General Accounting Office (GAO, now the Government Accountability Office). The results indicated that on average, three out of four landfills were located in majority-minority communities. In 1987, a study [26] of landfill sites across the US would find that zip codes with these hazardous sites had, on average, double the percent minority of zip codes without landfills. In a seminal meta-analysis of 49 environmental justice studies between 1995 and 2005, Ringquist [27] concluded that race-based environmental inequality was ubiquitous. Numerous subsequent case studies of urban environmental injustices have found that spatial distribution of pollution hazards and socially vulnerable populations (minority and low income) cluster together in Detroit [28], Los Angeles [29–31], New York City [32], Portland [33], St. Louis [34], and Tampa Bay [35].

However, few of these studies examined how these inequalities developed over time. In 2000, Pellow [36] introduced a theory of Environmental Inequality Formation which "occurs when different stakeholders struggle for access to scarce resources within the political economy, and the benefits and costs of those resources become distributed unevenly". A series of seminal environmental justice studies in the nineties found that the siting of industrial hazards in neighbourhoods did not target minority and poor communities, but nonminority and more affluent residents moved away from environmental pollution while minorities stayed or moved in [37–40]. In a recent national study combining individual-level mobility data with neighborhood-level industrial hazard risks, Crowder and Downey [41] found that "black and Latino householders move into neighbourhoods with significantly higher hazard levels than do comparable whites". Conversely, in another national study [42] of residential mobility between 1991 and 2007, the odds of starting with low pollution exposure but ending with high levels of exposure was 38 percent higher for immobile black households than for immobile whites. Likewise, a longitudinal analysis of Los Angeles' riskscape of hazardous waste Transport, Storage, and Disposal Facilities (TSDFs) found that when "controlling for other factors, minorities attract TSDFs but TSDFs do not generally attract minorities" [43]. Reversing the causal path, Hamilton [44] found that between 1987 and 1992, zip code neighbourhoods where commercial hazardous waste facilities expanded operations had an average nonwhite population of 25 percent *versus* 18 percent in zip codes hosting facilities with no expansion. In sum, this literature established that both residential and industrial mobility can contribute to environmental inequality formation. Thus we expect both to be factors in Seattle's land use restructuring, its inequitable stratification of socioeconomic development, and uneven distribution of environmental burdens. However, we go further in this case study and consider how local planning policy may contribute to environmental injustice [45,46].

Our Seattle case study contributes to the more recent scholarly attention to environmental gentrification [47–55]. Banzhaf and Walsh were some of the first that pondered how "in a world where households sort in response to changes in environmental quality, the bulk of the benefits of a policy that successfully cleans up dirtier neighbourhoods where the poor live may actually be captured by rich households... This 'environmental gentrification' may actually more than offset the direct gain of the environmental improvement, so that the original residents are actually worse off" [56]. In short, environmental cleanups attract gentrification while it's repelled by industrial activity and pollution. Ley and Dobson [57] observed three decades of redevelopment in Vancouver British Columbia and found "that inner-city districts distant from environmental amenities and proximate to industrial land use will be much less attractive to the incursions of gentrification". Conversely, deindustrializing neighbourhoods and their cleaner environments were more attractive for gentrification. While it has many definitions [58–62], gentrification generally refers to the upward socioeconomic transformation of urban neighbourhoods by income, house values, education, and occupational levels.

Since Bell's 1973 book on the postindustrial society [63], many urban geographers have examined the land use conflict between industrial and non-industrial uses through a lens of gentrification beginning with Ley [64,65]. More recently, Curran [66–69] has published a series of qualitative studies on the displacement of small manufacturers via gentrification in the New York neighbourhoods of Greenpoint and Williamsburg. Other qualitative case studies on gentrification have covered the major Northwest cities of Portland [55], Vancouver, BC [48], and Victoria, BC [49]. However, few have utilized multivariate quantitative methods [70,71] and, according to Lees, Slater, and Wyly [72], gentrification researchers rarely integrate quantitative and qualitative methods. We join these collection of studies [73–75] with a mixed-methods case study of Seattle Washington.

Gentrification is a growing challenge for Seattle. The city's rising housing costs and the equity and social impacts of explosive land use growth are an increasing concern for many Seattleites. In 2006, the median home value in Seattle was 7.7 times more than the median household income [76]. According to one observer, "the ever-increasing concentration of wealth could mean Seattle will become more and more the gilded city of the upper-middle and upper classes" [77]. Journalists in the *Seattle Times* also reported in 2006 that "the only area in Seattle where median-income folks could afford the median-priced house was the residential/industrial/commercial swath south of downtown that includes Georgetown and South Park" [78]. More recently, news headlines question "Growth Gone Wild?" [79], "Priced Out? Growing numbers appear to be fleeing King County" [80] and explore how "In Georgetown, the Housing Is Affordable and the Air Unbreathable" [81]. Similar patterns are being identified in dozens of US cities by one research group where the "creative class" [82] divides urban geographies by outcompeting service and working class residents for proximity to urban amenities like transit lines, universities, and natural areas [83].

2. Case Study Methods

2.1. Study Area and Data Sources

Sometimes called the Emerald City, this study uses Seattle as a single case approach [84] to examine the competing hypotheses of environmental gentrification (cleanup and deindustrialization attracts gentrification while industrial areas repel gentrification) and industrial gentrification (gentrification displaces industrial activity). This Emerald City provides an intriguing opportunity to examine the industrial transformation of a major West Coast city, which has over the last 30 years reinvented itself as a major hub for the United State's "tech" sector [85]. Seattle now follows San Jose in terms of the percentage of its workforce engaged in advanced industries.

The 1990s were a key time period in this transition as Microsoft became the center of an information technology cluster, followed by Amazon and startups Expedia and RealNetworks [86]. Between 1990 and 2009, of the 95,992 additional jobs created in Seattle, 28,614 new jobs (30 percent of the job growth) were in the professional, scientific and technical services sector. The job growth in this sector was second only to Arts, entertainment, recreation, accommodation and food service, which comprised 33 percent of the job growth. Consistent with the overall trend in the United States of manufacturing job loss, manufacturing employment as a percentage of total employment in Seattle between 1990 and 2009 has fallen from 15 to 8 percent (over a 50 percent decline). Seattle's manufacturing sector share of employment is now less than the national average [87].

The demand for knowledge workers by the region's tech firms has stimulated population growth. Like many other cities, Seattle has experienced rapid urban growth over the last 20 years, with the population growing by 15 percent between 1990 and 2010. In 2012, Seattle had the fastest rate of growth of the 50 most populous major cities [88]. Much of this population growth is the result of in-migration, both from domestic and foreign populations, which is consistent with Seattle's role as a regional employment and growth center.

Such growth pressures led Washington State to adopt growth management policies, a method of comprehensive land use planning significantly impacting Seattle's land use development. The city's comprehensive plan aimed to direct growth into four zones: (1) urban centers; (2) urban villages; (3) residential urban villages; and (4) manufacturing industrial centers [89]. A recent review of Seattle's planning strategy found that between 1994 and 2014, 75 percent of Seattle's total population growth (residential and employment) had been directed into these designated urban villages [90]. Notably, the area targeted for the largest employment gains (19 percent of the City's employment growth and 8 percent of its residential growth) was the South Lake Union (SLU) neighborhood [91], our first area of focus in this case study of Seattle's industrial land use planning.

Targeted for redevelopment from an industrial area to a hub urban village in the 1994 comprehensive plan (see Figure 1 below), change in SLU has occurred at a rapid pace, with over 90 percent of the City's projected employment gains between 2004 and 2024 having already been met by 2013 [92]. The redevelopment predominately consists of commercial office and residential towers, with supporting retail development. Major new tenants in the area are representative of Seattle's economic transition and include Amazon, the Gates Foundation, a number of medical research facilities, and somewhat ironically, the city's Museum of History and Industry.

After the transformation of the South Lake Union area, manufacturing in the City is being consolidated into the other two areas in our case study: the Ballard and Interbay Manufacturing/Industrial Center (BINMIC) and the Greater Duwamish Manufacturing/Industrial Center (GDMIC). While celebrated elsewhere, this conversion of Seattle's urban industrial zones to non-industrial development has generated local concerns. The city's planning department projected that "As non-industrial projects proliferate in industrial zones, it increases speculation that zoning may be changed to accommodate even larger non-industrial projects or even allow residential uses" [93]. In fact, city policy statements in 2005 aimed to "limit in manufacturing/industrial areas those commercial or residential uses that are unrelated to the industrial function, that occur at intensities posing short- and long-term conflicts for industrial uses, or that threaten to convert significant amounts of land to non-industrial uses" [94]. These concerns are now the focus of redevelopment pressures emerging in the BINMIC area.

Local news outlets reported the plans of online travel agency Expedia's impending move to the Interbay waterfront. One local TV website announced "Expedia's move likely to prompt changes to Seattle Interbay" [95] while an online blog noted that "the Interbay neighborhood has piqued the interest of developers and investors alike" [96]. City documents also noted that in BINMIC, "recent retail, residential, and office developments are pushing up property values, which could potentially squeeze out industrial uses and jobs. Some property owners would like more flexibility to develop their property. Enhanced transit service is yet another reason to rethink the future of this area" [97]. The restructuring of industrial land use patterns and their conflicts in SLU, BINMIC and GDMIC are therefore the focus of our case study (see Figure 1 below).

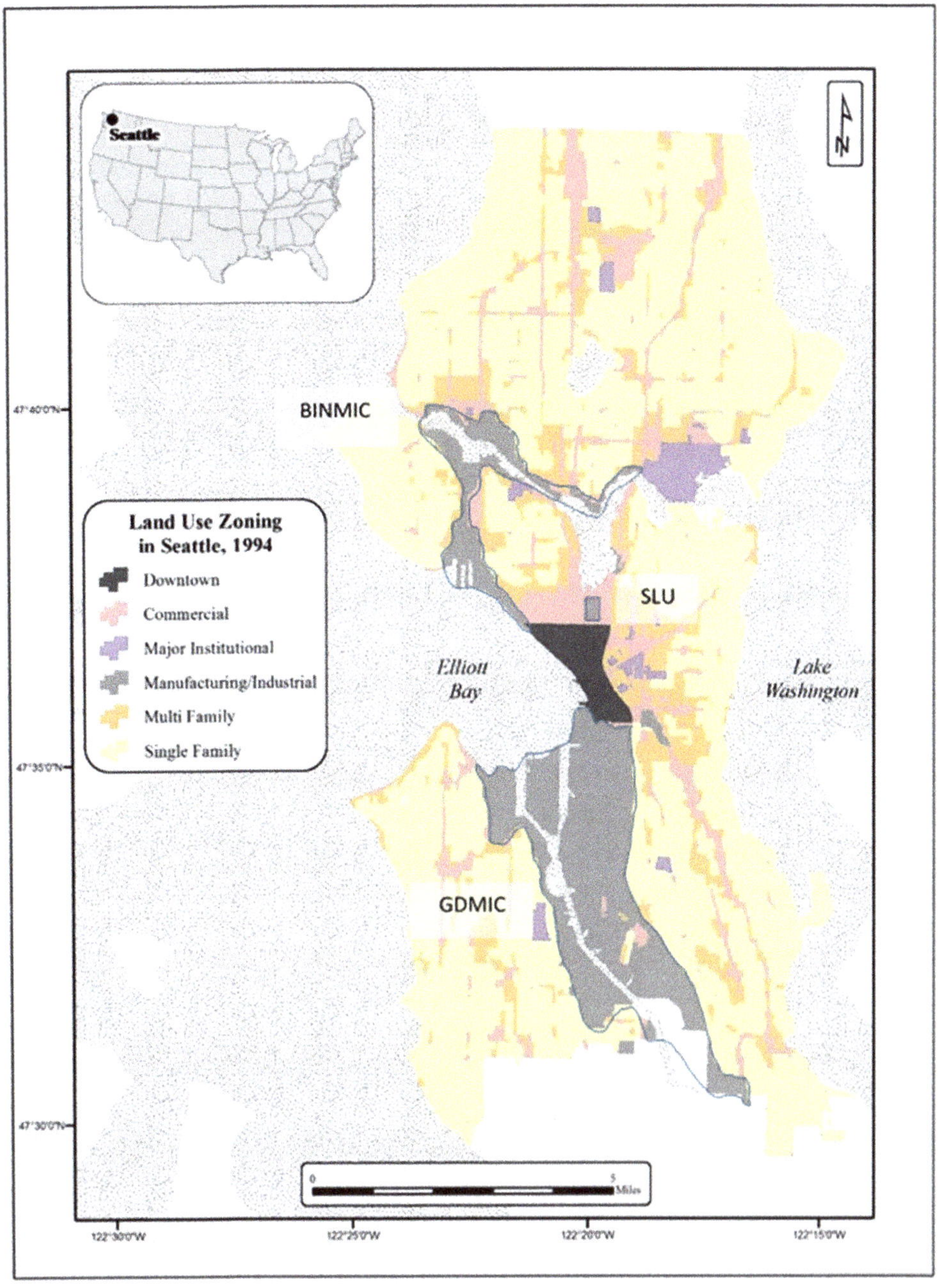

Figure 1. Seattle's zoning in 1994 and our industrial study areas outlined in red: (1) Ballard and Interbay Manufacturing/Industrial Center (BINMIC); (2) South Lake Union (SLU); and (3) the Greater Duwamish Manufacturing/Industrial Center (GDMIC).

BINMIC is located in the northwest part of the City of Seattle in Figure 1 and hosts wharfs, shipyards, and rail yards. The northern portion of BINMIC is home to Fisherman's Terminal, one of the largest commercial fishing terminals in the Northwest and one of Seattle's major railroad yards. At Interbay's south end are Terminal 91 (a large general cargo terminal complex) and Pier 86 (a Port of Seattle export grain terminal). This portion of BINMIC contains approximately 3 percent of the City's employment and the major industry sectors are Services (43%), Manufacturing (22%), and Wholesale, Transportation and Utilities (14%) [98].

GDMIC is located south of downtown Seattle in the Duwamish Valley and is one of the Pacific Northwest's largest and most intensely developed manufacturing/industrial areas. It is a major intermodal transportation hub, receiving and distributing goods via roadway, water, rail and air. GDMIC is home to Port of Seattle's primary marine shipping area, distribution and warehouse facilities, oil and petroleum storage facilities, major rail yards, and approximately 13 percent of the City's employment. The major industry sectors are Services (26%), Manufacturing (23%), and Wholesale, Transportation and Utilities (23%) [99], reflective of its base as a transportation, warehouse, distribution and manufacturing hub for the City and region. Consistent with citywide manufacturing job losses and the broader post-industrial trends of the US economy, both areas have experienced declining industrial employment over the last 20 years.

2.2. Data and Mixed-Methods

In the first part of our analysis, we analyze the stratification of Seattle's socioeconomic groups across the city with census data in 1990, 2000, and 2009 with measures and methods commonly used by urban geographers to characterize gentrification. We compared demographic information on Seattle's 568 Census Block Groups (CBGs) from the 1990 and 2000 censuses normalized by 2000 geographic boundaries [100] plus data from the 2009 American Community Survey (ACS) with a combination of Principal Components Analysis (PCA) and cluster analyses. PCA is designed to systematically reduce a large number of variables into a smaller, more conceptually coherent set of factors, dimensions, or components [101]. Component scores from the three PCAs are then used as independent variables in a cluster analysis of similar CBGs. While factor analysis results in groupings of variables, cluster analysis groups objects based on characteristics of interest to an analyst [102]. We used Ward's method of cluster analysis and its minimum distance hierarchical technique to differentiate relatively homogenous CBGs. This was the most appropriate technique for this research because of its maximization of between-group differences and minimization of within-group differences [103].

We then describe the spatial distribution of industrial environmental hazards and their shifting spatial locations in the second part of our analysis with two datasets from the US Environmental Protection Agency (EPA) over four time periods: (1) 1995 to 2000; (2) 2001 to 2005; (3) 2006 to 2010; and (4) 2011–2015. First, Seattle's largest industrial air polluters are compiled from the EPA's Toxics Release Inventory (TRI) along with their relative inhalation exposure risks derived from the Risk-Screening Environmental Indicators (RSEI) modeling program [104]. Second, we also plot the locations of facilities reporting to the EPA's Biennial Reporting System (BRS) to widen the characterization of Seattle's riskscape [105]. According to Atlas [106], no published study until his 2002 paper had examined the distribution of hazardous waste generators reporting to the BRS database and these kinds of facilities accounted for 90 percent of all industrial hazardous waste.

Finally, we complement the quantitative socioeconomic and environmental hazard assessment methods with a case study of Seattle's industrial area land use dynamics. Using policy documents and permit data on land use changes and building construction, we compare the trajectories of industrial land conversions in Seattle's remaining industrial areas. The key planning and policy documents included: (1) Seattle's Comprehensive Planning documents published in 1994 and 2005 [1,94]; (2) Industrial land studies released in 2007 and 2013 [93,107]; and (3) equity analyses from 1993 and 2015 [108,109]. We obtained the land use [110] and building construction permits [111] from SDPD's online permitting database.

These seemingly disparate methods are integrated through the perspective of sustainability as a conflicted process where affluent residents out-compete less affluent ones for neighbourhoods with fewer environmental hazards on the one hand. On the other hand, our methods offer an empirical application of the varied theoretical developments around environmental gentrification as a sociohistorical process and addressed several critical gaps in sustainability research including its inattention to the performance of cities on equitable development. It's the "creation and maintenance of economically and socially diverse communities that are stable over the long term, through means that generate a minimum of transition costs that fall unfairly on lower income residents" [112]. Equitable development requires, according to one urban practitioner, "...the promotion and management of economic growth that maximizes benefits for residents of low-income communities throughout metropolitan regions and assures their voice in the development process" [113]. Conversely, gentrification often involves inequitable development.

3. Results and Discussion

Since gentrification is considered to encompass change in any number of combinations of indicators, we first compiled and factor analyzed 12 variables from the broad categories of population, socioeconomic and housing measures to better understand the change in socioecological structure in each of the 568 CBGs in Seattle for the 1990, 2000, and 2009 time periods (see Table 1). In the first dataset, a three factor solution which explained about 73% of the variance in Seattle CBGs was obtained representing socioeconomic status, race/ethnicity, and household structure. The socioeconomic factor produced high loadings on percentage college graduates, percentage of professional occupations, median household income, median contract rent, and median house value. In the second factor, racial divides are manifest with the percentage of White alone inversely related to Black or Asian alone and the percentage at or below the poverty level. A third factor highlights a divide between traditional home-owning families and younger, unrelated residents who value urban living and amenities. The percentage of population age 25–34 loads together with non-family households while both are inversely correlated to median household income and homeownership rates.

The 2000 PCA results produced just two factors reflecting an urban structure shaped mostly by socio-racial status and household structure. Accounting for 65% of the variance in the arrangement of Seattle's urban landscape, the first factor represents a convergence of socioeconomic status and ethnicity.

Factor one produced positive loadings for the percentage of college graduates, percentage of professional employment, median household income, median contract rent, median house value, and percentage White. Variables with negative loadings on factor one included percentage Black alone, Asian alone, and residents at or below the poverty line. This structure reflected an emerging divide between the labor forces, yet the individual importance of median household income, median contract rent, and median house value was superseded by college graduates and professional occupation. Factor two is identical to factor three from 1990 and indicated that urban amenities as viewed by different types of households continue to be significant in the city's structural form.

The third PCA of 2009 ACS data accounted for 59% of the variance among the selected variables across two components reflecting socio-racial status and household structure. The first factor produced positive loadings for the percentage of college graduates, percentage of professional employment, median household income, median house value, and percentage White. Variables with negative loadings on factor one included percentage Black alone, Asian alone, and residents at or below the poverty line. This structure reflected a deepening divide between White and non-White populations in terms of professional status and affluence. The second factor again is nearly identical to factor three in 1990, and factor two in 2000 indicating that urban amenities for varying households continue to be significant in the city's structural form. The percentage of population age 25–34 loads negatively with non-family households and both are inversely correlated with median household income and rates of homeownership. In fact, the relative importance of the proportion of population age 25–34, median household income, and homeownership rates increased since 1990 indicating a significant disparity between traditional home-owning families and younger residents. These multi-year PCA results of seven factors were then utilized in a hierarchical cluster analysis to identify similar groups of CBGs.

3.1. Seattle's Gentrifying and Transitioning Urban Industrial Clusters

The results of the cluster analysis yielded groupings of CBGs with similar values on the seven factors derived from our PCA analyses. Like Morrill [114], we explored multiple cluster solutions and found a 15-cluster solution to be the most coherent ordering of Seattle's urban structure considering quantitative relationships as well as historical geographies of locally recognized neighbourhoods (See Figure 2 and Supplemental Materials). A sixteenth cluster was excluded from statistical estimations because it encompassed the industrial district of Harbor Island at the mouth of the Duwamish River containing no residences. To classify gentrification and transition clusters, we examined trends in income, house value, education and occupation. If the rate of increase in 3 out of 4 of these indicators and the value of each variable at the end of the study period was higher than the city average, the cluster was classified as gentrifying. Transition clusters were those agglomerations that may have increased on 3 out of 4 variables, but the values in 2009 remained below the Seattle average. A closer look at the clusters in and near Seattle's three industrial land use zones (3, 5, 13 and 15) in Table 2, Figure 2, and Table 3 illuminated some key trajectories in the city's restructuring riskscape.

Cluster 3 sprawls across northern and southern parts of the city in three different agglomerations while covering a small portion of Ballard and the BINMIC industrial area. This cluster saw a small increase in its Black population and the percentage of residents in poverty while Asian residents and those aged 25–34 remained relatively stable. Median household incomes tracked the city average but median contract rent surpassed the average in 2000 and then fell below it in 2009. Cluster 3 attracted an increasing percentage of college graduates and management professionals both up from less than one-third of the area's residents in 1990 to nearly half by 2009 suggesting an area growing in renters new to the workforce.

Table 1. Results of principal components analysis.

Category & Variable Name	1990			2000		2009	
	Socioeconomic Status	Racial Polarization	Household Structure	Socio-Racial Status	Household Structure	Socio-Racial Status	Household Structure
Population							
Population Age 25–34			0.746		−0.808		−0.789
Percent White alone		0.927		0.918		0.917	
Percent Black alone		−0.839		−0.678		−0.687	
Percent Asian alone		−0.635		−0.780		−0.708	
Socioeconomic							
College graduates	0.837			0.873		0.838	
Professional/Managerial	0.823			0.862		0.741	
Median Household Income	0.713		−0.551	0.647	0.638	0.551	0.670
Poverty		−0.562	0.488	−0.597		−0.476	
Housing							
2^+ Person Non-family Households			0.734		−0.729		−0.616
Median Contract Rent	0.627			0.555			0.493
Median House Value	0.821			0.559		0.507	
Owner-occupied			−0.794		0.843		0.834
Percent Variance	29.30	23.20	20.86	41.94	23.23	34.96	23.63
Cumulative Variance	29.30	52.50	73.36	41.94	65.17	34.96	58.59

Notes: Loadings −0.45 to +0.45 not shown. Varimax rotation with Kaiser normalization.

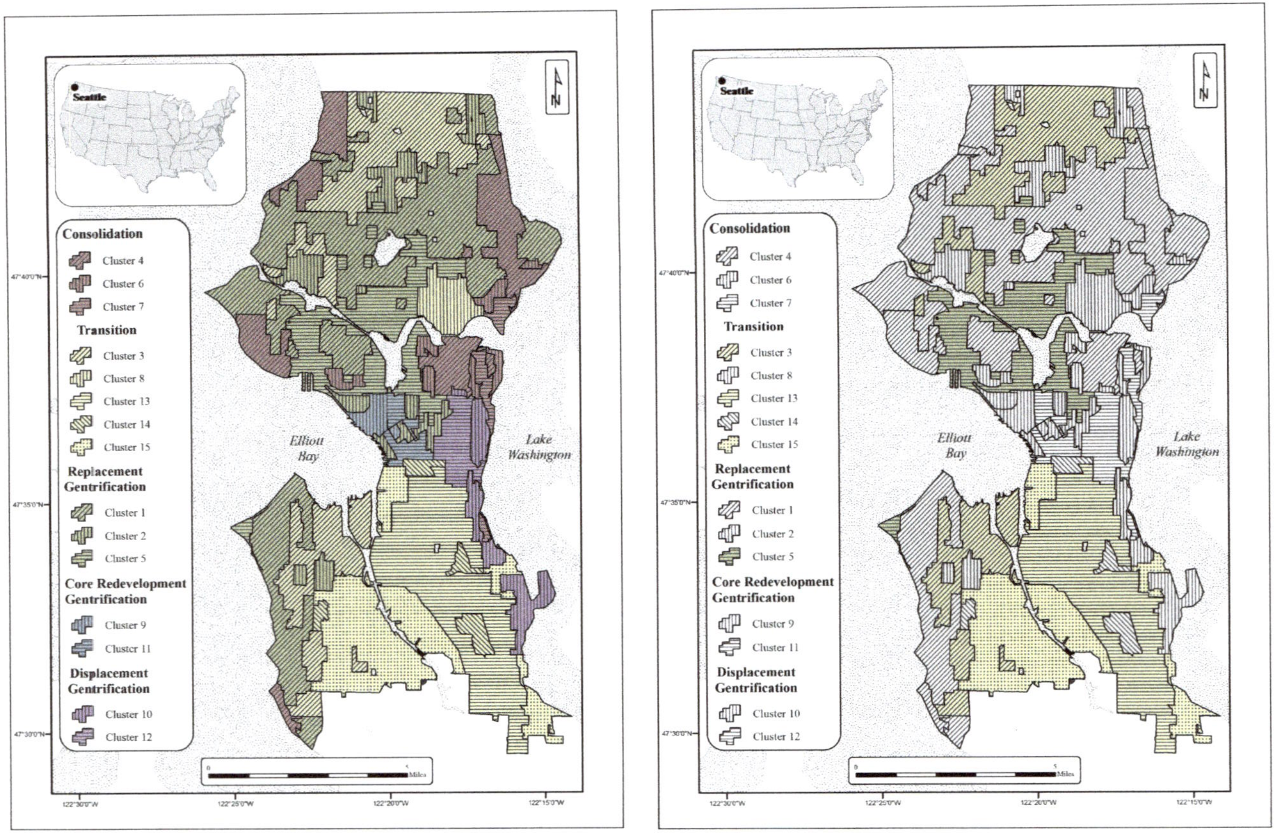

(a) (b)

Figure 2. Factorial Ecology of Seattle's Development trajectories in 1990, 2000, 2009 for (**a**) all clusters and (**b**) industrial areas.

Table 2. Gentrification typology for clusters in and near Seattle's industrial areas.

3	Transition cluster with increasing social status; middle income; young, non-families; increasing minorities
13	Transition cluster with Asian influx; little change in social status (working-class); above average home ownership
5	Replacement cluster with increased social status; above average incomes; young, non-families; above average home values; primarily renters
15	Transition cluster with Minority mixing; little change in social status (working-class); above average home ownership

Cluster 5 on the other hand stands out for its compactness and near alignment with the industrial areas on the edges of Lake Union and the BINMIC zone. This cluster saw an increase in young and nonfamily households while its residents also moved from below the average median household income in 1990 to above it in 2000 and 2009. Cluster 5 had increasing percentages of college graduates while also seeing one of Seattle's biggest surges (nearly 20 percent) of residents with professional and managerial occupations. This replacement gentrification cluster had a significant class shift with residents paying consistently higher rents then in Clusters 3, 13, and 15. Cluster 5 also saw a decrease in Black alone residents with a slight increase in Asian alone population.

Conversely, Cluster fifteen's trajectory in the center of the GDMIC industrial area was vastly different. Median household income in Cluster 15 remained below the city average in 1990, 2000, and 2009. While poverty declined between 1990 and 2000 in this cluster, it climbed from 12.3% to 15.8% between 2000 and 2009. Median house values increased at a higher rate than for the rest of the city (131.6% and 86.8%) but remained more than one hundred thousand dollars below the median value of a Seattle home. Moreover, Cluster 15's gap of college graduates compared to the rest of the city increased between 1990 and 2009. Seventeen percent of Cluster 15 held a college degree in 1990 while 37.9% was the city average (20.9% difference). In 2009, 30.5% of Cluster 15 residents held a college degree while the city average was 54.3% (23.8% difference).

Table 3. Changes in Race and Age, Social Status (adjusted to 2000 dollars), and Professional Status for clusters encompassing and contiguous to industrial zones.

Cluster (n)	Percent White Alone			Percent Black Alone		
	1990	2000	2009	1990	2000	2009
3 (83)	85.80	78.52	78.99	2.58	3.92	4.73
5 (69)	90.14	86.11	86.54	2.98	2.37	1.32
13 (51)	25.12	21.35	25.82	28.43	21.42	21.56
15 (31)	63.56	49.39	57.39	14.86	15.48	14.91
Seattle city	75.96	71.78	71.63	10.21	8.44	7.68
Cluster (n)	**Percent Asian alone**			**Percent age 25–34**		
	1990	2000	2009	1990	2000	2009
3 (83)	8.30	9.67	8.61	21.75	21.07	21.51
5 (69)	4.81	5.93	6.55	31.50	35.22	35.07
13 (51)	43.23	45.77	40.81	18.31	16.74	16.90
15 (31)	16.17	19.70	16.63	19.22	16.97	17.85
Seattle city	11.19	12.55	12.79	21.73	21.71	21.73
Cluster (n)	**Median household income**			**Median household income**		
	1990	2000	2009	1990	2000	2009
3 (83)	53,047	59,363	60,579	8.01	9.41	9.46
5 (69)	51,557	64,828	64,854	10.51	8.79	9.28
13 (51)	46,649	54,139	50,176	16.37	14.06	15.96
15 (31)	50,133	58,404	60,029	13.15	12.25	15.77
Seattle city	56,463	58,862	60,843	12.38	11.79	12.24
Cluster (n)	**Median contract rent**			**Median house value**		
	1990	2000	2009	1990	2000	2009
3 (83)	806	934	849	204,521	267,538	387,558
5 (69)	817	978	966	280,153	372,908	491,893
13 (51)	660	771	794	148,402	231,095	330,622
15 (31)	741	874	926	146,134	218,852	338,416
Seattle city	815	928	1024	239,198	334,105	446,900
Cluster (n)	**Percent college graduates**			**Percent professional/managerial**		
	1990	2000	2009	1990	2000	2009
3 (83)	28.35	38.58	49.40	30.23	42.50	48.69
5 (69)	52.27	63.01	68.85	42.41	57.30	60.56
13 (51)	15.81	21.93	27.11	17.91	27.20	30.48
15 (31)	16.99	25.43	30.48	19.61	30.91	37.99
Seattle city	37.91	47.19	54.31	36.27	48.41	52.29

3.2. Seattle's Shifting Riskscape

At the beginning of our study period, industrial toxic releases in the two clusters near BINMIC and SLU were not far behind the number of industrial facilities and risk levels seen in the two clusters near GDMIC (see Table 4). Clusters 3 and 5 hosted 19 TRI facilities in 1990 that produced a risk characterization above 13 million. In Clusters 13 and 15, 28 TRI facilities and their toxic emissions resulted in an exposure risk characterization of 17 million. By 2000, the BINMIC and SLU clusters exceeded the industrial emission and risk levels of GDMIC substantially. Nine industrial facilities in the industrial zones on the Northside produced nearly 78 percent of the air pollution exposure risk estimates while 16 facilities on the Southside of the city produced only 14 percent. By the end of our study period in 2009, Clusters 5 and 13 hosted only 3 TRI facilities, 4 were located in Cluster 3, and Cluster 15 hosted 11 industrial air polluters. More than 80 percent of the air pollution risk exposure estimates fell in the Southside's Cluster 15 compared to nearly 16 percent in the Northside's Cluster 3. Air emissions and their relative risk nearly disappeared in Clusters 5 and 13 as Seattle's industrial air pollution riskscape shifted south (see Figures 3 and 4).

Table 4. Seattle's air pollution exposure risk characterizations by cluster from 1990–2009.

Cluster (No. of TRIs)	Pounds	Risk Value	Total Risk, %	Cumulative Risk, %
		1990		
Cluster 3 (12)	330,224	1,418,360.74	4.65	4.65
Cluster 5 (7)	54,379	11,707,273.80	38.36	43.00
Cluster 13 (13)	140,700	15,368,208.89	50.35	93.35
Cluster 15 (15)	1,420,090	2,023,126.55	6.63	99.98
Top 10 facility totals	1,182,594	30,372,678.74	99.51	99.51
All facility totals (n = 58)	2,478,741	30,522,468.99	100.00	100.00
		2000		
Cluster 3 (6)	113,793	13,018.04	1.44	1.44
Cluster 5 (3)	10,492	691,922.98	76.42	77.85
Cluster 13 (4)	2317	87,022.83	9.61	87.46
Cluster 15 (12)	130,955	49,477.80	5.46	92.93
Top 10 facility totals	131,693	902,696.13	99.69	99.69
All facility totals (n=34)	285,737	905,478.23	100.00	100.00
		2009		
Cluster 3 (4)	15,604	5612.81	15.93	15.93
Cluster 5 (1)	10	0.83	0.00	15.93
Cluster 13 (2)	145	111.72	0.32	16.25
Cluster 15 (11)	47,296	29,218.24	82.92	99.17
Top 10 facility totals	60,716	34,662.91	98.37	98.37
All facility totals (n = 22)	68,580	35,237.07	100.00	100.00
		1990–2009		
Cluster 3 (21)	2,012,244	1,740,893.82	2.47	2.47
Cluster 5 (14)	337,601	17,161,948.51	24.33	26.80
Cluster 13 (22)	599,026	44,375,277.09	62.91	89.71
Cluster 15 (38)	4,938,664	6,625,084.58	9.39	99.10
Top 10 facility totals	2,630,549	68,299,288.08	96.82	96.82
All facility totals (n = 113)	8,978,347	70,539,262.74	100.00	100.00

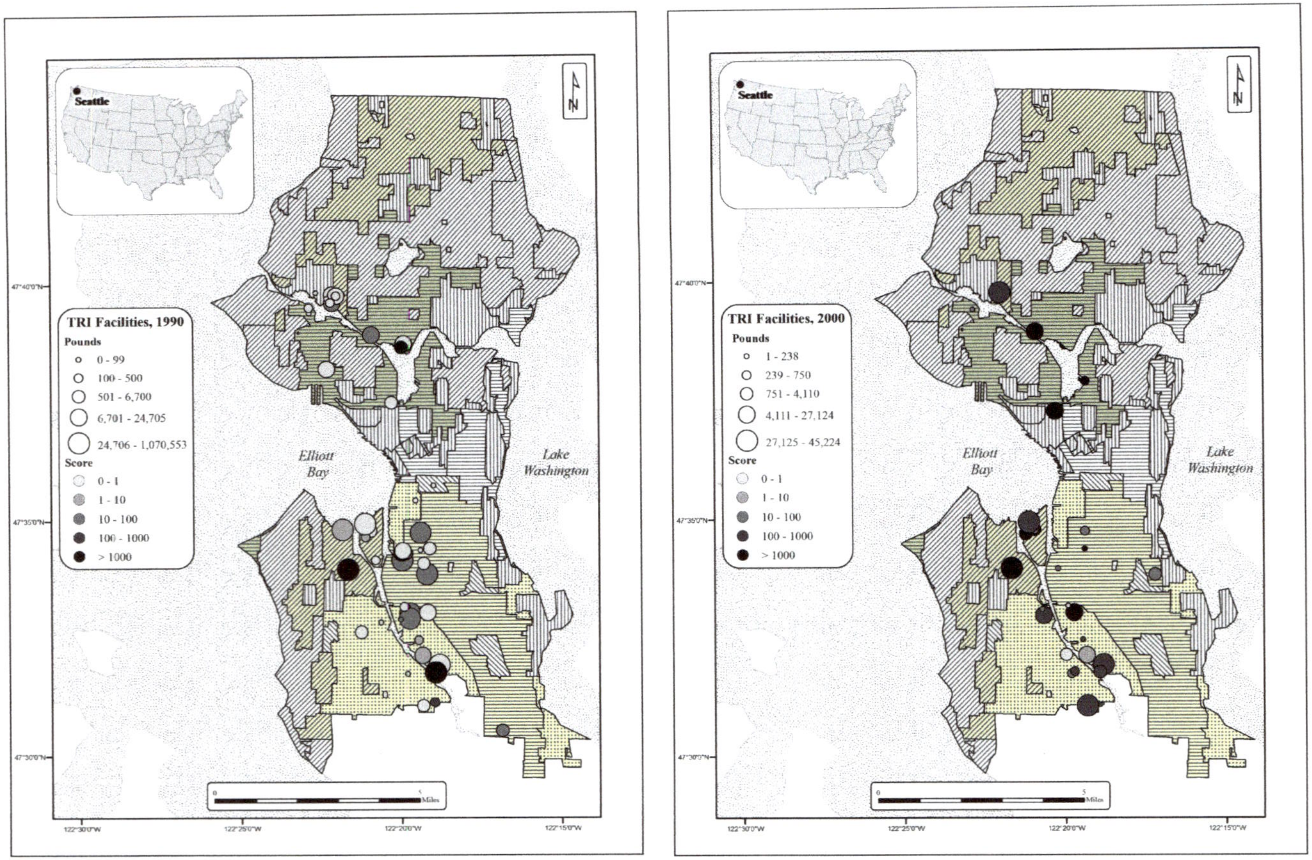

(a) (b)

Figure 3. (**a**) Seattle's Industrial Air Pollution Sources in the Toxics Release Inventory (TRI) in 1990 and (**b**) 2000.

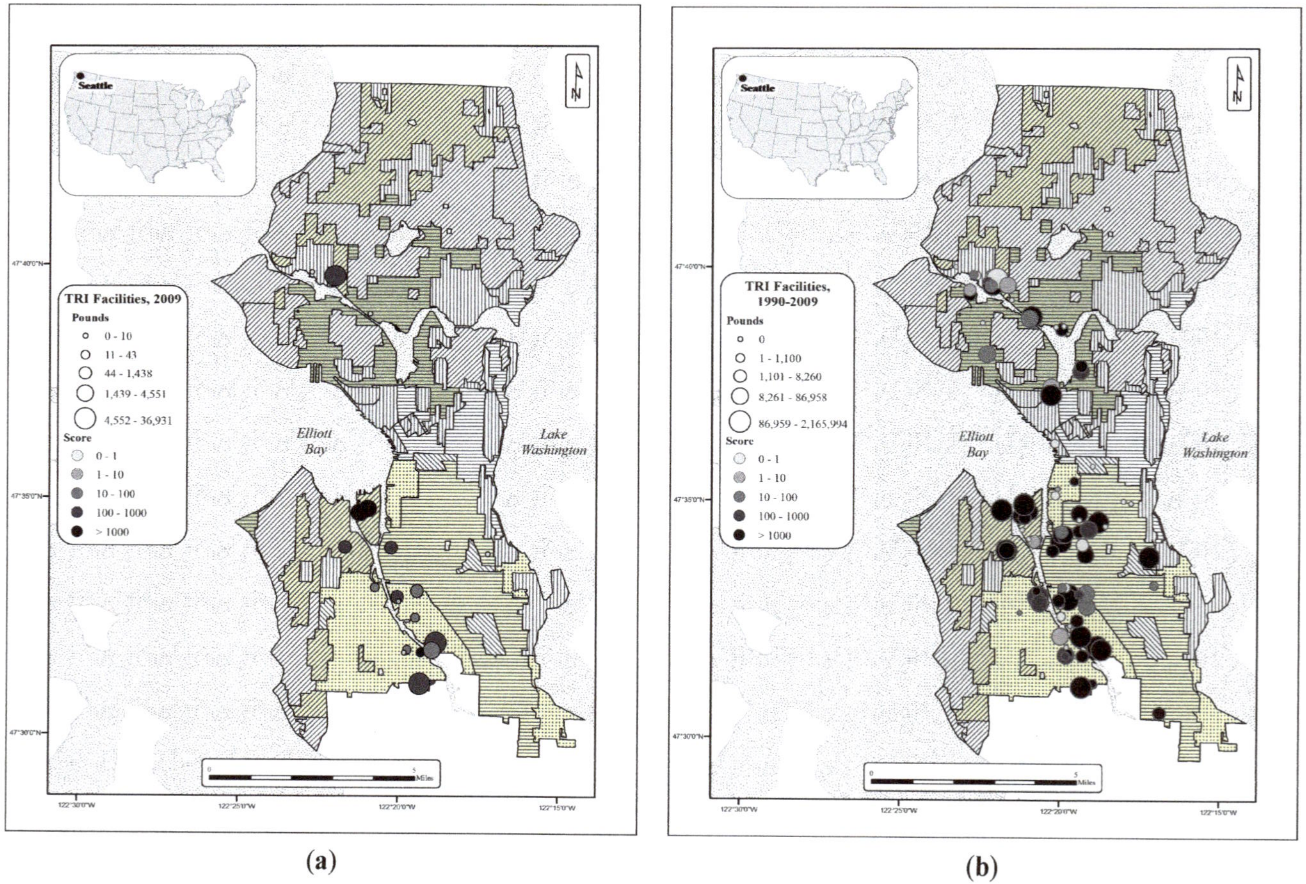

(a) (b)

Figure 4. (**a**) Seattle's industrial air pollution sources in the Toxics Release Inventory (TRI) in 2009 and (**b**) 1990–2009.

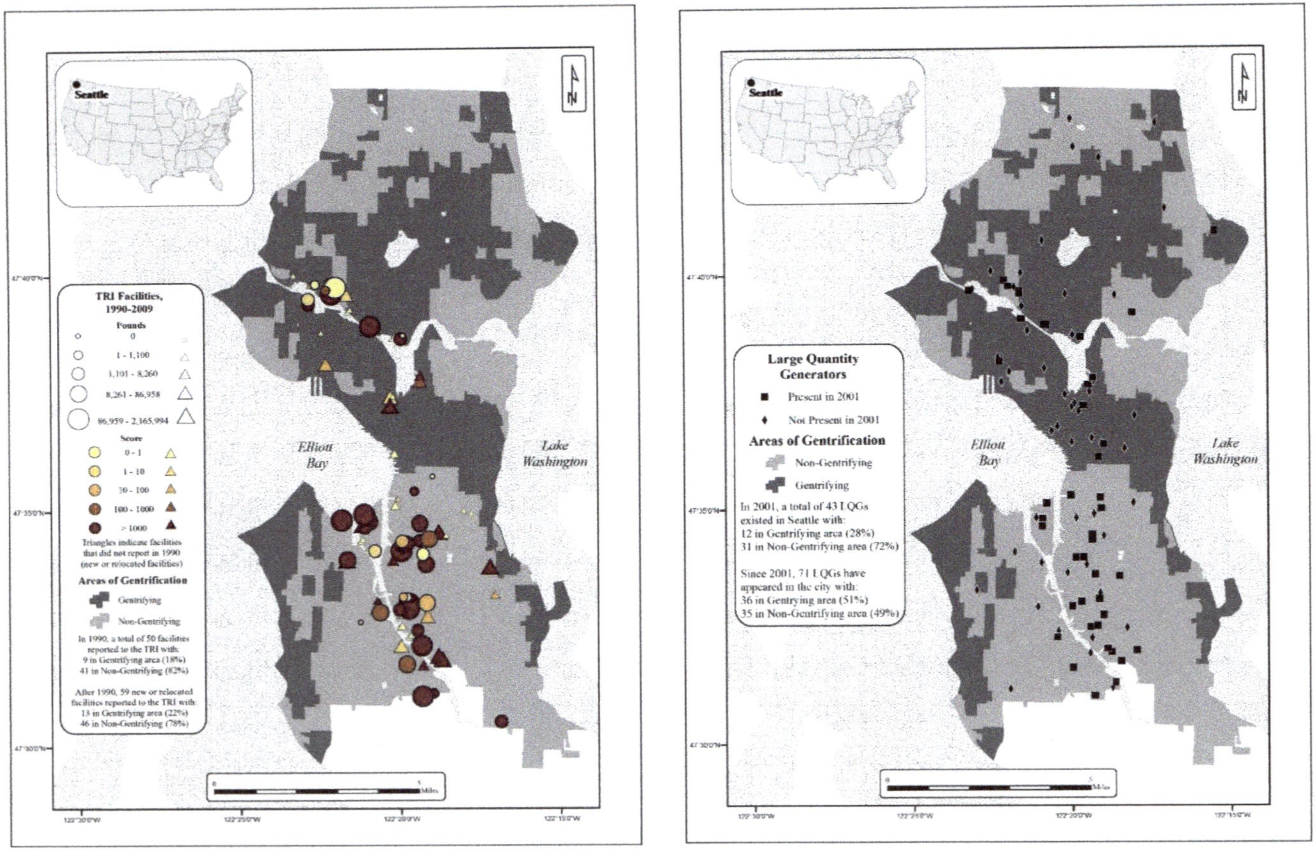

(a) (b)

Figure 5. (**a**) Seattle's shifting toxic air pollution sources, 1990–2009 and (**b**) large quantity hazardous waste generators.

To further explore whether gentrification may have influenced toxic industrial locations, we also plotted facilities appearing in the RSEI data in 1990 with a circle and those reporting emissions at a later date with triangles. In other words, triangles represent new toxic air polluters across Seattle's riskscape after 1990 (see Figure 5a). Likewise, and to expand on prior studies [22,23], we also obtained the spatial location and year of Large Quantity Generators (LQGs) of hazardous waste starting with 2001 represented by squares, and post-2001 sites with diamonds (see Figure 5b). We further dichotomized clusters into gentrifying and non-gentrifying CBGs with the former in a darker shade than the latter. In 1990, a total of 50 facilities reported toxic air emissions to the TRI with 18 percent (9) located in gentrifying areas while 82 percent (41) were in the non-gentrifying sections of Seattle. In the ensuing nineteen years, 59 new or relocated facilities reported air pollution emissions in the TRI (see Figure 5a). Only 22 percent (13) were in gentrifying clusters and 78 percent (46) in non-gentrifying areas. In 2001, there were 43 LQGs in Seattle with 28 percent (12) in gentrifying areas and 72 percent (31) in non-gentrifying areas. Between 2001 and 2013, 71 LQGs located to Seattle and a nearly equal number were present in gentrifying areas as non-gentrifying clusters (36 and 35 respectively). The distribution of Seattle's hazardous waste handlers were more evenly spread across the city than toxic air polluters.

3.3. Seattle's Industrial Zone Policy and Planning

In the early nineties, Seattle wasn't only breaking environmental ground because of its sustainability focused comprehensive plan in 1994. Four years earlier, the city embraced the Comparative Risk Assessment paradigm prominent at the time [115] and launched a two-year project to identify and assess Seattle's most pressing environmental challenges [116]. The committee's air team ranked transportation pollution sources, wood burning, and environmental tobacco smoke as high risk challenges for the city while medium-high risk problems included other indoor air pollution and industrial point sources. The air team then prioritized these risks for city action based on not only the relative risks, but also on the city's ability to address the problem. For example, the cross-media team ranked hazardous material use by business and industry in the highest relative-risk category but only ranked it in the second tier of priority actions "due to the effectiveness of existing regulations and the city's relatively limited role in further addressing the problem" [116]. Likewise, industrial air pollution sources were placed among the lowest category of action priorities even as it ranked as a medium-high risk problem. The same report noted that "the pollution from point sources tends to be localized, so that large areas or large numbers of people are not affected" [116]. Seattle's comparative risk assessment planning project would draw acclaim in a United Nations publication on the urban environment agenda [117].

A year later, the Mayor formulated and announced "Seattle's Environmental Action Agenda" in 1992 that influenced the city's next two budget cycles with 11 priorities including fostering stewardship, curbing cars, and protecting open space [118]. Moreover, the last section of the agenda titled "Improving Environmental Management and Coordination" identified several ideas for further action including the following. "Conduct a study of environmental equity in Seattle, and take appropriate measures" [118]. This recommendation was quickly implemented and in 1993, Seattle's Planning Department completed the city's first and only study of environmental equity [108].

For the two case studies: city planners performed two kinds of analyses: (1) an analysis of race and the location of commercial hazardous waste facilities; and (2) a descriptive study on the racial composition of communities with uncontrolled toxic waste sites. Interestingly, the authors of the environmental equity report noted a significant weakness in Seattle's Environmental Priorities Project that had led in part to the equity analysis. "The priorities were based on a citywide examination of environmental risks. No attempt was made to evaluate environmental problems on a site-by-site basis or across population groups. Some population groups may experience a set of localized environmental problems that would not be identified in a city wide analysis" [108].

The city's researchers identified 70 facilities that reported using EPA defined Extremely Hazardous Substances (EHS) with 11 in the BINMIC industrial area, seven around the Lake Union shoreline and the SLU industrial area and 25 across the GDMIC zone. Twelve were dispersed across nonindustrial areas of the city leaving 15 that either were not included on the map or were obscured in the symbology of overlapping points because of their close proximity to another facility. The report noted that "although the citywide proportion of non-white population groups is 24.7 percent, almost 35 percent of the people living in the same census blocks as the identified EHS facilities are persons of color". Also, "Hispanics appear to be disproportionately represented in the two impact areas, with percent differences of over 83 percent and 55 percent between the immediate and greater EHS facility areas and the citywide totals respectively". The report concluded in one section that "several population groups are, in fact, disproportionately represented in the impact areas". However, in the concluding summary section, the report also stated the following. "Given the general lack of information on the correlation between risk sources and risk burdens that exists in the scientific community, it does not seem possible to determine with any certainty whether environmental risks are inequitably distributed in Seattle" [108]. Expectedly, the cities attention to environmental inequities in subsequent planning efforts disappeared after this 1993 report.

One scholar recalled that "for a time in 1994, it appeared the city would be the first public institution to complete the cycle of comparative risk by updating the analysis and repeating the problem rankings. The departure of key staff people, however, may have stopped that initiative" [119]. Nonetheless, the city was under a state mandate to produce a new comprehensive plan to manage growth. With automobile pollution at the top of the city's environmental risk priorities, Seattle's 1994 comprehensive plan aimed to steer new residents and their jobs into denser urban centers and urban villages (See the Supplemental Materials for a map of these designations). "The goal that unifies all the elements of the Comprehensive Plan is to preserve the best qualities of Seattle's distinct neighbourhoods while responding positively and creatively to the pressures of change and growth. A key component of the City's plan to achieve this goal is the urban village strategy" [120]. The report plan proclaimed that "The City will continue to work with its residents, businesses, and institutions to promote conditions that will help each of its communities thrive, but will pay special attention to those areas where the majority of growth and change is expected". Moreover, the plan outlined different goals and policies for residential neighbourhoods, the new HUVs, and the city's industrial zones.

For example, in the South Lake Union (SLU) section of the 1994 plan, the first stated goal for this neighborhood's character aimed for "a mixed use neighborhood with an emphasis on small business and light industry" [120]. But for Manufacturing and Industrial Centers like BINMIC in the same plan, the goal was to "ensure that adequate accessible industrial land is available to promote a diversified employment base and sustain Seattle's contribution to regional high-wage job growth". Moreover, one of the seven policy statements for the city's industrial areas prescribed the following.

> Limit in industrial/manufacturing areas commercial or residential uses that are unrelated to the industrial function, that occur at intensities posing short- and long-term conflicts for industrial uses, or that threaten to convert significant amounts of industrial land to non-industrial uses [120].

However, the designation of SLU and its existing industrial area as a HUV embraced numerous non-industrial development possibilities. For instance, one HUV policy from the comprehensive plan aimed to "provide zoning to accommodate a wide range of housing types and retail and commercial services to support the business and residential population in the village, the surrounding community, and beyond". Such a contradiction was worrisome for industrial interests when the comprehensive plan draft was released two years earlier. A landowner in the industrial South Lake Union area was quoted in 1992 as saying "business people are not enthusiastic about the wholesale creation of a housing neighborhood" [121]. Likewise, in 1995, the encroachment of non-industrial development in BINMIC pitted the City of Seattle and retailer Fred Meyer against a Ballard group named "Save Our

Industrial Lands" (SOIL) in a court battle that lasted three years [122]. SOIL settled with the retailer in 1999 and the former site of Salmon Bay Steel Company became the home of a new Fred Meyer store [123].

3.4. *Non-Industrial Encroachments, Creeping Gentrification*

Industrial area zoning changes and conflicts continued between 1995 and 2006 with more nonindustrial proposals for BINMIC and nearby shoreline [124–129], a big shift on the north end of GDMIC [130–134], some concern for GDMIC's southern portion [135,136], but suspicion, cynicism and then resignation for SLU [137–143]. In the midst of these developments, the *Seattle Times* editorial board sounded an alarm in 2000 and called for the city to "Hold Fast on Zoning for Industrial Uses" [144]. A few months later, a city councilor described planning for the non-industrial developments. "We've done what we need to make sure the industrial base is protected. I don't think you can absolutely prevent encroachment. What you can do is try to set policies that encourage as much as possible of the commercial development toward the north" [145]. Then, in 2007, the seemingly fragmented and decade of creeping deindustrialization in Seattle's industrial zones spurred another city study.

Intent on understanding the key dilemmas with Seattle's manufacturing and industrial areas, the city's Department of Planning and Development (SDPD) performed a policy analysis of land use changes between 1996 and 2006 across BINMIC, GDMIC, and the other areas of industrial use like SLU [93]. SDPD analysts found that in the northern subareas of BINMIC next to the Ballard HUV, 20 percent of construction and land use permits proposed conversions from industrial to nonindustrial activity while 20 percent involved changes from one industrial use to another. Nonindustrial conversions also exceeded one out of every five permits (24 percent) in the industrial zones outside of the BINMIC and GDMIC areas like SLU and those along the shoreline of Lake Union. Conversely, in the Georgetown neighborhood of the Southeast section of GDMIC, only four percent of the permit activity over a decade involved nonindustrial conversions from industrial activity. In the southern section of GDMIC around the residential area of South Park, the analysis reported that 15 percent of permit changes were from one industrial use to another while 63 percent were industrial conversions from nonindustrial uses. This reflected, according to the report, "the area's vibrant industrial economy" [93].

We replicated and extended this analysis by analyzing and mapping change of use construction permits between 2008 and 2015 and land use permits between 2001 and 2015 (see Figure 6, Figure 7 and Supplemental Materials). Among construction permits that included a change of use, we identified a total of 153 for the Greater Duwamish (GDMIC) manufacturing zone and 50 for the Ballard and Interbay (BINMIC) industrial area. The largest share of construction permit activity in GDMIC was for deindustrializing activity at 45 sites (29 percent) with 20 locations changing from industrial to nonindustrial and 25 new nonindustrial. Industrial to industrial uses followed closely behind at 39 sites (26 percent) while industrializing sites were the third most frequent construction change of use (14 Nonindustrial to Industrial and 20 New Industrial). Nonindustrial to nonindustrial changes were the fourth most frequent change of use. Consistent with its manufacturing character, the majority of land use permits (42 or 51 percent) in GDMIC involved industrial activity, 43 percent (35) were commercial, and 5 percent institutional. Conversely, 51 percent (25) in BINMIC involved commercial activity while 41 percent (20) were industrial and 6 percent (3) institutional.

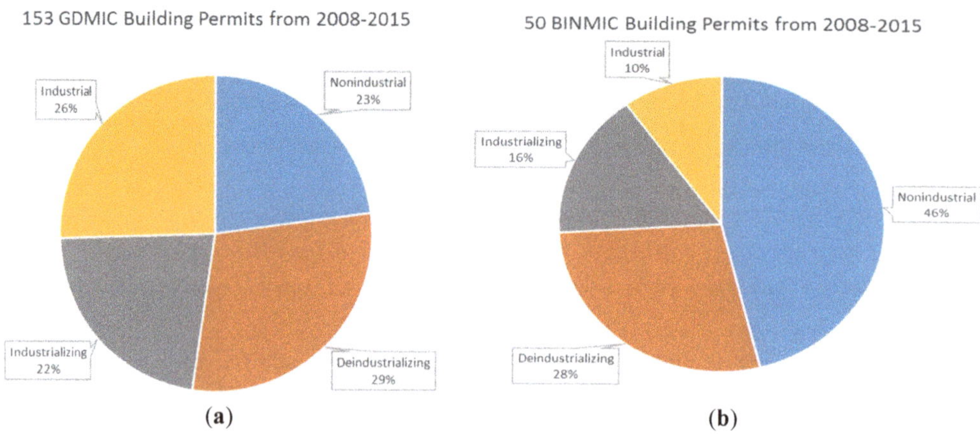

Figure 6. (**a**) 2008–2015 Construction Change of Use Permits Issued in GDMIC and (**b**) BINMIC.

In the BINMIC zones, the majority of construction permits with a change of use were non-industrial to non-industrial activity represented at 23 sites or 46 percent of permit activity. Half of those permits were clustered at the top of BINMIC near the location of the Ballard Hub Urban Village (See Supplemental Materials). Deindustrializing conversions (industrial to nonindustrial and new nonindustrial) were the second most frequent change of use permit activity at 12 sites representing 28 percent of BINMIC's change of use permit actions. Conversely, only 16 percent of BINMIC permits were for industrializing activity with 5 involving non-industrial to industrial conversions and 3 for new industrial development. In sum, 74 percent of BINMIC construction permits and 52 percent in GDMIC were for non-industrial uses. BINMIC also only saw 26 percent of its construction permit activity for industrial development while 48 percent of GDMIC's permits were for industrial uses. In other words, BINMIC was becoming a predominately nonindustrial development area while GDMIC saw nearly equal amounts of industrial and nonindustrial development.

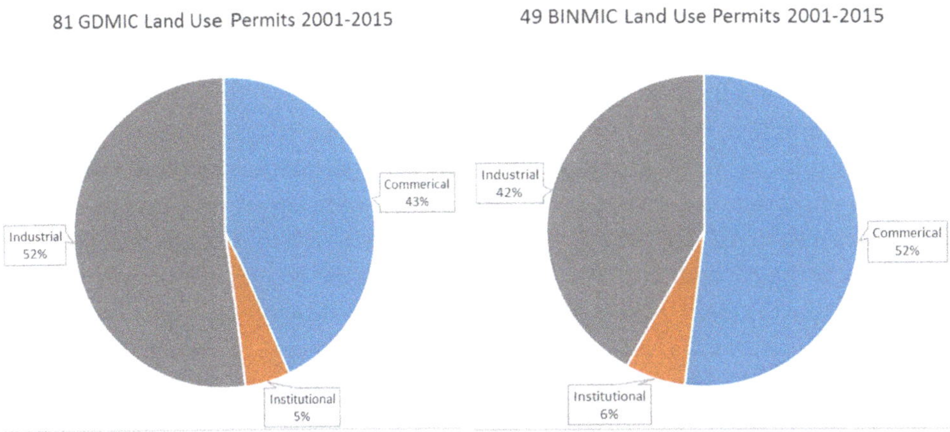

(a) **(b)**

Figure 7. (**a**) 2001–2015 Land Use Permits Issued in GDMIC and (**b**) BINMIC.

4. Discussion

Seattle experienced a significant transformation of its economic and urban geography through a dramatic post-industrial shift. The city lost a significant share of its industrial footprint as air pollution decreased 99 percent and the simulated inhalation exposure risk also dropped by 96 percent. The environmental dimension of Seattle's sustainability was dramatically better. Likewise, the economic indicators for the city suggested that performance on another sustainability dimension were also stellar. Median household incomes jumped by eight percent, poverty was down, and the city's minimum wage was boosted by local initiative to support a living wage. Seattle's share of its workforce in professional and managerial occupations also increased from thirty-six to fifty-two percent. Educational attainment surged for the city as its residents with a college degree grew from less than thirty-eight to fifty-four percent. Such transformations earned this Northwest city a third place rank among urban areas advancing a new economy [146] and the following kind of praise. "Seattle has completely transformed itself from a decaying old-economy provincial town into one of the world's preeminent innovation hubs. In the process, its residents have become some of the most creative and best-paid workers in the United States" [147]. However, our research joins those finding an increasingly divided city but also speaks to a larger shortcoming of sustainability research that overlooks the contested geographies of post-industrial and industrial development.

4.1. Environmental Gentrification and the Post-Industrial City

The conventional thesis in the relatively new literature on environmental gentrification "predicts that environmental quality improvement in poor communities may spur gentrification and the displacement of residents" [51]. Our research supports a different, and more complex process. We observed strong gentrification signals across Seattle over our entire study period in areas with and without active pollution sources. In particular, the replacement gentrification observed in cluster 5 completely encompassed the BINMIC and SLU industrial areas even though there were active, and some of the relatively riskiest air pollution sources in two of our three study period cross-sections. In 1990, the gentrifying cluster 5 that included industrial zoned land hosted 7 facilities that produced a simulated 38 percent of the inhalation exposure risk to Seattle residents from industrial air pollution. Ten years later, cluster 5 still hosted three of the most hazardous facilities producing 76 percent of the city's relative inhalation exposure risk from TRI facilities. Gentrification did not follow pollution cleanups and deindustrialization in the BINMIC area. Instead, industrial displacement through gentrification [66–69] better describes the process driving the restructuring of urban land use on Seattle's Northside.

For example, newspaper accounts between 1994 and 2005 describe significant nonindustrial redevelopment proposals by the Seattle's Port Authority and real estate investors for the BINMIC and SLU industrial lands. Nonindustrial redevelopment permits in these two areas also were much higher than in GDMIC revealing a more heated contest for land in North Seattle's industrial areas. City planners also targeted three different parts in or near BINMIC and SLU for denser redevelopment. These results suggest that rather than gentrification following cleanup, gentrification was occurring in parallel if not preceding environmental improvements around BNMIC and SLU.

4.2. *Planning for Environmental Discrimination?*

While the Seattle's industrial air pollution riskscape was spatially distributed between the northern and southern industrial areas in a relatively even pattern in both 1990 and 2000, it became extremely skewed by 2009. Several of the riskiest and many new TRI facilities were located in the southern half of the GDMIC industrial zone in the midst of cluster 15, one of the most socially vulnerable neighborhood grouping in the city. Conversely, the share of the city's relative risk exposure of facilities in replacement gentrification clusters dropped from 38 to less than one percent during our study period. The riskiest industrial facilities and lowest socioeconomic strata converged in the areas represented by Custer 15 surrounding two residential neighbourhoods called South Park and Georgetown. In the 98108 ZIP code containing both communities, more than 70% of the 5,070 residents are non-white minorities, including Asian, Pacific Islander, Hispanic, African, African-American, and Native American [148]. Forty-two percent of residents are foreign-born, and more than 20 languages are spoken. Nearly a third (32%) of residents live below 200% of the poverty level and 78% of children at the local elementary school qualify for free or reduced price lunch. The neighbourhoods also have a significantly higher percentage of elderly residents (>65) and children (<5) than the city average who are more vulnerable to air toxics pollution exposures.

Even as the city was dismissing environmental inequality in a 1993 planning study, other studies and community groups were raising concerns about inequitable pollution burdens particularly in South Seattle. In 1991, the US Environmental Protection Agency (EPA) cited the Duwamish industrial zone for violating federal standards for particulate pollution [149,150]. Three years later, a Seattle Times study in 1994 highlighted the concentration of environmentally hazardous facilities near South Park and reported the following quote from one South Park resident [151]. "I don't see a lot of environmental justice. I see a lot of environmental injustice". In a second newspaper story on the same date, same source, the reporter noted that "most sites in Seattle where toxic chemicals are released or hazardous waste is stored are in neighbourhoods whose populations are less white or less affluent than the city as a whole" [152]. In 1995, the city's mayor formed a task force to develop a city wide plan to address environmental justice and they released their report a year later [153]. However, there is little evidence that the mayor or other Seattle officials implemented any environmental justice action.

National research confirmed the relative shallow environmental justice efforts by Seattle. Warner [154] conducted a comparative study of 77 city's efforts to integrate environmental justice in their sustainability planning. Focusing on website content, only five cities were identified with substantial amounts of information about environmental justice including Seattle. However, only one policy statement committing to "compliance with applicable environmental justice regulations" from Seattle's Council on Airport Affairs was available on the website plus a link to the US EPA's environmental justice website. Thus, the City of Seattle's early symbolic efforts quickly faded as no policy or planning effort was sustained for five or more years even as other Seattle groups and agencies continued to identify environmental inequities.

In 1997, the city and county's Department of Public Health [155] documented higher respiratory hospitalization rates, decreased life expectancies, and higher mortality rates in South Seattle. A local environmental justice group identified more than forty industrial and waste facilities in 2001 within a one to five-mile radius of South Park homes [156]. In 2002, the Puget Sound Clean Air Agency (PSCAA) used monitored data to evaluate cancer and non-cancer risks in different Seattle locations and found the highest risk estimates in the Duwamish industrial valley neighborhood of Georgetown [157]. Likewise, a 2008-2009 study sampling over 100 different air toxics across four sites in both Seattle and the industrial port of Tacoma to the South found that the potential cancer risks from Diesel emissions were the highest at the monitoring site just north of the Georgetown and South Park neighbourhoods [158]. Another longitudinal air toxics monitoring program for Seattle published in 2011 found higher inhalation cancer risks from data collected at two South Seattle sites compared to four other locations across the city [159]. Yet, Seattle city agencies, planners, and politicians continued to ignore the growing number of studies and community concerns registering how environmental injustice remained unaddressed in South Seattle.

Twenty years after an environmental justice movement gained national policy attention and South Seattle's unequal environmental burdens were documented over and over, sections of one of the most sustainable cities in the world still do not enjoy the same environmental amenities as most other neighbourhoods in the city. Duwamish Valley residents instead have been burdened by a series of decisions over time that have seen Seattle promote nonindustrial development in Ballard, Interbay, and South Lake Union leaving new and relocating industrial development to concentrate in the southern section of GDMIC. In fact, the City's struggles with the intersections of environmental injustice and social stratification were acknowledged when the EPA in 2014 selected a Duwamish Valley Coalition as one of 10 communities across the nation for a two-year Environmental Justice Collaborative Problem-Solving (EJCPS) grant to focus on cleaning up air pollution in the Georgetown and South Park neighbourhoods [160].

4.3. A Tale of Two Emerald Cities and Beyond

A recent compilation of air pollution levels, health data, and simulations of excess cancer risk from air toxics documents that air pollution is on the rise in Seattle again, asthma hospitalization rates for Duwamish Valley residents are significantly higher than the county and city, and excess cancer risks were much higher in South Park and Georgetown census tracts [161]. Since its groundbreaking comprehensive plan for sustainability in 1994, the City of Seattle and its urban geography has underwent a significant, but skewed transformation. Our research illuminated how the Emerald City has fractured into two different urban development trajectories because of gentrification, an increasingly skewed riskscape, and inaction on the part of city officials. In an ironic moment of déjà vu on Earth Day 2015, Seattle's newest mayor recently announced a "first-of-its-kind" equity and environment initiative [162]. Two months later, the mayor then signed an executive order to reorganize Seattle's Department of Planning and Development into a new office of Planning and Community Development [163]. The move was spurred in part by a local study reporting that while the city has had some success in directing growth to designated urban villages, neighborhood investments had been uneven [164]. In July of this year, the mayor then announced the findings of the city's Task Force on Housing Affordability. He was quoted saying "my vision is where the people who work in Seattle can afford to live in Seattle. Without this plan, people will continue to be forced to live outside this city. We will continue to have a city of economic apartheid" [165]. But until Seattle's leaders recognize the connections between gentrification, zoning, affordable housing, and skewed air pollution exposures, the city's economic stratification and environmental injustice will continue to tarnish the Emerald City's brand of sustainability.

4.4. Integrating Quantitative and Qualitative Methods

Our mixed-method research design also provides an alternative approach to avoid the over simplistic presumptions behind the prominent approaches to examining environmental gentrification. National level and structural gentrification analyses using census tracts or larger scales of analysis result in too coarse of a resolution to capture significant spatial variations at the neighborhood level. Thus, it often provides little to no guidance for policy and planning strategies. Likewise, studies using individual level mobility data to assess gentrification are too granular. Both methods display an over commitment to either a structural model of gentrification processes or the pure agency of behavioral individualism. Our methods detailed here instead paint a picture of how micro-structural analysis can be used to screen for hot spots of both environmental risk sources and socially disruptive gentrification trends.

In sum, our study is more consistent with prior research concluding that the plans of city officials and real estate developers can contribute to industrial displacement that in turn often results in a cleaner environment for redevelopment areas. Such a dynamic is better illuminated with our combination of longitudinal geographic cluster analysis, relative-risk air pollution screening, and qualitative policy analysis. Our study reveals the risky business of Seattle's complicated yet intersecting trajectories of environmental inequality, socioeconomic stratification, and planning blind spots. Similar mixed-method and historical studies assessing the equity dimensions of urban land development, land use conflicts, and skewed riskscapes will be a crucial addition to the much larger research task assessing urban sustainability.

5. Conclusions

While Seattle has been heralded for its leadership in sustainability, we join the growing number of voices that critically interrogate this reputation. A livability and gentrification conflict marked twenty years of Seattle's neighborhood geography resulting in a city divided by class, and to a lesser extent, by race. This trajectory is characterized by a pattern of development clusters highly stratified by occupation, income and property values. In the broadest simplification, Seattle has divided between an affluent and highly educated postindustrial workforce centered on South Lake Union and encroaching on the city's northern industrial zones. A concentration of mixed, and predominately intermediate industrial and service workforce are increasingly locating in south central Seattle. Finally, the remaining burdens of the city's remaining industrial economy are being relegated to the Duwamish Valley. Parts of Seattle fared poorly in each of sustainability's three dimensions. A more just sustainability for this Emerald City and other metropolitan regions will require more research like ours that can inform the political and policy attention needed to confront the trajectories of inequitable development and environmental injustice to mitigate obscured industrial land use conflicts.

Acknowledgments: The research was supported by Western Washington University's Research and Sponsored Programs, the Department of Environmental Studies graduate program, and the Peninsulas Program of Huxley College of the Environment. We also thank the librarians staffing the Seattle room of the Seattle Public Library, Western Washington University's Elissa Mowry, John Wilkens, and the editor of this special issue, Yehua Dennis Wei.

Author Contributions: Troy D. Abel originated the study, prepared the relative risk screening exposure modeling, conducted the qualitative analysis, and led the writing of the article. Jonah White designed and conducted the factor and cluster analysis and contributed to the writing of the article. Stacy Clauson collected and analyzed permit data and contributed to the writing of the article.

Conflicts of Interest: The authors declare no conflict of interest.

References and Notes

1. Seattle City Council. *The City of Seattle Comprehensive Plan: Toward a Sustainable Seattle: A Plan for Managing Growth, 1994–2014*; Seattle City Council: Seattle, WA, USA, 1994.
2. Beatley, T. Sustainability in Planning: The Arc and Trajectory of a Movement, and New Directions for the Twenty-First-Century City. In *Planning Ideas the Matter: Livability, Territoriality, Governance, and Reflective Practice*; MIT Press: Cambridge, MA, USA, 2012; pp. 91–124.
3. World Commission on Environment and Development. *Our Common Future*; Oxford University Press: Oxford, UK, 1987.
4. Dobson, A. *Justice and the Environment: Conceptions of Environmental Sustainability*; Oxford University Press: New York, NY, USA, 1998.
5. Dobson, A. Social Justice and Environmental Sustainability: Ne'er the Twain Shall Meet. In *Just Sustainabilities: Development in An Unequal World*; Julian Agyeman, R.D.B., Evans, B., Eds.; MIT Press: Cambridge, MA, USA, 2003.
6. Agyeman, J. *Sustainable Communities and the Challenge of Environmental Justice*; New York University Press: New York, NY, USA, 2005.
7. Connelly, S. Mapping sustainable development as a contested concept. *Local Environ.* **2007**, *12*, 259–278.
8. McManus, P. Contested terrains: Politics, stories and discourses of sustainability. *Environ. Polit.* **1996**, *5*, 48–73.
9. Williams, C.C.; Millington, A.C. The diverse and contested meanings of sustainable development. *Geogr. J.* **2004**, *170*, 99–104.

10. Campbell, S. Green Cities, Growing Cities, Just Cities? Urban Planning and the Contradictions of Sustainable Development. *J. Am. Plan. Assoc.* **1996**, *62*, 296–312.
11. Godschalk, D.R. Land Use Planning Challenges: Coping with Conflicts in Visions of Sustainable Development and Livable Communities. *J. Am. Plan. Assoc.* **2004**, *70*, 5–13.
12. AtKisson, A. Developing indicators of sustainable community: Lessons from sustainable Seattle. *Environ. Impact Assess. Rev.* **1996**, *16*, 337–350.
13. Karlenzig, W. What makes today's green city? In *Growing Greener Cities: Urban Sustainability in the Twenty-First Century*; Birch, E., Wachter, S., Eds.; University of Pennsylvania Press: Philadelphia, PA, USA, 2008; pp. 346–363.
14. Portney, K. *Taking Sustainable Cities Seriously: Economic Development, the Environment, and Quality of Life in American Cities*; MIT Press: Cambridge, MA, USA, 2003.
15. Karlenzig, W. *How Green is Your City? The Sustainlane US City Rankings*; New Society Publishers: Gabriola Island, BC, USA, 2007.
16. Krueger, R.; Agyeman, J. Sustainability schizophrenia or "actually existing sustainabilities?" toward a broader understanding of the politics and promise of local sustainability in the US. *Geoforum* **2005**, *36*, 410–417.
17. Berke, P.R.; Conroy, M.M. "Are We Planning for Sustainable Development: An Evaluation of 30 Comprehensive Plans". *J. Am. Plan. Assoc.* **2000**, *66*, 21–33.
18. Sanders, J.C. *Seattle and the Roots of Urban Sustainability: Inventing Ecotopia*; University of Pittsburgh Press: Pittsburgh, PA, USA, 2010.
19. Abel, T.D.; White, J. Skewed riskscapes and gentrified inequities: Environmental exposure disparities in Seattle, Washington. *Am. J. Public Health* **2011**, *101*, 246–254.
20. Abel, T.; White, J. Gentrified sustainability: Inequitable development and Seattle's skewed riskscape. *Interdiscip. Environ. Rev.* **2015**, *16*, 124–157.
21. Dierwechter, Y. The Spaces That Smart Growth Makes: Sustainability, Segregation, and Residential Change across Greater Seattle. *Urban Geogr.* **2014**, *35*, 691–714.
22. Dooling, S. Ecological Gentrification: A Research Agenda Exploring Justice in the City. *Int. J. Urban Reg. Res.* **2009**, *33*, 621–639.
23. Klingle, M. *Emerald City: An Environmental History of Seattle*; Yale University Press: New Haven, CT, USA, 2008.
24. Bullard, R. Solid Waste Sites and the Black Houston Community. *Sociol. Inq.* **1983**, *53*, 273–288.
25. United States Government Accounting Office (GAO). *Siting of Hazardous Waste Landfills and Their Correlation with Racial and Economic Status of Surrounding Communities*; United States Government Accounting Office: Washington, DC, USA, 1983.
26. United Church of Christ (UCC). Commission for Racial Justice. In *Toxic Wastes and Race in the United States*; United Church of Christ: New York, NY, USA, 1987.
27. Ringquist, E.J. Assessing Evidence of Environmental Inequities: A Meta-Analysis. *J. Policy Anal. Manag.* **2005**, *24*, 223–247.
28. Downey, L. Environmental racial inequality in Detroit. *Soc. Forces* **2006**, *85*, 771–796.

29. Su, J.G.; Morello-Frosch, R.; Jesdale, B.M.; Kyle, A.D.; Shamasunder, B.; Jerrett, M. An index for assessing demographic inequalities in cumulative environmental hazards with application to Los Angeles, California. *Environ. Sci. Technol.* **2009**, *43*, 7626–7634.
30. Sadd, J.L.; Pastor, M.; Morello-Frosch, R.; Scoggins, J.; Jesdale, B. Playing it safe: Assessing cumulative impact and social vulnerability through an environmental justice screening method in the south coast air basin, California. *Int. J. Environ. Res. Public Health* **2011**, *8*, 1441–1459.
31. Marshall, J.D. Environmental inequality: Air pollution exposures in California's South Coast Air Basin. *Atmos. Environ.* **2008**, *42*, 5499–5503.
32. Sze, J. *Noxious New York: The Racial Politics of Urban Health and Environmental Justice*; MIT Press: Cambridge, MA, USA, 2007.
33. Elliott, J.R.; Frickel, S. The Historical Nature of Cities: A Study of Urbanization and Hazardous Waste Accumulation. *Am. Sociol. Rev.* **2013**, *78*, 521–543.
34. Abel, T.D. Skewed riskscapes and environmental injustice: A case study of metropolitan St. Louis. *Environ. Manag.* **2008**, *42*, 232–248.
35. Chakraborty, J.; Bosman, M.M. (Eds.) *Spatial and Environmental Injustice in an American Metropolis: A Study of Tampa Bay, Florida*; Cambria Press: Amherst, MA, USA, 2010.
36. Pellow, D.N. Environmental inequality formation toward a theory of environmental injustice. *Am. Behav. Sci.* **2000**, *43*, 581–601.
37. Been, V. Locally undesirable land uses in minority neighborhoods: Disproportionate siting or market dynamics? *Yale Law J.* **1994**, *103*, 1383–1422.
38. Been, V.; Gupta, F. Coming to the Nuisance or Going to the Barrios-A Longitudinal Analysis of Environmental Justice Claims. *Ecol. Law Q.* **1997**, *24*, 1–56.
39. Oakes, J.M.; Anderton, D.L.; Anderson, A.B. A longitudinal analysis of environmental equity in communities with hazardous waste facilities. *Soc. Sci. Res.* **1996**, *25*, 125–148.
40. Yandle, T.; Burton, D. Reexamining environmental justice: A statistical analysis of historical hazardous waste landfill siting patterns in metropolitan Texas. *Soc. Sci. Q.* **1996**, *77*, 477–492.
41. Crowder, K.; Downey, L. Inter-neighborhood migration, race, and environmental hazards: Modeling micro-level processes of environmental inequality. *Am. J. Sociol.* **2010**, *115*, 1110–1149.
42. Pais, J.; Crowder, K.; Downey, L. Unequal Trajectories: Racial and Class Differences in Residential Exposure to Industrial Hazard. *Soc. Forces* **2013**, *92*, 1189–1215.
43. Pastor, M.; Sadd, J.; Hipp, J. Which Came First? Toxic Facilities, Minority Move-in, and Environmental Justice. *J. Urban Aff.* **2001**, *23*, 1–21.
44. Hamilton, J.T. Testing for environmental racism: Prejudice, profits, political power? *J. Policy Anal. Manag.* **1995**, *14*, 107–132.
45. National Academy of Public Administration (NAPA). *Addressing Community Concerns: How Environmental Justice Relates to Land Use Planning and Zoning*; NAPA: Washington, DC, USA, 2003.

46. National Environmental Justice Advisory Council (NEJAC). *Unintended Impacts of Redevelopment and Revitalization Efforts in Five Environmental Justice Communities*; NEJAC: Washington, DC, USA, 2006.
47. Cameron, T.A.; McConnaha, I.T. Evidence of environmental migration. *Land Econ.* **2006**, *82*, 273–290.
48. Quastel, N. Political ecologies of gentrification. *Urban Geogr.* **2009**, *30*, 694–725.
49. Dale, A.; Newman, L.L. Sustainable development for some: Green urban development and affordability. *Local Environ.* **2009**, *14*, 669–681.
50. Checker, M. Wiped out by the "greenwave": Environmental gentrification and the paradoxical politics of urban sustainability. *City Soc.* **2011**, *23*, 210–229.
51. Eckerd, A. Cleaning up without clearing out? A spatial assessment of environmental gentrification. *Urban Aff. Rev.* **2011**, *47*, 31–59.
52. Gamper-Rabindran, S.; Timmins, C. Hazardous waste cleanup, neighborhood gentrification, and environmental justice: Evidence from restricted access census block data. *Am. Econ. Rev.* **2011**, *101*, 620–624.
53. Pearsall, H. Moving out or moving in? Resilience to environmental gentrification in New York City. *Local Environ.* **2012**, *17*, 1013–1026.
54. Wolch, J.R.; Byrne, J.; Newell, J.P. Urban green space, public health, and environmental justice: The challenge of making cities "just green enough". *Landsc. Urban Plan.* **2014**, *125*, 234–244.
55. Goodling, E.; Green, J.; McClintock, N. Uneven development of the sustainable city: Shifting capital in Portland, Oregon. *Urban Geogr.* **2015**, *36*, 504–527.
56. Banzhaf, S.H.; Walsh, R.P. *Do People Vote with Their Feet? An Empirical Test of Environmental Gentrification*; Resources for the Future: Washington, DC, USA, 2006.
57. Ley, D.; Dobson, C. Are there limits to gentrification? The contexts of impeded gentrification in Vancouver. *Urban Stud.* **2008**, *45*, 2471–2498.
58. Glass, R. Introduction: Aspects of Change. In *London: Aspects of Change*; Glass, R., Ed.; MacKibbon and Kee: London, UK, 1964.
59. Hamnett, C. The blind men and the elephant: The explanation of gentrification. *Trans. Inst. Br. Geogr.* **1991**, *16*, 173–189.
60. Smith, N. Toward a theory of gentrification: A back to the city movement by capital, not people. *J. Am. Plan. Assoc.* **1979**, *45*, 538–548.
61. Smith, N. Gentrification. In *The Dictionary of Human Geography*, 4th ed.; Johnston, R.J., Gregory, D., Pratt, G., Watts, M., Eds.; Blackwell: Oxford, UK, 2000.
62. Ley, D. *The New Middle Class and the Remaking of the Central City*; Oxford University Press: Oxford, UK, 1996.
63. Bell, D. *The Coming of the Post-Industrial Society*; Basic Books: New York, NY, USA, 1973.
64. Ley, D. Liberal ideology and the postindustrial city. *Ann. Assoc. Am. Geogr.* **1980**, *70*, 238–258.
65. Ley, D. Alternative explanations for inner-city gentrification: A Canadian assessment. *Ann. Assoc. Am. Geogr.* **1986**, *76*, 521–535.

66. Curran, W.; Hamilton, T. Just green enough: Contesting environmental gentrification in Greenpoint, Brooklyn. *Local Environ.* **2012**, *17*, 1027–1042.
67. Curran, W. Gentrification and the nature of work: Exploring the links in Williamsburg, Brooklyn. *Environ. Plan. A* **2004**, *36*, 1243–1258.
68. Curran, W.; Hanson, S. Getting globalized: Urban policy and industrial displacement in Williamsburg, Brooklyn. *Urban Geogr.* **2005**, *266*, 461–482.
69. Curran, W. "From the Frying Pan to the Oven": Gentrification and the Experience of Industrial Displacement in Williamsburg, Brooklyn. *Urban Stud.* **2007**, *44*, 1427–1440.
70. O'Sullivan, D. Toward micro-scale spatial modeling of gentrification. *J. Geogr. Syst.* **2002**, *4*, 251–274.
71. Hwang, J. Gentrification in changing cities: Immigration, new diversity, and racial inequality in neighborhood renewal. *Ann. Am. Acad. Polit. Soc. Sci.* **2015**, *660*, 319–340.
72. Lees, L.; Slater, T.; Wyly, E.K. *Gentrification*; Routledge/Taylor & Francis Group: New York, NY, USA, 2008.
73. Lees, L. Super-gentrification: The case of Brooklyn heights, New York City. *Urban Stud.* **2003**, *40*, 2487–2509.
74. Ley, D. Artists, aestheticisation and the field of gentrification. *Urban Stud.* **2003**, *40*, 2527–2544.
75. Smith, N.; DeFilippis, J. The reassertion of economics: 1990s gentrification in the Lower East Side. *Int. J. Urban Reg. Res.* **1999**, *23*, 638–653.
76. Cohen, A. Home, wage gap widens. *Seattle Post Intelligencer*, 11 September 2007.
77. McGee, H.W. Seattle's Central District, 1990—2006: Integration or Displacement? *Urban Lawyer* **2007**, *39*, 167–256.
78. Rhodes, E.; Mayo, J. Only 9 areas in King County left for middle-income buyers. *Seattle Times*, 16 July 2006; A1.
79. Westneat, D. Growth Gone Wild? Website Map Pins it down. Available online: http://www.seattletimes.com/seattle-news/growth-gone-wild-website-map-pins-it-down (accessed on 5 April 2015).
80. Balk, G. Priced out? Growing Numbers Appear to Be Fleeing King County. Available online: http://www.seattletimes.com/seattle-news/data/priced-out-has-the-king-county-diaspora-begun (accessed on 5 April 2015).
81. Bernard, S. In Georgetown, the Housing Is Affordable and the Air Unbreathable. *Seattle Weekly*. Available online: http://www.seattleweekly.com/home/960984-129/in-georgetown-the-housing-is-affordable (accessed on 24 October 2015).
82. Florida, R. *The Rise of the Creative Class—Revisited: Revised and Expanded*; Basic books: New York, NY, USA, 2012.
83. Florida, R.; Matheson, Z.; Adler, P.; Brydges, T. *The Divided City and the Shape of the New Metropolis*; Martin Prosperity Institute, Rotman School of Management, University of Toronto: Toronto, ON, Canada, 2014.
84. Yin, R.K. *Case Study Research: Design and Methods, Applied Social Research Methods Series*, 3rd ed.; Sage Publications: Thousand Oaks, CA, USA, 2002; Volume 5.

85. Muro, M.; Rothwell, J.; Andes, S.; Fikri, K.; Kulkarni, S. America's Advanced Industries: What Are They, Where They Are, and Why They Matter. Available online: http://www.brookings.edu/~/media/Research/Files/Reports/2015/02/03-advanced-industries/final/AdvancedIndustry_FinalFeb2lores.pdf?la=en (accessed on 30 August 2015).
86. Brookings Institute; JP Morgan Chase. The 10 Traits of Globally Fluent Metro Areas. Available online: http://www.brookings.edu/~/media/Multimedia/Interactives/2013/tentraits/Seattle.pdf (accessed on 30 August 2015).
87. Seattle Jobs Initiative. Understanding King County's Manufacturing Sector: Preparing Our Workforce for Good Job Opportunities. Available online: http://www.seattlejobsinitiative.com/wp-content/uploads/SJI_ManufacturingReport_vFINAL.pdf (accessed on 12 September 2015).
88. Balk, G. Census: Seattle Is the Fastest-Growing Big City in the U.S. Available online: http://blogs.seattletimes.com/fyi-guy/2014/05/22/census-seattle-is-the-fastest-growing-big-city-in-the-u-s/ (accessed on 12 September 2015).
89. Dierwechter, Y. *Urban Growth Management and Its Discontents*; Palgrave Macmillan: New York, NY, USA, 2008.
90. Steinbrueck, P.; Winter, M.; Patterson, M.; Williamson, S.; Barbe, D.; Greaney, Y. Seattle Sustainable Neighbourhoods Assessment Project. Available online: http://www.seattle.gov/dpd/cs/groups/pan/@pan/documents/web_informational/p2233677.pdf (accessed on 15 September 2015).
91. Seattle Department of Planning and Development. South Lake Union Urban Center Neighborhood Plan. Available online: http://www.seattle.gov/dpd/cs/groups/pan/@pan/documents/web_informational/dpds021899.pdf (accessed on 15 September 2015).
92. City of Seattle Department of Planning and Development (SDPD). Urban Center/Village Employment Growth Report. Available online: http://www.seattle.gov/dPd/cs/groups/pan/@pan/documents/web_informational/dpds022046.pdf (accessed on 15 September 2015).
93. City of Seattle Department of Planning and Development (SDPD). *Seattle's Industrial Lands Background Report*; City of Seattle: Seattle, WA, USA, 2007.
94. City of Seattle Department of Planning and Development (SDPD). *Toward a Sustainable Seattle: A Plan for Managing Growth 2004–2024*; City of Seattle: Seattle, WA, USA, 2005.
95. Expedia's move likely to prompt changes to Seattle Interbay and Bellevue. Available online: http://www.kirotv.com/news/news/expedias-move-likely-prompt-changes-seattle-interb/nkmCX/ (accessed on 8 July 2015).
96. All Eyes On Interbay As Expedia Plans Move. Available online: http://blog.seattlepi.com/seattlerealestate/2015/04/15/all-eyes-on-interbay-as-expedia-plans-move/ (accessed on 8 July 2015).
97. Seattle Department of Planning and Development. Envisioning Ballard to Interbay-What & Why 2015. Available online: http://www.seattle.gov/dpd/cityplanning/completeprojectslist/ballardinterbay/whatwhy/ (accessed on 31 August 2015).

98. Puget Sound Regional Council. 2013 Regional Centers Monitoring Report: Ballard-Interbay MIC. Available online: http://www.psrc.org/assets/265/mic-profile-Seattle-Ballard-Interbay.pdf (accessed on 12 September 2015).
99. Puget Sound Regional Council. 2013 Regional Centers Monitoring Report: Duwamish MIC. Available online: http://www.psrc.org/assets/270/mic-profile-Seattle-Duwamish.pdf (accessed on 12 September 2015).
100. GeoLytics. *CensusCD 1990 Long Form in 2000 Boundaries*; GeoLytics: East Brunswick, NJ, USA, 2008.
101. Dunteman, G.H. *Principal Components Analysis*; Sage: Newbury Park, CA, USA, 1989.
102. Aldenderfer, M.S.; Blashfield, R.K. *Cluster Analysis*; Sage Publications: Beverly Hills, CA, USA, 1984.
103. Lattin, J.M.; Carroll, J.D.; Green, P.E. *Analyzing Multivariate Data*; Thomson Brooks/Cole: Pacific Grove, CA, USA, 2003.
104. Manufacturing facilities are required to report the release of any toxic chemical into the environment with EPA's TRI program if (1) it is in the manufacturing, metal mining, coal mining, electrical utilities, hazardous waste treatment and disposal facilities, chemical plants, petroleum plants and terminals, solvent recovery services, or federal facilities industrial sectors; (2) has 10 or more full-time employees; and (3) manufactures or processes more than 25,000 pounds or otherwise uses more than 10,000 pounds of any listed chemical during the calendar year. Using self-reported TRI air release estimates, RSEI simulates a steady-state Gaussian plume of downwind pollutant concentrations from both stack and fugitive emission levels. RSEI than estimates a surrogate inhalation dose for neighboring census populations and produces a unit-less indicator value that can be used to rank relative impacts by geography, industry, and facility. See Risk-Screening Environmental Indicators (RSEI). Available online: www.epa.gov/oppt/rsei (accessed on 8 July 2015). and Schmidt, Charles W. The risk where you live. *Environ. Health Perspect.* **2003**, *111*, A404.
105. The BRS is another EPA information disclosure program that requires Large-Quantity Generators (LQGs) to report their hazardous waste generation if they manage at least 1000 kilograms monthly. See United States Environmental Protection Agency (EPA). *Biennial Report: Reportable and Non-Reportable Wastes*; Office of Resource Conservation and Recovery: Washington, DC, USA, 2013.
106. Atlas, M. Few and far between? An environmental equity analysis of the geographic distribution of hazardous waste generation. *Soc. Sci. Q.* **2002**, *83*, 365–378.
107. City of Seattle Department of Planning and Development (SDPD). *Duwamish M/IC Policy and Land Use Study*; City of Seattle: Seattle, WA, USA, 2013.
108. City of Seattle Planning Department. *Two Case Studies of the Spatial Relationship between Environmental Risks and Population Groups in Seattle*; City of Seattle: Seattle, WA, USA, 1993.
109. City of Seattle Department of Planning and Development (SDPD). *Growth and Equity: Analyzing Impacts on Displacement and Opportunity Related to Seattle's Growth Strategy*; City of Seattle: Seattle, WA, USA, 2015.

110. Land Use Permits. Available online: https://data.seattle.gov/Permitting/Land-Use-Permits/uyyd-8gak (accessed on 6 July 2015).
111. Building Permits: Current. Available online: https://data.seattle.gov/Permitting/Building-Permits-Current/mags-97de (accessed on 6 July 2015).
112. Kennedy, M.; Leonard, P. *Dealing with Neighborhood Change: A Primer on Gentrification and Policy Choices*; Brookings Institution Center on Urban and Metropolitan Policy and PolicyLink: Washington, DC, USA, 2001.
113. Blackwell, A.G. Promoting equitable development. *Indiana Law Rev.* **2000**, *34*, 1273–1290.
114. Morrill, R. Gentrification, Class, and Growth Management in Seattle, 1990–2000. In *Global Perspectives on Urbanization*; Pomeroy, G., Webster, G., Eds.; University Press of America: Lanham, MD, USA, 2008.
115. United States Environmental Protection Agency (EPA). *Reducing Risk: Setting Priorities and Strategies for Environmental Protection*; Science Advisory Board: Washington, DC, USA, 1990.
116. City of Seattle. *Environmental Risks in Seattle: A Comparative Assessment*; Office for Long-range Planning: Seattle, WA, USA, 1991.
117. United Nations Centre for Human Settlements (HABITAT). *Implementing the Urban Environmental Agenda*; United Nations Environment Programme: Nairobi, Kenya, 1997.
118. City of Seattle. *Mayor's Recommended Environmental Action Agenda: Environmental Stewardship in Seattle*; Planning Department: Seattle, WA, USA, 1992.
119. Davies, J.C. *Comparing Environmental Risks: Tools for Setting Government Priorities*; Routledge: London, UK, 1996.
120. City of Seattle. *Seattle's Comprehensive Plan toward a Sustainable Seattle: A Plan for Managing Growth 1994–2014*; Strategic Planning Office: Seattle, WA, USA, 1994.
121. Lilly, D. Rice's Idea for Urban Growth Draws Praise. *Seattle Times*, 20 March 1992; C1.
122. Kokmen, L. Plot Thickens as Fred Meyer seeks to Build Store—An Industrial *vs.* Retail Saga. *Seattle Times*, 8 April 1996; E1.
123. Stubbs, D. Fred Meyer Finally Comes to Ballard—Store Opening Closes Years of Controversy. *Seattle Times*, 10 November 1999; E5.
124. Kraft, G. Commons Property is Better Site for Immunex Development. *Seattle Times*, 3 February 1996; A13.
125. Solomon, C. Change is Coming to North Waterfront—but Slowly—Zoning is an Obstacle. *Seattle Times*, 13 December 1996; D1.
126. Vinluan, F. Study eyes options at terminal if fisherman vacate some space. *Seattle Times*, 8 February 2002; C1.
127. McOmber, J.M. Tenants on Lake Getting the Boot—Makeover in Works for Wallingford Waterfront. *Seattle Times*, 3 August 2002; C1.
128. McOmber, J.M. Port's Big Plans for Terminal 91–82 Acres of Prized but Unused Land Eyed for Immense New Development. *Seattle Times*, 11 October 2003; A1.
129. Young, B.; Scott, A. City Council Says No to Housing on Seattle Port's North Bay Tract. *Seattle Times*, 3 May 2005; B5.
130. Boyer, T. Working Waterfront Losing a Mainstay. *Seattle Times*, 25 June 2005; A1.

131. Boyer, T. North Bay Plan is Ok'd by Port—Multiple Uses Development Proposal Faces Obstacles. *Seattle Times*, 29 June 2005; E1.
132. Lane, P. High-tech, Retail Complex Proposed—Development Linking Waterfront, Sports Arenas would require a Change from Industrial Zoning. *Seattle Times*, 18 February 2000; D1.
133. Seven, R. How goes Sodo?—Beyond the Trucks and Trains, It's a Work in Progress. *Seattle Times*, 19 October 2003; 18.
134. McOmber, J.M. Cargo or Condos? The Future of Terminal 46. *Seattle Times*, 4 April 2004; E1.
135. Stockton, P. Heralding the New Seattle—Is Blue-Collar going Extinct? Creeping Gentrification has South Park, Georgetown, and Sodo Residents Wondering if they're the Next Fremont. *Seattle Times*, 1 October 2000; L1.
136. Stockton, P. Long Ordeal Ends with Company Leaving. *Seattle Times*, 10 August 2002; A1.
137. Bjorhus, J. Allen Raises Rents, Dander, Suspicions—Technology Corridor South of Lake Union? *Seattle Times*, 4 April 1997; B1.
138. Editorial. Beware Taller Buildings near South Lake Union. *Seattle Times*. 20 December 2000, p. B6. Available online: http://search.proquest.com/docview/383345652?accountid=15006 (accessed on 13 July 2015).
139. Editorial. A New South Lake Union, warts and all. *Seattle Times*. 18 June 2001, p. B4. Available online: http://search.proquest.com/docview/383319082?accountid=15006 (accessed on 13 July 2015).
140. McOmber, J.M.; Young, B. Paul Allen's Makeover offers High Risks, Huge Payoffs. *Seattle Times*, 27 April 2003; A1.
141. Young, B. Compton: Free Trip on Allen Jet may have Left "Unfortunate Impression"—Disclosure Comes as Council Takes Up Developer's Lake Union Plans. *Seattle Times*, 27 August 2003; A1.
142. Young, B. Panel Oks Taller Buildings for Biotechs at S. Lake Union. *Seattle Times*, 4 December 2003; B1.
143. Young, B. Coalition Gives up on Vulcan Talks—South Lake Union Development. *Seattle Times*, 9 March 2005; B2.
144. Editorial. Hold Fast on Zoning for Industrial Uses. *Seattle Times*, 20 March 2000; B4.
145. Conlin, R. So Long to Blue-Collar Zone?—New City Plan Leaves Ample Room for Offices and Other Development in Once-Industrial Stadium Area. *Seattle Times*, 5 May 2000; A1.
146. Atkinson, R.D.; Gottlieb, P.D. *The Metropolitan New Economy Index*; Progressive Policy Institute and Center for Regional Economic Issues: Washington, DC, USA, 2001.
147. Moretti, E. *The New Geography of Jobs*; Houghton Mifflin Harcourt: New York, NY, USA, 2012.
148. United States Census. American Fact Finder. Available online: http://factfinder.census.gov/ (accessed on 28 July 2014).
149. United States Environmental Protection Agency (EPA). Approval and Promulgation of Implementation Plans and Designation of Areas for Air Quality Planning Purposes: Washington. *Fed. Regist.* **2001**, *66*, 14492–14500.

150. Ortega, B. Paying the Price for Cleaner Air—Drivers Expected to Foot the Bill. *Seattle Times*, 21 November 1991; B1.
151. Pryne, E. South Park—We're Seattle's Dumping Ground. *Seattle Times*, 24 February 1994; A1.
152. Priyne, E. The Search for Environmental Justice—Tackling a Toxic Issue. *Seattle Times*, 24 February 1994; A1.
153. City of Seattle. *Environmental Justice Task Force Report*; Mayor's Office: Seattle, WA, USA, 1996.
154. Warner, K. Linking local sustainability initiatives with environmental justice. *Local Environ.* **2002**, *7*, 35–47.
155. Seattle King County Department of Public Health. *Hospitalization Rates for Respiratory Diseases by Age*; Seattle King County Department of Public Health: Seattle, WA, USA, 1997.
156. Community Coalition for Environmental Justice. *Global Struggles Local Struggles: Indigenous People, People of Color and the Working Class in Washington State and Beyond Fight for Social and Environmental Justice*; Community Coalition for Environmental Justice: Seattle, WA, USA, 2001.
157. Keill, L.; Maykut, N. *Puget Sound Air Toxics Evaluation*; Puget Sound Clean Air Agency (PSCAA): Seattle, WA, USA, 2002.
158. Yost, M.; Gilroy, M.; Strange, K.H. *Tacoma and Seattle Area Air Toxics Evaluation*; Puget Sound Clean Air Agency (PSCAA): Seattle, WA, USA, 2010.
159. Wu, C.; Liu, L.J.S.; Cullen, A.; Westberg, H. Williamson, John. Spatial–Temporal and cancer risk assessment of selected hazardous air pollutants in Seattle. *Environ. Int.* **2011**, *37*, 11–17.
160. United States Environmental Protection Agency Newsroom. EPA Awards 2014 Environmental Justice Collaborative Problem-Solving Cooperative Agreements. Available online: http://yosemite.epa.gov/opa/admpress.nsf/8b770facf5edf6f185257359003fb69e/d4e4de8dbee4ef5585257d5500485609!OpenDocument (accessed on 16 September 2014).
161. Clauson, S.; Abel, T.D. A Tale of Two Emerald Cities: Environmental Inequality Formation and Participatory Geography in Action. In Proceeding of the Annual Meeting of the Association of American Geographers, Chicago, IL, USA, 21–25 April 2015.
162. City of Seattle. Mayor Murray Launches First-Of-Its-Kind Equity and Environment Initiative. Available online: http://murray.seattle.gov/mayor-murray-launches-first-of-its-kind-equity-environment-initiative/#sthash.2LlW8hCz.5rMKORLF.dpbs (accessed on 16 July 2015).
163. Beekman, D. New City Office would Oversee Growth, Quality-of-Life Issues. *Seattle Times*, 24 June 2015; B1.
164. Beekman, D. A Mixed Report on City's "Urban Villages". *Seattle Times*, 30 January 2015; B1.
165. Beekman, D. Mayor: Build Affordable Housing or Pay Fees. *Seattle Times*, 14 July 2015; B1.

How Does Land Development Promote China's Urban Economic Growth? The Mediating Effect of Public Infrastructure

Xianwei Fan, Dan Zheng and Minjun Shi

Abstract: Although substantial studies emphasized the close relationship among land development, public infrastructure, and urban economic growth, the mediating effect of public infrastructure remains unexplored. Using panel data of 253 prefecture-level Chinese cities from 1999 to 2012, we empirically conduct a mediating effect analysis to examine how land development promotes urban economic growth. It is found that land development has a positive impact on public infrastructure, whereas the construction of public infrastructure is positively related with urban economic growth. Therefore, land development exerts a positive influence on urban economic growth through one important mediator: public infrastructure. It is also found that the mediating effect of public infrastructure is partial. The estimation results are robust to various specifications and sensitivity analysis.

Reprinted from *Sustainability*. Cite as: Fan, X.; Zheng, D.; Shi, M. How Does Land Development Promote China's Urban Economic Growth? The Mediating Effect of Public Infrastructure. *Sustainability* **2016**, *8*, 279.

1. Introduction

China's land policy has experienced dramatic changes in the past decades. Local governments are authorized with monopolistic power in land acquisition and deposition in land developing. Since the reform of the tax sharing system in 1994, central fiscal revenue surged whereas local government revenue decreased rapidly. The unbalanced administrative rights and financial power made local officials have little option but to take on the role of land developers in China's primary land market [1]. Developing and managing land has become a major business for many local governments [2,3]. From 1999 to 2012, the area of urban construction land increased by 119%, from 2.08×10^4 km^2 to 4.57×10^4 km^2. Existing studies reveal that land development has two kinds of impacts on urban growth [4,5]. The first one is the expansion of urban construction land and the second one is the substantial increase in land revenue. The interrelationship between urban spatial expansion and economic growth has been well discussed [6–8]. By comparison, our research focuses on land revenue which local officials acquired during the land development. According to the different obtaining ways, we divide land revenue into land conveyance revenue and land finance revenue.

With China's booming real estate market, urban housing and land price has grown rapidly [9]. Land gradually became the greatest immobile asset controlled by local government [1,10]. Local officials' land revenue, especially land conveyance revenue, skyrocketed [11,12]. From 2009 to 2013, the total land conveyance revenue reached 12.94 billion yuan. Most land revenues are spent on the construction of urban public infrastructure [13,14]. For example, during the period 2008–2012, a total of 2.59 trillion yuan of land conveyance revenue was spent on public infrastructure. Such huge spending greatly improved the level of China's urban infrastructure. The total length of city road was increased from 2.6 million km in 2000 to 3.27 million km in 2012. In addition, the massive construction of infrastructure greatly promoted economic growth [15–17]. Therefore, it can be inferred that infrastructure has played a mediating role. It is not surprising that the new phenomenon of "local government-driven land-based development", which can simply be summarized as land development, land revenue, public infrastructure, urban economic growth, was very prevalent for local government in the past decades.

So far, the important role of land development in China's urban growth has been stressed by many studies from different perspectives. Nevertheless, how land development promotes China's urban economic growth remains unexplored, and the mechanism and pathway from land development and land revenue to urban economic growth was overlooked. From the point view of land revenue, this paper aims to answer the above question by conducting a thorough empirical analysis on the mediating effect of public infrastructure. The adoption of mediating effect analysis allows us to obtain robust estimation results and verifies the validity of public infrastructure as a mediator. It is found that land development has a positive impact on public infrastructure, whereas the construction of public infrastructure is positively related with urban economic growth. Therefore, land development exerts a positive influence on urban economic growth through one important mediator: public infrastructure. It is also found that the mediating effect of public infrastructure is partial. The estimation results are robust to various specifications and sensitivity analysis.

The paper is structured as follows. Section 2 briefly reviews studies of land development, public infrastructure and urban growth. In Section 3, we demonstrate the analytical framework and propose three hypotheses in relation with the mediating effect of public infrastructure. Section 4 describes methodology and data used in this paper. Section 5 reports and discuss the estimation results, and conduct some sensitive analysis. Section 6 concludes the paper and provides some policy suggestions.

2. Literature Review

The relationship between land development and urban economic growth has been well discussed [18]. The prevailing view took land use change as the outcome of urban economic growth [7,13,19]. Economic growth has generated enormous demand for urban construction land converted from agricultural land [20,21]. As an important production factor, some studies empirically verified the positive impact of urban land expansion on economic growth [6]. The supply of urban construction land has been proven to be a growth engine in China [1,22,23]. He *et al.* [8] investigated the interrelationship between urban spatial expansion and economic growth. They found that urban construction land expansion was not only the consequence of economic growth but also its direct and indirect driver. For the consequence of substantial increasing in land revenue, Liu *et al.* [24] confirmed that leasing land for industrial and commercial use is beneficial to local budget revenue growth by generating a stream of future revenues, land leasing significantly contributes to local GDP growth in the current and following years. The substantial increase in land revenue caused by land leasing greatly stimulated urban economic growth [1].

To attract both domestic and foreign investment, substantial revenues accumulated from land development have been used to invest in urban public infrastructure [25–27]. Zheng *et al.* [28] investigated the self- reinforcing mechanism of China's financing and investment channel for urban infrastructure. They found that, on one hand, the municipal governments who have rich land revenue are able to invest heavily on urban public infrastructure; On the other hand, investment on urban public infrastructure will increase land prices through improving economic productivity and quality of life. The role of infrastructure in stimulating economic growth has been well documented theoretically and empirically. In the theoretical literature, infrastructure is modeled as an important source which could generate external economies. As an intermediate input, it can have spillover externalities, such as enhancing local amenities and attracting external investment [15–17]. However, empirical findings on this issue are inconsistent and contrary to each other. For example, Fedderke *et al.* [29] failed to find any strong linkage between infrastructure investment and output in South Africa by utilizing VAR technique. But, for the same country, Fedderke *et al.* [30] found strong effects of infrastructure on output per worker as well as on TFP. As for East Asian countries, Young [31] and Hsieh [32] also found different conclusions.

In summary, existing studies paid more attention to the direct impact of land development on urban growth, but overlooked how land development influences urban economic growth. Although substantial studies emphasized the close relationship among land development, public infrastructure, and urban economic growth, to the best of our knowledge, the role of public infrastructure in land development or urban growth has not been systemically investigated. Our analysis contributes to existing studies in this aspect by empirically examining the mediating effect of public infrastructure.

3. Analytical Framework and Hypotheses Development

3.1. Analytical Framework

As one of the main consequences of land development, land revenue, which was the local officials' biggest extra-budgetary revenue, has skyrocketed during the past decades [6,33]. Substantial empirical studies have directly verified the positive impact of land revenue on economic growth. But except the directly positive impact, land revenue also has an indirect impact on economic growth. Previous research and statistical data analysis shows that most of local land revenue has been used for urban public infrastructures [34,35]. The non-budgetary revenues generated from urban land leasing hold the promise to empower local governments' fiscal capacity and flexibility in financing various urban public infrastructures, such as transportation, water, and gas supplies and green spaces, aiming to enhance local amenities and attract external investment. Thus, urban public infrastructure, as a mediator, has proven its positive effects on urban economic growth (Figure 1).

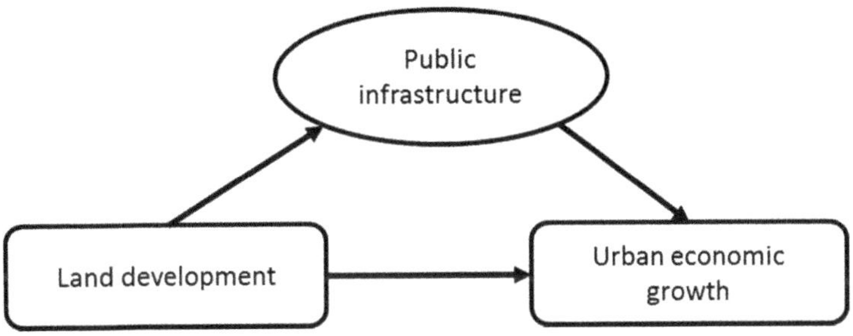

Figure 1. Analytical framework of the mediating effect of public infrastructure.

3.2. Hypothesis of Land Development, Land Revenue, and Public Infrastructure

Since the early 1990s, local officials have turned land development into a money-making business and acquired a significant amount of land revenues [1,3]. Land conveyance revenue, which was historically set mainly by negotiation but is increasingly set by auction (pai mai), public tender (zhao biao), and quotation (gua pai) subject to competitive bidding land supply system, increased by more than 55 times, from 51.43 billion yuan in 1999 to 2.85 trillion yuan in 2012 (Figure 2). The proportion of total land conveyance revenue to local governments' fiscal revenue was more than 30% nationwide after 2004. For land finance revenue, which includes land mortgage revenue and quasi-municipal bonds (chengtouzai) based on the credit of land. In 2013, the net land mortgage revenues reached 1.77 trillion yuan and the total mortgage revenues reached 7.76 billion yuan from 2009 to 2013 (Figure 3).

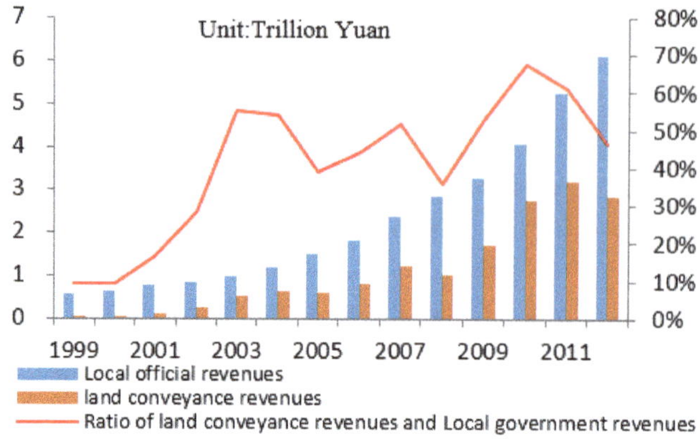

Figure 2. Growth rate and proportion of land conveyance revenue. Sources: China City Statistical Yearbook [36], China Statistical Yearbook of Land and Resources [37] and author's calculations.

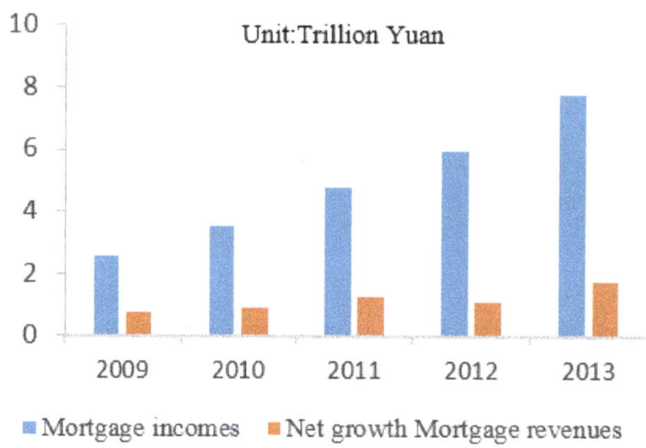

Figure 3. 84 major cities' land mortgage revenue. Source: China Land and Resources bulletin [38].

Substantial land revenue accumulated from land development has been used to finance the provision and improvement of urban public infrastructure, such as transportation, water and gas supplies, and green spaces, *etc.* [25], so as to attract both domestic and foreign investment [26,27]. Before 2007, the expenditure of land revenue in China was under little scrutiny and local officials can freely use them for any purpose. With the state-owned land conveyance revenue and expenditure management approach becoming effective, the expenditure of land conveyance revenue was limited in land acquisition and relocation compensation, land development, urban infrastructure construction, *etc.* in 2007. About 58.97% (369.22 billion yuan) of land conveyance revenue was spent on urban public infrastructure after deducting the cost of land compensation in 2012 (Table 1). Liu and Jiang [39] investigated the investment sources of local officials' infrastructure in several southeast coastal counties in 2005. They found that only 10% investment came from fiscal revenue, while the remaining 90% was related to land development, among which 30% came from land conveyance revenue, 60% came from land mortgage revenue.

Table 1. Net incomes of land leasing and its expenditure.

	2008	2009	2010	2011	2012
Net revenue of land leasing	456.31	686.88	221.60	942.32	626.14
Urban infrastructure expenditure	336.23	377.41	855.34	655.80	369.22
Other expenditure	121.93	169.38	264.41	237.57	210.47
Ratio of urban infrastructure expenditure	73.69%	54.95%	70.02%	69.59%	58.97%

Unit: Billion Yuan; Source: China Land and Resources bulletin [38].

We proposed Hypothesis 1 as follows:

Hypothesis 1: The construction of urban infrastructure is positively related with urban land development.

3.3. Hypothesis of Public Infrastructure and Urban Economic Growth

Existing studies indicated that as investment expenditure, the construction of public infrastructure can stimulate economic growth directly, while also enhancing local amenities and attracting external investment [15–17]. Thus public infrastructure has a spillover effect and provision of public infrastructure is one of the effective means by which governments can promote economic growth [40,41].

We proposed Hypothesis 2 as follows:

Hypothesis 2: The construction of urban public infrastructure has a positive impact on urban economic growth.

Although most land revenue has been invested in urban public infrastructure, there is also a certain amount dedicated to education and compensation expropriation. Table 2 indicates that this amount reaches proximately 30% to 50% of the total land revenue. The education and compensation expropriation also has an impact on economic growth. While on the other hand, analysis shows that the urban construction land expansion is also a result for land developing activities [33]. Urban construction land as a production factor, can stimulate economic growth directly and in the same time can be successfully used as a tool to attract foreign investments and to sustain infrastructure investments, indirectly triggering economic growth [4,8]. Thus, the improvement of infrastructure as a mediator is only a partial reason for economic growth. Land development promotes urban economic growth through comprehensive channels besides this mediator.

Table 2. Sources and descriptive statistics of variables.

Variables	Sources	Mean	S.D.	Min.	Max.
ln GDP	China City Statistical Yearbook [36]	6.118	1.117	−4.605	9.804
ln PI	Author calculation	−2.479	0.545	−6.023	−0.622
ln Lc	China Statistical Yearbook of Land and Resources [37]	0.141	2.071	−4.605	6.478
ln Lfc	China Urban Construction Statistics Yearbook [42]	1.945	2.192	−4.605	7.350
ln Li	Author calculation	2.248	2.002	−4.605	7.496
ln Lac	China City Statistical Yearbook [36]	0.797	1.469	−4.605	4.551
ln $pFDI$	China City Statistical Yearbook [36]	2.887	2.160	−4.605	8.073
ln Psi	China City Statistical Yearbook [36]	3.832	0.260	2.197	4.511

Taking Hypothesis 1 and Hypothesis 2 together, we proposed Hypothesis 3 as follows:

Hypothesis 3: Public infrastructure mediates the impact of land development and urban economic growth. Moreover, the mediating effect of public infrastructure is partial.

4. Methodology and Data

4.1. Mediating Effects Analysis Method

Mediating effect models are frequently used in the research of social science. Mediating indicates that the effect of an independent variable (x) on a dependent variable (y) is transmitted through a third variable (m), and m is called "mediator". Using the mediating effects model, we can identify the mediator's effect and verify how land development influence urban economic growth.

According to Baron and Kenny's causal steps approach [43], in order to examine how land development influences urban economic growth, we estimate three regression models to examine the mediating effects. Model 1 examines the impact of land revenues on urban economic growth, Model 2 examines the impact of land revenues on urban public infrastructure, and Model 3 examines the impact of land revenues on urban economic growth under the control of public infrastructure. If all of the three effects are significant, the mediating effects of public infrastructure are proven to exist, which means the urban public infrastructure is a mediator. Furthermore, if the land revenues' coefficient in Model 3 is significant and the coefficient is smaller than that in Model 1, urban public infrastructure is identified as a partial mediator, otherwise, urban public infrastructure is identified as a completed mediator.

4.2. Model Specification

According to the mediating effects analysis method, we established three models to exam three hypotheses proposed in Section 3.

4.2.1. Model 1: Economic Growth Model

We introduced the government's land revenue into an economic growth model to exam the impact of land revenue on urban economic growth. Based on the Cobb-Douglas production function, after controlling for human capital, foreign direct investment (FDI), and the output of secondary industry, we introduced land revenue into urban economic growth estimation model (See Equation (1)):

$$GDP_{it} = \alpha_0 + \alpha_1 Li_{it} + \alpha_2 X_{it} + f_i + \varepsilon_{1,\ it} \tag{1}$$

where GDP_{it} is the Gross Domestic Product of city i in year t; Li_{it} is the land revenue of city i in year t, which includes local government land conveyance revenue (Lc), land finance revenue (Lfc) and total land revenue; X_{it} is a vector of control variables, including human capital (Lac), FDI per capita ($pFDI$) and the proportion of secondary industry output (Psi).

4.2.2. Model 2: Urban Public Infrastructure Model

We introduced the government's land revenues into an urban public infrastructure model to exam the effect of land revenue on urban public infrastructure (See Equation (2)):

$$PI_{it} = \beta_0 + \beta_1 Li_{it} + \beta_2 X_{it} + f_i + \varepsilon_{2,\ it} \tag{2}$$

where PI_{it} are the public infrastructure level of city i in year t and the definition of Li_{it} is the same as Model 1. X_{it} is also a vector of control variables, the variables should the same as Model 1 according to mediating effects analysis method.

4.2.3. Model 3: Urban Economic Growth Model under the Control of Public Infrastructure

Under the control of public infrastructure, we also established an urban economic growth model to examine the direct effect of land revenue on economic growth (See Equation (3)):

$$GDP_{it} = \gamma_0 + \gamma_1 Li_{it} + \gamma_2 PI_{it} + \gamma_3 X_{it} + f_i + \varepsilon_{3,\ it} \tag{3}$$

where the definition of GDP_{it}, Li_{it} and the control variables (X_{it}) are the same as Model 1, the definition of PI_{it} are the same as Model 2.

4.3. Data and Variable Measurements

Our empirical study aims at investigating how land development influences urban economic growth and verifies the mediating effect of public infrastructure. We use a panel of time-series cross-section data and restrict our analysis to a time period beginning in 1999, the first year in which reliable land conveyance revenue data is available, and ending in 2012. Given that the data of some prefecture-level cities are missing or hard to find, we exclude these cities and restrict our attention to the 253 cities (including four municipalities). The main economic data comes from the China City Statistical Yearbook, the land revenue data comes from China Statistical Yearbook of Land and Resources, and the public infrastructure data comes from China Urban Construction Statistics Yearbook (See Table 2).

So far, there is no systematic statistic data of urban public infrastructure in the existing yearbook. In order to measure urban public infrastructure, we establish a public infrastructure evaluation index system, which contains four second class indicators and nine third class indicators. Table 3 reports the weight of each indicator, which is calculated with expert scoring. Then, after standardizing the original data, we measure urban public infrastructure of 253 cities from 1999 to 2012.

Table 3. Urban public infrastructure index system and the weight.

Urban infrastructure level	Living Facilities (25%)	Length of heating pipeline (8.33%)
		Length of water supply pipeline (8.33%)
		Length of gas pipeline (8.33%)
	Road Facilities (25%)	Public transport vehicles of per million people (12.5%)
		Road area of per people (12.5%)
	Ecological Facilities (25%)	Green area of per people (12.5%)
		Green area coverage rate (12.5%)
	Sanitation Facilities (25%)	Length of sewage drainage pipeline (12.5%)
		Number of sewage treatment station (12.5%)

Since systematic statistic data on land finance revenue are lacking in the existing yearbook, following Zheng *et al.* [28], we use the sum of domestic loans and bonds of urban infrastructure investment funding as the amount of urban land finance revenue. Land revenues are the sum of land conveyance incomes and land finance revenue.

In addition, we make logarithmic transformation to both the dependent and independent variables.

5. Estimation Results and Sensitive Analysis

5.1. Estimation Results

Table 4 reports the estimation results of all models. Because Hausman test indicates rejecting the hypothesis of no correlation at a 5% significance level, we estimated all specifications of these equations using fixed-effect panel data regression.

Column Model 1 in Table 4 shows that land revenue has a direct positive impact on urban economy. Increasing land revenue by 1% will result in a 0.23% increase in urban GDP. The result is consistent with most direct study of the relationship between land revenue and economic growth [44]. Column Model 2 reports that land revenue has a positive impact on public infrastructure. The coefficient of Model 2 shows that if we increase land revenue by 1%, urban public infrastructure level will increase by 0.06%. It is consistent with the finding of Zheng *et al.* [28] and verified Hypothesis 1. The construction of urban infrastructure is positively related with urban land revenue, which is gained by the local officials during the land development.

From Column Model 3, we can find that urban public infrastructure has a positive impact on urban economic growth. Increasing urban public infrastructure level by 1% will result in a 0.27% increase in urban GDP. This result consistent with the finding of most empirical studies on the relationship between public infrastructure and economic growth, such as Bronzini *et al.* [41]. It also verified Hypothesis 1.The construction of urban public infrastructure has a positive impact on urban economic growth.

According to coefficients in column Model 3, we know that urban public infrastructure is a partial mediator since the coefficient in Model 3 is significant at a 1% level and is smaller than the coefficient in Model 1. Therefore, Hypothesis 3 is verified. This result means that the effect of local government's land revenues on urban economic growth is not only transmitted through public infrastructure. Public infrastructure is not the only mediator and other factors or channels may also impact economic growth.

Then we replaced land conveyance revenue and land finance revenue to the Model 1 to Model 3 respectively to investigate their influence on urban economic growth and to examine the mediating effect of public infrastructure. Table 4 reports the estimation results. It is shown that the mediating effect exists. Both land conveyance revenue and land finance revenue have a positive impact on public infrastructure, the construction of public infrastructure is positively related with urban economic growth, and urban public infrastructure level is an effective mediator.

Table 4. Estimation result of mediating effects.

Dependent Variable	Model 1			Model 2			Model 3		
	GDP			Urban Infrastructure Level			GDP		
ln PI							0.420 *** (0.033)	0.294 *** (0.028)	0.274 *** (0.028)
ln Lc	0.045 *** (0.007)			0.026 *** (0.004)			0.034 *** (0.007)		
ln Lfc		0.214 *** (0.009)			0.041 *** (0.006)			0.202 *** (0.009)	
ln Li			0.231 *** (0.010)			0.059 *** (0.006)			0.215 *** (0.010)
ln Lac	0.341 *** (0.010)	0.257 *** (0.009)	0.240 *** (0.010)	0.100 *** (0.006)	0.089 *** (0.006)	0.078 *** (0.006)	0.299 *** (0.010)	0.231 *** (0.009)	0.218 *** (0.010)
ln $pFDI$	0.142 *** (0.007)	0.083 *** (0.006)	0.084 *** (0.006)	0.091 *** (0.004)	0.088 *** (0.004)	0.085 *** (0.004)	0.104 *** (0.007)	0.057 *** (0.007)	0.061 *** (0.007)
ln Psi	0.366 *** (0.049)	0.297 *** (0.041)	0.291 *** (0.041)	0.213 *** (0.027)	0.235 *** (0.025)	0.228 *** (0.025)	0.276 *** (0.048)	0.228 *** (0.041)	0.229 *** (0.041)
_Cons	4.0721 *** (0.183)	4.162 *** (0.152)	4.089 *** (0.153)	−3.626 *** (0.101)	−3.769 *** (0.093)	−3.776 *** (0.092)	5.596 *** (0.215)	5.272 *** (0.183)	5.123 *** (0.200)
R-sq:	0.668	0.716	0.706	0.5740	0.567	0.573	0.695	0.715	0.716
Number of obs.	2882	3334	3347	2882	3334	3347	2882	3334	3347

Note: Standard errors are reported in parentheses; *** denotes significance higher than 0.01.

5.2. Sensitive Analysis

In this section, we conduct some sensitive analysis to check the robustness of estimation results of Models 1–3.

Firstly, we divided the total sample into three sub-samples: east, middle, and west. The east sub-sample contains 113 cities, the middle sub-sample contains 72 cities, and the west sub-sample contains 68 cities. According to the result of Hausman test of specification, all models are estimated with fixed-effect panel data regression. Table 5 reports the estimation results of the sensitive analysis. Note that three hypotheses we proposed still hold in the east, middle, and west sub-samples. As a result, public infrastructure is an effective mediator in all of the three areas. In addition, land revenues have greatly promoted urban economic growth in each area. In particular, increasing land revenue by 1% will result in a 0.20%, 0.09%, and 0.20% increase in urban GDP in the east, middle, and west areas, respectively.

Table 5. Estimation results of mediating effects in different regions.

Sub-Sample	Dependent Variable	GDP	Public Infrastructure	GDP
East sub-sample	ln *PI*			0.4114 *** (0.042)
	ln *Li*	0.197 *** (0.013)	0.0480 *** (0.007)	0.1769 ***(0.012)
	ln *Lac*	0.307 *** (0.013)	0.0982 *** (0.007)	0.2665 *** (0.013)
	ln *pFDI*	0.108 *** (0.009)	0.0850 *** (0.006)	0.0731 *** (0.010)
	ln *Psi*	0.310 *** (0.058)	−0.0237 (0.035)	0.3200 *** (0.057)
	_Cons	4.129 *** (0.212)	−2.756 *** (0.126)	5.2631 *** (0.236)
	R-sq:	0.805	0.638	0.819
	Number of obs.	1535	1535	1535
Middle sub-sample	ln *PI*			0.0707 * (0.047)
	ln *Li*	0.0891 *** (0.019)	0.029 ** (0.013)	0.087 *** (0.019)
	ln *Lac*	0.3626 *** (0.015)	0.046 *** (0.010)	0.3593 *** (0.015)
	ln *pFDI*	−0.0493 **(0.011)	0.071 *** (0.008)	−0.0543 ***0.012)
	ln *Psi*	0.1473 *** (0.060)	0.358 *** (0.041)	0.1220 * (0.063)
	_Cons	5.0916 *** (0.228)	−4.162 *** (0.156)	5.3858 *** (0.301)
	R-sq:	0.650	0.477	0.656
	Number of obs.	977	977	977
West sub-sample	ln *PI*			0.122 *** (0.052)
	ln *Li*	0.203 *** (0.023)	0.111 *** (0.016)	0.189 *** (0.024)
	ln *Lac*	0.170 *** (0.020)	0.076 *** (0.014)	0.161 *** (0.021)
	ln *pFDI*	0.031 ** (0.012)	0.057 *** (0.008)	0.024 *** (0.013)
	ln *Psi*	0.424 *** (0.079)	0.464 *** (0.053)	0.368 *** (0.082)
	_Cons	3.594 *** (0.303)	−4.782 *** (0.204)	4.176 *** (0.391)
	R-sq:	0.526	0.508	0.531
	Number of obs.	835	835	835

Note: Standard error are reported in parentheses. ***, ** and * denotes significance higher than 0.01, 0.05 and 0.1, respectively.

Secondly, to control for time-specific effect, we divided the whole time period into two periods: period 1999–2004, and period 2005–2012. We estimate Models 1–3, respectively in these two periods and reports the results in the following Table 6. It is clear that the signs and significance of each variable are similar to that of the whole-period. Taken together, we conclude that our estimation results obtained by fixed-effect panel data regression are robust to various specification tests as well as change of sample size and period.

Table 6. Estimation results of mediating effects in different time periods.

Dependent Variable	Model 1		Model 2		Model 3	
	GDP		Urban Infrastructure Level		GDP	
Time Period	1999–2004	2005–2012	1999–2004	2005–2012	1999–2004	2005–2012
ln PI					0.234 *** (0.039)	0.304 *** (0.040)
ln Li	0.203 *** (0.013)	0.258 *** (0.015)	0.051 *** (0.009)	0.065 *** (0.008)	0.192 *** (0.013)	0.238 *** (0.015)
ln Lac	0.248 *** (0.013)	0.230 *** (0.013)	0.085 *** (0.009)	0.074 *** (0.007)	0.228 *** (0.013)	0.208 *** (0.014)
ln $pFDI$	0.095 *** (0.008)	0.073 *** (0.009)	0.080 *** (0.006)	0.088 *** (0.005)	0.076 *** (0.009)	0.047 *** (0.009)
ln Psi	0.306 *** (0.059)	0.283 *** (0.056)	0.173 *** (0.041)	0.268 *** (0.031)	0.265 *** (0.059)	0.202 *** (0.056)
_Cons	4.041 *** (0.218)	4.083 *** (0.211)	−3.606 *** (0.151)	−3.921 *** (0.117)	4.888 *** (0.257)	5.275 *** (0.260)
R-sq:	0.685	0.651	0.46	0.552	0.695	0.663
Number of obs.	1329	2018	1329	2018	1329	2018

Note: Standard errors are reported in parentheses. *** denotes significance higher than 0.01.

6. Conclusions

In recent decades, developing and managing land has become a major business for Chinese local officials. Although substantial studies emphasized the close relationship among land development, public infrastructure, and urban economic growth, the mediating effect of public infrastructure remains unexplored. Using panel data of 253 prefecture-level Chinese cities from 1999 to 2012, we empirically conducted a mediating effect analysis to examine how land development promotes urban economic growth. It is found that land development has a positive impact on public infrastructure, whereas the construction of public infrastructure is positively related with urban economic growth. Therefore, land development exerts a positive influence on urban economic growth through one important mediator: public infrastructure.

Our paper can be viewed as the first attempt to empirically investigate the mechanism and path way from land development to urban economic growth, through public infrastructure. As discussed in Zheng *et al.* [28], the interrelationship among land development, land revenues, and urban economic growth is complicated. The self-reinforced mechanism between land revenues and urban public infrastructure and the cumulative causation effect among land development, land revenues and urban economic growth may exist. Though the relationship between land development and urban economic growth have been widely discussed, how the mechanism that urban economic growth effect land revenues are rarely studied.

Therefore, it is worthwhile to extend our empirical analysis to investigate land revenues and urban public infrastructure's self-reinforced mechanism and the cumulative causation effect among land development, land revenues, and urban economic growth.

Although the paper verified the prevailing phenomenon that land development promotes urban economic growth greatly through mediate effectors of public infrastructure in the past decade, there is a critical issue for us to rethink: is this kind of land-dependent development mode sustainable in the future? Since the financial crisis, China's land price and land revenue have experienced a downward trend. In addition, it is emphasized in China's new urbanization plan (2014–2020) that central government will strictly control the scale of urban land development. Therefore, urban construction land will not grow as fast as it has in the past decades. The "local government-driven land-based development" mode will not be sustainable in the future. In this regard, it is highly suggested that local government should transform their extensive growth model from now on.

Acknowledgments: The authors would like to acknowledge funding of National Natural Science Foundation of China (71173212, 71503241, 41401197) and financial support from the project-sponsored by Scientific Research Foundation for the Returned Overseas Chinese Scholars, State Education Ministry.

Author Contributions: Minjun Shi, Dan Zheng, and Xianwei Fan contributed to research design. Xianwei Fan performed research and analyzed the data. Xianwei Fan, Dan Zheng, and Minjun Shi wrote the paper. All authors have read and approved the final manuscript.

Conflicts of Interest: The authors declare no conflict of interest.

References

1. Tao, R.; Su, F.; Liu, M.; Cao, G. Land leasing and local public finance in China's regional development: Evidence from prefecture-level cities. *Urban Stud.* **2010**, *47*, 2217–2236.
2. Zheng, H.; Wang, X.; Cao, S. The land finance model jeopardizes China's sustainable development. *Habitat Int.* **2014**, *44*, 130–136.
3. Wu, Q.; Li, Y.L.; Yan, S.Q. The incentives of China's urban land finance. *Land Use Policy* **2015**, *42*, 432–442.
4. Lichtenberg, E.; Ding, C. Local officials as land developers: Urban spatial expansion in China. *J. Urban Econ.* **2009**, *66*, 57–64.
5. Wang, L.; Li, C.; Ying, Q.; Cheng, X.; Wang, X.; Li, X.; Hu, L.; Liang, L.; Yu, L. China's urban expansion from 1990 to 2010 determined with satellite remote sensing. *Chin. Sci. Bull.* **2012**, *57*, 2802–2812.
6. Ding, C. Policy and praxis of land acquisition in China. *Land Use Policy* **2007**, *24*, 1–13.
7. Deng, X.; Huang, J.; Rozelle, S.; Uchida, E. Growth, population and industrialization, and urban land expansion of China. *J. Urban Econ.* **2008**, *63*, 96–115.
8. He, C.; Huang, Z.; Wang, R. Land use change and economic growth in urban China: A structural equation analysis. *Urban Stud.* **2014**, *51*, 2880–2898.

9. Pan, J.N.; Huang, J.T.; Chiang, T.F. Empirical study of the local government deficit, land finance and real estate markets in China. *China Econ. Rev.* **2015**, *32*, 57–67.
10. Tian, L.; Ma, W. Government intervention in city development of China: A tool of land supply. *Land Use Policy* **2009**, *26*, 599–609.
11. Li, J. Land sale venue and economic growth path: Evidence from China's urban land market. *Habitat Int.* **2014**, *41*, 307–313.
12. Ye, L.; Wu, A.M. Urbanization, land development, and land financing: Evidence from Chinese cities. *J. Urban Aff.* **2014**, *36*, 354–368.
13. Lin, G.C.S. Reproducing spaces of Chinese urbanization: New city-based and land-centred urban transformation. *Urban Stud.* **2007**, *44*, 1827–1855.
14. Clarke, A.P.; Peterson, G.E. *Financing Cities: Fiscal Responsibility and Urban Infrastructure in Brazil, China, India, Poland and South Africa*; World Bank: Washington, DC, USA, 2007.
15. Lucas, R.E., Jr. On the Mechanics of Economic Development. *J. Monet. Econ.* **1988**, *22*, 3–42.
16. Aschauer, D.A. Is public expenditure productive? *J. Monet. Econ.* **1989**, *23*, 177–200.
17. Munnell, A.H. Infrastructure investment and economic growth. *J. Econ. Perspect.* **1992**, *6*, 189–198.
18. Ioppolo, G.; Cucurachi, S.; Salomone, R.; Saija, G.; Shi, L. Sustainable Local Development and Environmental Governance: A Strategic Planning Experience. *Sustainability* **2016**.
19. Gibson, J.; Li, C.; Boe-Gibson, G. Economic Growth and Expansion of China's Urban Land Area: Evidence from Administrative Data and Night Lights, 1993–2012. *Sustainability* **2014**, *6*, 7850–7865.
20. Lin, G.C.S.; Ho, S.P.S. The state, land system, and land development processes in contemporary China. *Ann. Assoc. Am. Geogr.* **2005**, *95*, 411–436.
21. Gao, J.; Wei Y, D.; Chen, W.; Yenneti, K. Urban Land Expansion and Structural Change in the Yangtze River Delta, China. *Sustainability* **2015**, *7*, 10281–10307.
22. Ding, C.; Lichtenberg, E. Land and Urban Economic Growth in China. *J. Reg. Sci.* **2011**, *51*, 299–317.
23. Bai, X.; Chen, J.; Shi, P. Landscape urbanization and economic growth in China: Positive feedbacks and sustainability dilemmas. *Environ. Sci. Technol.* **2011**, *46*, 132–139.
24. Liu, M.; Tao, R.; Yuan, F.; Cao, G. Instrumental land use investment-driven growth in China. *J. Asia Pac. Econ.* **2008**, *13*, 313–331.
25. Lin, G.C.S.; Yi, F. Urbanization of capital or capitalization on urban land? Land development and local public finance in urbanizing China. *Urban Geogr.* **2011**, *32*, 50–79.
26. Yew, C.P. Pseudo-urbanization? Competitive government behavior and urban sprawl in China. *J. Contemp. China* **2012**, *21*, 281–298.
27. Chen, W.Y.; Hu, F.Z.Y. Producing nature for public: Land-based urbanization and provision of public green spaces in China. *Appl. Geogr.* **2015**, *58*, 32–40.

28. Zheng, S.Q.; Sun, W.Z.; Wu, J.W. Infrastructure Investment, Land Leasing and Real Estate Price: A Unique Financing and Investment Channel for Urban Development in Chinese Cities. *Econ. Res. J.* **2014**, *8*, 14–27. (In Chinese).
29. Fedderke, J.W.; Perkins, P.; Luiz, J.M. Infrastructural investment in long-run economic growth: South Africa 1875–2001. *World Dev.* **2006**, *34*, 1037–1059.
30. Fedderke, J.W.; Bogetić, Ž. Infrastructure and growth in South Africa: Direct and indirect productivity impacts of 19 infrastructure measures. *World Dev.* **2009**, *37*, 1522–1539.
31. Young, A. The Tyranny of Numbers: Confronting the Statistical Realities of the East Asian Growth Experience. *Q. J. Econ.* **1995**, *110*, 641–680.
32. Hsieh, C.T. Productivity growth and factor prices in East Asia. *Am. Econ. Rev.* **1999**, *89*, 133–138.
33. Li, H.; Wei, Y.H.D.; Huang, Z. Urban Land Expansion and Spatial Dynamics in Globalizing Shanghai. *Sustainability* **2014**, *6*, 8856–8875.
34. Tao, R. Status and risk of land financing. *Land Resour. Her.* **2013**, *84*, 26–30. (In Chinese).
35. Jiang, S.X.; Liu, S.Y. Land reform and national economic growth. *Manag. World* **2007**, *9*, 1–9. (In Chinese).
36. National Bureau of Statistics. *China City Statistical Yearbook*; Statistical Press of China: Beijing, China, 2000–2013. (In Chinese)
37. Ministry of Land and Resources. *China Statistical Yearbook of Land and Resources*; Geological Publishing House: Beijing, China, 2000–2013. (In Chinese).
38. Ministry of Land and Resources. China Land and Resources bulletin, 2008–2012. Available online: http://www.mlr.gov.cn/zwgk/tjxx/ (accessed on 16 March 2016). (In Chinese).
39. Liu, S.Y.; Jiang, S.X. Financial Risks of Land Financing by Local Governments—Case Study of a Developed Area in East China. *China Land Sci.* **2005**, *19*, 3–9. (In Chinese).
40. Kelejian, H.H.; Robinson, D.P. Infrastructure productivity estimation and its underlying econometric specifications: A sensitivity analysis. *Pap. Reg. Sci.* **1997**, *76*, 115–131.
41. Bronzini, R.; Piselli, P. Determinants of long-run regional productivity with geographical spillovers: The role of R & D, human capital and public infrastructure. *Reg. Sci. Urban Econ.* **2009**, *39*, 187–199.
42. Ministry of Housing and Urban-Rural Development. *China Urban Construction Statistics Yearbook*; China Architecture& Building Press: Beijing, China, 2000–2013. (In Chinese).
43. Baron, R.M.; Kenny, D.A. The moderator-mediator variable distinction in social psychological research: Conceptual, strategic, and statistical considerations. *J. Personal. Soc. Psychol.* **1986**, *51*, 1173–1182.
44. Du, X.; Huang, Z.; Wu, C. China Land Revenue and Economic Growth. *Financ. Trade Econ.* **2009**, *1*, 60–64. (In Chinese).

Natural Disasters, Economic Growth and Sustainable Development in China—An Empirical Study Using Provincial Panel Data

Ji Guo, Hui Liu, Xianhua Wu, Jiong Gu, Shunfeng Song and Yinshan Tang

Abstract: Using a newly developed integrated indicator system with entropy weighting, we analyzed the panel data of 577 recorded disasters in 30 provinces of China from 1985–2011 to identify their links with the subsequent economic growth. Meteorological disasters promote economic growth through human capital instead of physical capital. Geological disasters did not trigger local economic growth from 1999–2011. Generally, natural disasters overall had no significant impact on economic growth from 1985–1998. Thus, human capital reinvestment should be the aim in managing recoveries, and it should be used to regenerate the local economy based on long-term sustainable development.

Reprinted from *Sustainability*. Cite as: Guo, J.; Liu, H.; Wu, X.; Gu, J.; Song, S.; Tang, Y. Natural Disasters, Economic Growth and Sustainable Development in China—An Empirical Study Using Provincial Panel Data. *Sustainability* **2015**, *7*, 16783–16800.

1. Introduction

In recent decades, natural disasters have become international concerns. The Richter magnitude 9.0 earthquake and the subsequent tsunami in Japan in 2011, the Indian Ocean tsunami in 2004, Hurricane Katrina in 2005, and the earthquakes in Haiti and Chile in 2010 resulted in substantial losses in terms of finance and human lives. In China, many disasters occurred during a period of rapid economic growth. Recent events include the catastrophic earthquake in Sichuan Province in 2008, the severe droughts in Yunnan, Guizhou, and Sichuan Provinces in 2009 and 2010, another earthquake in Yushu County, Qinghai Province in 2010, and an unprecedented storm in Beijing on 21 July 2012. The causes of the high frequency of severe natural disasters have become popular research topics, and the findings have improved the understanding of the relationship between human activities and the environment. However, the exploration into the relationship between natural disasters and long-run economic growth in China has been minimal. This paper fills this gap by investigating the economic effects of post-disaster rebuilding activities in the current Chinese economic environment.

Published works have divided the long-run economic effects into three different groups, namely, negative, positive, and inconclusive effects. In defining the negative effects, frequent natural disasters are linked to low economic growth rates (Benson and Clay [1] and Yu Xiao [2]) and enduring consequences (Gall *et al.* [3], Coffman and Noy [4] and Okuyama [5]). Natural disasters reduce household consumption over time without any sign of economic recovery (Dercon [6] and Mechler [7]). Furthermore, natural disasters reduce real gross domestic product (GDP) per capita in the long term (Raddatz [8]), reduce national welfare (Arndt *et al.* [9,10]), increase socioeconomic instability (Porfiriev [11]), and create poverty traps (Hallegatte and Dumas [12]). Area-specific studies in Central America and the Caribbean regions have shown that less democratic countries suffer more losses on imports and exports after natural disasters (Gassebner *et al.* [13], Strobl [14]). Evidence shows that natural disasters increase poverty by 1.5%–3.6%, and has a significant negative effect on the Human Development Index in Mexico (Rodriguez-Oreggia *et al.* [15]). A negative effect on knowledge spillovers is also observed between developing and developed countries (Cuaresma *et al.* [16]), although a "creative destruction" process could occur in developed countries (Skidmore and Toya. [17]). With regard to positive effects, natural disasters, through the rebuilding and recovery process, promote economic growth, improve agricultural and construction outputs and capital formation, and address fiscal and trade deficits (Albala-Bertrand [18,19]; Dacy and Kunreuther [20]; Otero and Marti [21]; Tol and Leek [22]). Such positive effects are evidenced by the accumulation of human capital, growth of GDP per capita, and total factor productivity (Skidmore and Toya [17]; Dacy and Kunreuther [20]; Wang [23]; Kim [24]; Bradley *et al.* [25]). For instance, Dacy and Kunreuther [20] pointed out that the GDP of the damaged area would increase after the disaster. That's because natural disasters may lead to casualties and economic losses, while on the other hand, new technology may be developed for post-disaster reconstruction, thus promoting economic growth. Skidmore and Toya [17], based on the Schumpeter's creative destruction, proposed that natural disasters would create opportunities for the rejuvenation of capital and encourage people to adopt new techniques. They also pointed out that natural disasters would lower the expected return to physical capital, which would drive people to invest in human capital, a key factor in economic growth (Lucas [26]). Bradley *et al.* [25] found that the reconstruction of commercial and residential housing and infrastructure in Oklahoma City after the 3 May 1999 tornado stimulated the labor market and increased the average employment rate. Older facilities are more vulnerable to damage, and thus the reinvestment in such facilities will have a positive effect on overall economic growth and productivity in the long run (Okuyama [27]; Okuyama *et al.* [28]). Different authors have also reported the inconclusive relationships between natural disasters and economic growth. From a macroeconomic perspective, no links have been established between the Kobe earthquake in Japan in 1995 and capital stock (Horwich [29]). A negative effect on short-term economic growth is found, but not on long-term growth (Zou [30]).

Companies in regions hit by floods show a higher growth in total assets and employment with a negative impact on productivity than those in unaffected regions (Leiter *et al.* [31]). Different disasters have different impacts in different regions of the world (Loayza *et al.* [32]). Shabnam [33] also reported a similar classification in his recent review.

The resulting regional economic status after a natural disaster can be influenced heavily by how the recovery activities are managed. Management of these activities is a process that is also linked to the national wealth and political system of a country. In relation to these considerations, China, a newly developed economic superpower, possesses distinctive culture, ethical characteristics as well as a different governing system. China's situation is distinct because of three aspects, namely, the relatively frequent disasters, increasing innovation abilities, and powerful rebuilding supports from the government. With these characteristics, will natural disasters have positive effects on the long-term economic growth in China? What are the mechanisms for such impacts? What can be done to promote a long-term sustainable development, particularly in the places of China where natural disasters have become more frequent in recent years? The purpose of this research is to answer these questions.

2. Data, Disaster Indicator and Hypotheses

In this section, we describe the data used in the paper, construct the indicator that measures the effects of natural disasters, and propose hypotheses on the relationship between natural disasters and economic growth.

2.1. Data

Emergency Events Database (EM-DAT) provides comprehensive and updated information on natural disasters around the world (Guha-Sapir and Below [34]). In this study, disaster related data were collected from EM-DAT, which is maintained by the Center for Research on the Epidemiology of Disasters (CRED) at the Catholic University of Louvain, Belgium. Natural disasters (NDs) are separated into two different groups, namely, meteorological disasters and geological disasters. Meteorological disasters (MDs) are those created by deviations in the normal water cycle and atmospheric processes, including floods, landslides (the classification of EM-DAT, where landslide is categorized into meteorological disaster, is adopted in this paper), cyclones, hurricanes, tornadoes, and typhoons and storms, including ice snow storms. Geological disasters (GDs) are those caused by geological processes, including volcanic eruptions, natural explosions, avalanches, and earthquakes. MDs occur more frequently and are more predictable and regularly occurring than geological disasters. GDs are more difficult to predict, limiting the effectiveness of evacuation procedures. MDs are a reasonable proxy of risk to physical capital, and

GDs may be perceived as a threat to both people's lives and properties (Skidmore and Toya [17]).

A limitation of the EM-DAT data is the highlight on a region instead of an individual province. If an earthquake affects many provinces, the EM-DAT provides damage records of this group of provinces, instead of an individual province. Thus, the data on the losses of each province are collected from *China Meteorological Disasters Book* (Wen, K.G. [35])and previous literature (Disaster Information Department of China National Disaster Reduction Center [36], The State Flood Control and Drought Relief Headquarters Office [37]). It is necessary to point out that *China Meteorological Disasters Book* (Wen, K.G. [35]) only records the data before 2000, thus those after 2000 are mainly from the previous literature in *Disaster Reduction in China* (Disaster Information Department of China national Disaster Reduction Center [36], The State Flood Control and Drought Relief Headquarters Office [37]) except a small part from China's Civil Affairs Network or China Seismic Information. The proportions of each province's losses are calculated and then the total losses of each disaster in EM-DAT are allocated to each province in proportion (an example is given in Table A1). This study used the data collected from 30 provinces between 1985 and 2011. All data related to the economy, education, and population were derived from the *China Statistical Yearbook* (China's National Bureau of Statistics [38]) and *China Compendium of Statistics 1949–2008* (The National Economy Comprehensive Statistics Division of the National Bureau of Statistics [39]).

2.2. Disaster Indicator

How to evaluate the disaster comprehensively is an important content of disaster losses evaluation. Researchers have developed a number of indicators to evaluate the losses of disasters, such as disaster frequency (number of disasters) (Skidmore and Toya [17]; Wang [23]; Raddatz [9]; Kim [24]; Kellenberg and Mobarak [40]), disaster index based on maximum sustained wind velocity for hurricanes (Strobl [14]), economic losses (Cuaresma [16]; Li [41]), mortality (the number of deaths/total population), and loss rate (disaster losses/GDP) (Noy and Nualsri [42]). Although losses are effectively linked to the hazard-formative factors, hazard-formative environments, and hazard-affected bodies, limitations exist in these indicators. Zhao [43] suggested that two disaster events with the same intensity could have significantly different effects if one occured in a sparsely populated and less developed region, whereas the other occured in a densely populated and developed area. A magnitude 6.0 earthquake in China may produce the same level of losses as an 8.0 earthquake in Japan because Japan has more experiences and advanced technologies to prevent and reduce the losses. Therefore, we need an aggregative indicator to estimate the disaster. The frequency and the extent of damages are the most important in the selection of evaluation index of natural disasters. The frequency is often referred to as the number of disasters. The extent of damages includes two aspects, namely, death toll and economic losses. Actually, only the data on the number of disasters, death toll and economic losses can be found in all databases for natural disasters (EM-DAT, *China Meteorological Disasters Book* (Wen, K.G. *et al.* [35]) and other databases). Thus, the variables selected for the synthesized disaster indicator include the number of disasters, death toll, and economic losses.

What method to be adopted for integrating multiple indicators is also a problem. Thus, the entropy method, where a new aggregative indicator is obtained by weighting the number of disasters with the extent of damages, is adopted in this paper. The entropy method is an objective weighting method which holds that the more ordered a system, the smaller the information entropy. Therefore, the information entropy serves as a measurement of disorder degree in a system. The data size is a determinant of the reliability and accuracy of evaluation. In other words, the smaller the information entropy of an indicator, the more information the indicator offers. Then, the indicator will have a higher weight in the synthetical evaluation (Tang *et al.* [44] and Zhang *et al.* [45]).As argued by Burg [46], Darbellay and Wuertz [47] and Li *et al.* [48], synthesizing the indexes with entropy method has four advantages. First, adopting the entropy method to determine the index weight can avoid human interference factors and thus obtain more realistic assessment. Second, by calculating the entropy value of each indicator, it measures the amount of information of each indicator objectively. Third, by considering the correlation among samples and determining the weight based on multiple sample points, the entropy method reduces the impact of outliers on the assessment. Fourth, the entropy method enjoys a better adaptability because it can be applied to any situation that needs to calculate the weight. Therefore, a synchronous indicator with entropy weight has been suggested to improve the measure of severity of disasters (Tang *et al.* [44]. Zhang *et al.* [45]).

The steps for obtaining the synthetical value through entropy method are as follows: (1) calculate the proportion of disaster indicator; (2) calculate the value of information entropy of the disaster indicator; (3) calculate the weight of the disaster indicator; and (4) obtain the synthetical value. First, we calculated the proportion of disaster indicator (j) in region (i):

$$y_{ij} = x_{ij}/\sum_{i=1}^{n} x_{ij}, (i = 1, 2, \ldots, n; j = 1, 2, \ldots, m) \tag{1}$$

where i represents 30 Chinese provinces and central municipalities, j is an indicator of NDs (the number of disasters, death toll, and economic losses), and x_{ij} is the value of the disaster indicator (j) in region (i). Then, we calculated the value of information entropy of the disaster indicator (j):

$$e_j = -K\sum_{i=1}^{n} y_{ij}\ln y_{ij} \tag{2}$$

where K is a constant, $K = 1/\ln n$, n is the number of regions used in this paper, and $n = 30$. The information utility value of disaster indicator (j) is:

$$d_j = 1 - e_j \tag{3}$$

The evaluation is more vital because the information utility value is larger. The weight of the disaster indicator (j) is:

$$w_j = d_j/\sum_{j=1}^{m} d_j \tag{4}$$

Using a weighted summation method, we obtained:

$$U = \sum_{j=1}^{m} x_{ij}w_j \tag{5}$$

where U is the synthetic value calculated from the three natural disaster indicators. Finally, we used a weighted model to obtain a new indicator to measure the degree of NDs. In the following empirical analysis, the new disaster indicators used are ND (degree of total natural disasters), GD (degree of geological disasters), and MD (degree of meteorological disasters).

2.3. Hypotheses

As mentioned in the introduction, many researchers held that the correlations between economic growth and different natural disasters vary. Skidmore and Toya [17] and Zou [30] believed that there was a positive correlation between the MDs of some countries and regions and economic growth as the occurrence of MDs is evident and easy to forecast. Wang [23] and Li [41] held that there was a negative correlation between the GDs of most countries and regions and economic growth because the occurring regularity of GDs is hard to find out and the effect of MDs is wider than that of GDs in regions. As there is little research on the correlation between MDs (or GDs) and economic growth in China, this paper attempts to obtain some new findings by verifying the correlation with China's data and analyze the possible reasons for the correlation. To determine the role of different NDs in regional economic growth, this paper included all three measures of NDs (*ND*, *MD*, and *GD*) in the analytical model. Specifically, *ND* or *MD* and *GD* were factored into the classical growth model.

H_1: Meteorological and geological disasters have different effects on economic growth.

NDs may affect investment decisions on production, and thus affect the economic growth. Leiter [31], based on difference-in-difference (DID) approach, found that physical capital accumulation was significantly higher in regions experiencing a major flood-event. Albala-Bertrand [18] proposed that NDs might have brought a bad influence to the economy in the short run, however, the demands expanded by the capital inflow for post-disaster constructions would definitely promote economic growth eventually. Kunreuther and Kleffner [49] and Wisner [50] held that the productivity restoration of post-disaster economy was directly related to the government's expenditure on disaster prevention and reduction which is conducive to the national security and public welfare. Moreover, the government's investment in disaster prevention and reduction is a key factor for promoting economic growth. Though physical capital investment may be reduced because of the temporary impact of disasters, the government will rapidly increase physical capital investment afterward because China has strong rescue system. A higher saving rate in the short run may promote the accumulation of physical capital. In addition, Okuyama [27] pointed out that NDs might have damaged the previous production system, while old facilities destroyed might also need to be reinstalled or upgraded so as to resume production, in which case NDs was a catalyst for capital investment and industry upgrade, thus, increasing physical capital investment. This assumption leads to the second hypothesis.

H_2: The occurrence of NDs leads to faster accumulation of physical capital.

We also analyzed the impact of NDs on human capital investment. Wang [23] found that NDs had a huge impact on macro-economic activities mainly through affecting the investing behavior of human capital. Li [41] held that when an economy suffered from NDs, new equipment and technologies would be introduced and adopted, which might improve the efficiency of the laborers. Furthermore, Skidmore and Toya [17] believed that people might choose investment between human capital and physical capital in the endogenous growth framework. When people are threatened by NDs, they may reduce the investment in physical capital and opt to invest in human capital. The economy may allocate more resources to improve the human capital. Moreover, Bradley [25] believed that post-disaster reconstruction increased employment rate which would accelerate the accumulation of human capital. Accordingly, the third hypothesis is proposed.

H_3: The occurrence of NDs leads to a faster accumulation of human capital.

3. The Empirical Analysis

According to the number of disasters in the EM-DAT, we summarized the frequency and distribution of NDs in China. From 1985 to 2011, the average number of MDs is 17 each year. The lowest frequency is nine in 1989 and the highest is 30 in 2006. Compared with the number of MDs, the number of GDs is much less frequent. The average number of GDs is four every year. The lowest frequency is one in 1986 and 2007 and the highest is 11 in 2003. The remaining years have less than eight GDs.

All provinces in China suffer from MDs. The highest number of MDs is 95 in Guangdong and the least is three in Tianjin and Ningxia. Examining the most to least frequent occurrences revealed that the number of MDs is bigger than 30 in 11 provinces, including Guangdong, Sichuan, Hunan, Fujian, Zhejiang, Hubei, Jiangxi, Guangxi, Yunnan, Anhui, and Guizhou. Most of them are located in the central and southeast coastal regions of China. Among them, 21 provinces are located in the inland regions of China, and have suffered from GDs during the period from 1985–2011. These provinces are Yunnan (with a frequency of 38), Xinjiang (19), Sichuan (17), Gansu (8), Qinghai (5), Shanxi (5), Hebei (4), Inner Mongolia (4), Tibet (4), Guizhou (3), Jiangxi (3), Hubei (2), Guangdong (2), Liaoning (2), Ningxia (2), Fujian (1), Guangxi (1), Henan (1), Hunan (1), Shannxi (1), and Zhejiang (1).

Figure 1 shows the number of NDs over the period from 1985–2011 (We attempted to analyze the impacts of natural disasters on economic growth based on the data from 30 provinces during the period from 1985 to 2011. There are no statistically significant results for disaster variables and some control variables. In addition, due to the lack of data and changes of statistics caliber on relevant variables over the years, we are unable to collect comparable data before 1985). An upward shift in the frequency around 1998 is revealed. The average frequency of NDs from 1999–2011 is 1.62 times of that from 1985–1998. Since the catastrophic flood in 1998, the Chinese government has established observation networks for NDs by constructing more facilities for the disaster prevention and resilience, on the other hand, the institutional construction has been enforced by issuing laws and regulations on disaster prevention and damage reduction. The disaster prevention and damage reduction has stepped into a new stage since 1998 (Jiang [51]).Therefore, 1998 can be seen as a demarcation point for the impact of natural disasters on economic growth. To verify if 1998 is also a structural break point regarding the relationship between NDs and economic growth, we performed the Chow test to the data using the per capita GDP growth rate as the dependent variable and the frequency of NDs as the independent variable. At the 1% significance level, the null hypothesis of no breakpoint in 1998 is rejected. Therefore, the study period can be statistically divided into the first period (from 1985 to 1998) and the second period (from 1999 to 2011). Accordingly, we summarized the empirical evidence for these two sub-periods.

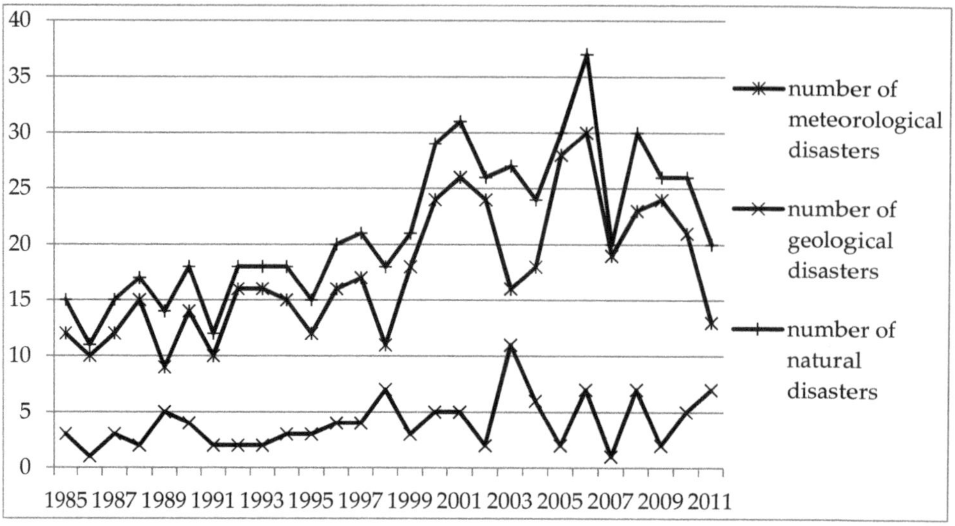

Figure 1. Number of disasters from 1985 to 2011.

To empirically investigate the relationship between NDs and economic growth, after reviewing the studies conducted by Skidmore and Toya [17], Wang [23], Kim [24], Gassebner *et al.* [13], Guo [52], and Liu *et al.* [53], a regression analysis based on production function was used with the data of different disasters and a number of economic variables. Let's us say that y is production, K is capital input, and H is labor input, then the expression of Cobb–Douglas production function is written as:

$$y = f(K, H) \tag{6}$$

According to the research of Skidmore and Toya [17], by bringing the endogeneity of NDs into model, we get:

$$y = f(K, H, ND) \tag{7}$$

where y is measured by per capita GDP growth rate, K is measured by investment growth rate, and H is measured by Education degree. After reviewing the studies of Barro and Lee [54], Guo [52], and Lin *et al.* [55], we chose control variables that include openness (total volume of foreign trade/GDP), SOE (the number of state-owned enterprises), and birth rate.

From the viewpoint of factor inputs, the role of investment in economic growth can be concluded in two aspects, the pulling effects of investment demand and the driving effects of investment supply. Extensive investment demands can produce positive pulling effects to the economic growth and social consumption, whereas adequate investment supply can inject new production factors directly to the social reproduction process, increase the supply of production materials, and provide the material conditions for expanded reproduction, which will directly promote the growth of the gross national product. A larger percentage of output used as investment will stimulate fast economic growth (Kim [24]) (this finding is consistent with the Say Law, *i.e.*, supply creates demand). The growth rate in fixed-asset investment is used as the control variable to measure investment.

Human capital is a key factor for economic growth (Lucas [26]). Generally, either the average education years based on expenditure approach or the average wage based on income approach is used for evaluating human capital. In this paper, the average education years is adopted and the formula of it is written as below (in this paper, the average education years is adopted for calculation instead of educational expenditure because the data of the former is easier to get (Wang [23]). Moreover, as the statistical caliber of average wage changed in 1998 (only includes the wages of on-post staffs), thus, it is quite hard to obtain the comparable data of average wage.):

$$E = \sum P_h \times N_h / \sum P_h, (h = 1, 2, \ldots, 5) \tag{8}$$

where P_h is the population of each education level, N_h is the number of years in each education level, and h is the education level comprising illiteracy, primary education, lower secondary education, upper secondary education, and tertiary education.

Openness is also a determinant of economic growth. Openness is an important channel to introduce advanced technological knowledge from abroad, which serves as a proxy for the local international trade level and the degree of competitiveness to a certain degree. Good economic openness can introduce advanced technology and equipment, which will greatly improve the labor productivity and save development costs. Consequently, economic growth is promoted. Openness can also provide new opportunities in relation to the overcapacity of domestic market and further stimulate domestic production capacity. Accordingly, the total volume of foreign trade/gross provincial product (GDP) is used as the control variable for openness.

With the efficiency loss and soft budget constraint, Kornai [56] argued that the government cannot bail out the loss-making state-owned enterprises, and state-owned enterprises (SOEs), which may disrupt the development of private enterprises and decelerate economic growth (Liu *et al.* [53]). Soft budget constraints lead to efficiency loss, which can further induce the creation of moral hazard of local governments and SOEs, leading to low efficiency, lack of innovation, and economic shortage (Lin *et al.* [55]). Soft budget constraints can also cause excessive bad debts, the portfolio of risk assets, and financial crises. The number of SOEs was used as the control variable in our calculations (due to the lack of data, we are unable to use data on production value of state-owned enterprises; nevertheless, to a certain degree, the number of state-owned enterprises reflects the condition of state-owned capital in the economic structure).

Birth rate was also used as a control variable. According to parents' utility functions with characteristics of altruism provided by Becker and Barro [57], high birth rates can raise discount rates of future consumption and reduce the degree of altruism of each child for parents, which will not be conducive to human capital investment on their children. A high birth rate will increase the burden on societies and families, reduce the level of investment per capita, and create a negative impact on education, and thus affect the rate of economic growth because of the large population in China (Guo [52]). To consider the induced effect of birth rate on the economy, we chose this factor as a control variable (in our analysis, we tried to include other variables such as foreign direct investment and household consumption; however, they were excluded in this paper because these variables suffered too many missing observations at the provincial level before 1997).

Unit root tests with ADF and PP methodologies were performed to avoid spurious regression. Table A2 shows that parts of some variables were rough at the initial level. After the first difference, they returned to smoothness.

Based on the test results in Table A2, we used the following panel model to examine the determinants of economic growth rate. After correlation analyses among economic variables, we found that only the education degree and birth rate were highly correlated (their correlation coefficient is –0.781857 (p = 0.0000)). To avoid the multi-collinearity, the birth rate is eliminated in Equations (9) and (10):

$$G_{it} = c_1 + c_2 ND_{it} + c_3 I_{it} + c_4 E_{it} + c_5 O_{it} + c_6 S_{it} + \varepsilon_{it}, \quad (i = 1,2,\ldots,30; t = 1,2,\ldots) \tag{9}$$

where G_{it} is the per capita GDP growth rate in region (i) in period (t) (the base year is1985), ND_{it} is the degree of NDs in region (i) in period (t), I_{it} is the investment (investment in the fixed assets) growth rate in region (i) in period (t) (the base year is 1985), E_{it} is the education degree in region (i) in period (t) in region (i) in period (t), O_{it} is the degree of openness (total volume of foreign trade/GDP) in region (i) in period (t) (O_{it} is rough in the second period, and then replaced with ΔO_{it}, the degree of openness at the first difference level), S_{it} is the number of SOEs (the number of state-owned and state-holding industrial enterprises) in region (i) in period (t), $c_z (z = 1,2,\ldots,6)$ is the regression coefficient, and ε_{it} is the error term.

Column 1 of Table 1 reports Panel Generalized Least Squares (GLS) results (the random-effect model was used in this paper, and we eliminated heterocedasticy with Generalized Least Squares (GLS) and autocorrelation with cluster robust standard error). The estimated coefficient of ND is statistically insignificant. Therefore, when MD and GD are grouped together, ND neither hinders nor promotes economic growth. The estimated coefficients of openness and SOE are statistically insignificant. Both education degree and investment has a positive effect on per capita GDP growth rate.

Table 1. Panel data regression results (Equations (9) and (10)).

Dependent Variable	Per Capita GDP Growth Rate			
Period	The First Period (1985–1998)		The Second Period (1999–2011)	
C	0.095202 *** (0.0000)	0.092598 *** (0.0000)	0.085283 *** (0.0000)	0.084266 *** (0.0000)
Degree of natural disasters *ND*	0.00000126 (0.8169)		0.00000136 (0.1481)	
Degree of meteorological disasters *MD*		0.000000837 (0.8791)		0.0000164 ** (0.0267)
Degree of geological disasters *GD*		0.000248 (0.1044)		0.00000142 (0.3085)
Growth rate of investment *I*	0.228303 *** (0.0000)	0.229056 *** (0.0000)	0.174843 *** (0.0000)	0.168970 *** (0.0000)
Education degree *E*	0.005103 * (0.0930)	0.005430 * (0.0740)	0.005968 ** (0.0114)	0.006027 ** (0.0102)
Openness *O*	−0.00000382 (0.2591)	−0.00000380 (0.2218)		
D (openness) ΔO			0.000135 *** (0.0005)	0.000132 *** (0.0007)
SOE *S*	−0.00000214 (0.2591)	−0.00000219 (0.2475)	−0.0000241 *** (0.0000)	−0.0000246 *** (0.0000)
R^2	0.311052	0.315356	0.261843	0.269959
Obs	420	420	360	360

Note: a. Probs are reported in brackets; b. The symbols *, **, and *** stand for 10%, 5%, and 1% significance, respectively.

Column 3 of Table 1 presents the GLS results in the second period. Similarly, the estimated coefficient of *ND* is statistically insignificant. All other control variables show the expected results. Education degree, investment and international trade promote economic growth, whereas SOE hinder economic growth.

To further examine the different impacts of different disasters on economic growth, we separated ND into MD and GD. Accordingly, Equation (9) is expressed as:

$$G_{it} = c_1 + c_2 MD_{it} + c_3 GD_{it} + c_4 I_{it} + c_5 E_{it} + c_6 O_{it} + c_7 S_{it} + \varepsilon_{it}, \quad (i = 1, 2, \ldots, 30; t = 1, 2, \ldots) \tag{10}$$

where MD_{it} and $GD_{it}GD_{it}$ are new independent variables, MD_{it} is the degree of *MD* in region (*i*) in period (*t*), and GD_{it} is the degree of GD in region (*i*) in period (*t*). Other variables are the same as those in Equation (6).

The results in Column 2 of Table 1 indicate the absence of significant evidence of any correlation between the disaster variables (*MD* and *GD*) and GDP growth. The results of other control variables are similar to those obtained by Equation (9). In the second period, the results from GLS are summarized in Column 4 of Table 1. *MD* has marginally positive significant estimated coefficient (at 5% level), whereas *GD* has no significant effect on per capita GDP growth rate. This finding suggests that MDs can marginally promote economic growth, whereas GDs can not. The results of other control variables are similar to those obtained by Equation (9).

A few possible reasons may help to explain why MDs marginally affected economic growth in the second period from 1999 to 2011 in stead of the first period from 1985 to 1998. First, the annual average frequency of MDs in the first period (13.21) was much smaller than that in the second period (21.85). Second, the volatility of MD frequency in the first period (sd = 2.66) was much lower than that that in the second period (sd = 4.90). Third, prevention and reconstruction were relatively weak in the first period. Since 1998, the Chinese government has systematically formulated and promulgated a series of laws, rules, and regulations, including *Disasters Mitigation Plan of the People's Republic of China (1998–2010)* (The State Council of the People's Republic of China [58]), *Earthquake Disaster Mitigation Law of the People's Republic of China* (The State Council of the People's Republic of China [59]) and *National Disaster Relief Emergency Plan* (China Flood Control and Drought Relief [60]).They help the Chinese government improve the modern system of disaster relief and prevention (Jiang [51]), which reduced the impact of disasters on economic growth.

We further investigated the mechanism of how ND affected economic growth in the second period. We analyzed the relationship between NDs and physical capital. We chose investment growth rate as the dependent variable to measure physical capital and employed the following panel model:

$$\begin{aligned} I_{it} &= c_1 + c_2 MD_{it} + c_3 GD_{it} + c_4 G_{it} + \varepsilon_{it}, \\ &(i = 1, 2, \ldots, 30; t = 1, 2, \ldots) \end{aligned} \tag{11}$$

The GLS results in Column 1 of Table 2 show that neither MDs nor GDs are significantly correlated with investment growth rate. Thus, MDs and GDs may not have affected economic growth through increasing physical capital investment.

Skidmore and Toya [17] reported that meteorological disasters may promote investment in human capital. Suppose a society can choose the level of investment for factors of production, human capital investment will be more attractive because society can invest more in human capital when physical capital is destroyed by MDs. Human capital investment mainly includes investment in education, health, training, and labor force flow. Human capital investment can improve the abilities of workers to use new technologies and equipment, which can enhance the output efficiency of physical capital, and thus promote economic growth. In return, economic development can help workers gain a larger human capital investment and further promote the accumulation of human capital. Therefore, human capital and economic growth are inseparable.

Table 2. Panel data regression results in the second period (1999–2011) (Equations (11) and (12)).

Dependent Variable	Growth Rate of Investment	Education Degree
C	0.086913 *** (0.0000)	9.743637 *** (0.0000)
Degree of meteorological disasters MD	0.00000811 (0.5846)	0.000177 * (0.0534)
Degree of geological disasters GD	−0.000000397 (0.9561)	−0.000000613 (0.6564)
Per capita GDP growth rate G	0.876742 *** (0.0000)	
Birth rate B		−0.148292 *** (0.0000)
D (income level) ΔIN		18.97086 *** (0.0001)
R^2	0.218076	0.235808
Obs	390	360

Note: a. Probs are reported in brackets; b. The symbols *, **, and *** stand for 10%, 5%, and 1% significance, respectively.

We used the degree of education to measure human capital because education is one of the main ways to form human capital. Human capital is greater when the levels of education of individual workers are higher. According to the household production function model, birth rate is negatively related to economic development. Higher birth rates will increase the burden on society and family, and thus affect the education level (Guo [52]). The income level is also a key factor because income is the foundation of education investment. Lu *et al.* [61] have pointed out that low income will lead to credit constraints, which will further reduce education levels. We used education degree as the dependent variable and birth rate and income level as control variables to examine the relationship between human capital and NDs.

$$E_{it} = c_1 + c_2 MD_{it} + c_3 GD_{it} + c_4 B_{it} + c_5 \Delta IN_{it} + \varepsilon_{it}, \quad (i = 1, 2, \ldots, 30; t = 1, 2, \ldots) \tag{12}$$

where ΔIN_{it} is the income level (general budget revenue/GDP) at the first difference level in region (i) in period.

The results from GLS are shown in column 2 of Table 2. Two conclusions can be derived from the results. First, our empirical results confirm the finding of Skidmore and Toya [17] which holds that MDs promote investment in human capital. Second, GDs have no significant relationship with education level. These two conclusions suggest that MDs promote economic growth not through increasing physical capital investment, but by improving human capital, whereas GDs do not promote economic growth because these events neither increase physical capital investment nor improve human capital. One possible reason is the much smaller number of GDs recorded during the study period. Most GDs occurred in less developed western parts of China. In less developed regions, the post-disaster relief tends to focus more on "disaster recovery" (such as repairing the houses and making arrangement for the victims' life) than on "economic development" (such as large-scale construction of road, post-disaster migration, educational training and further improvement on the system of disaster prevention and reduction). Therefore, reliefs received after disasters may have only been used for short-term recovery instead of long-term development.

To further verify the robustness of the estimates, this paper has, based on the experience of Skidmore and Toya [17], conducted regression estimation after including five control variables in Equations (10) and (12), respectively. The five variables are listed below: (1) Control variable: population density. Disasters with the same extent differ from each other in damages. The higher the population density, the more serious the damages (Ma [62]). (2) Dummy variable: whether the province is along the Yangtze River Basin or the Yellow River Basin (yes = 1; no = 0). According to the data from EM-DAT between 1985 and 2011, flood is the most frequent and serious disaster both in the Yangtze River Basin and the Yellow River Basin (Hong [63]). (3) Dummy variable: whether the province is along coastal areas (yes = 1, no = 0). Compared with inland areas, the economy in coastal areas is more prosperous and the people here have a better educational background (Wei [64]). (4) Control variable: longitude and latitude. The frequency of disasters may also be partly determined by geographical factors (Skidmore and Toya [17]). As shown in Tables 3 and 4 after these five control variables are included, the degree of meteorological disasters *MD* and the degree of geological disasters *GD* in Equation (10) and Equation (12) remain the same. In addition, to reduce possible endogeneity of the disaster data with respect to economic performance (growth), according to Coffman and Noy(2011) [4], and Noy *et al.* (2007) [42], setting different lagged dependent variables will eliminate endogeneity of the disaster data with respect to economic performance (growth), and the econometric analysis carried out by us also shows that the results have no significantly changes.

Table 3. Per capita GDP growth rate and Disasters: Robustness Tests in the second period (1999–2011).

Variable	Per Capita GDP Growth Rate				
Degree of meteorological disasters *MD*	0.0000157 ** (0.0285)	0.0000173 ** (0.0196)	0.0000130 * (0.0780)	0.0000172 ** (0.0270)	0.0000174 ** (0.0161)
Degree of geological disasters *GD*	0.00000138 (0.23212)	0.00000130 (0.3529)	0.00000125 (0.3650)	0.00000141 (0.3121)	0.00000130 (0.2565)
Population density	−0.0000224 *** (0.0003)				
Coasts		−0.010654 * (0.0658)			
Rivers			0.016849 *** (0.0031)		
Latitude				0.000135 (0.7627)	
Longitude					−0.000548 * (0.0848)
Obs	360	360	360	360	360
R^2	0.298119	0.277010	0.287935	0.270151	0.275212

Note: The symbols *, **, and *** stand for 10%, 5%, and 1% significance, respectively. Other explanatory variables used in Table 1 are included but not reported here.

Table 4. Education degree and Disasters: Robustness Tests in the second period (1999–2011).

Variable	Education Degree				
Degree of meteorological disasters *MD*	0.000183 ** (0.046)	0.000177 * (0.0562)	0.000178 * (0.0534)	0.000183 ** (0.0481)	0.000177 * (0.0535)
Degree of geological disasters *GD*	−0.00000022 (0.8752)	−0.00017700 (0.732)	−0.00000056 (0.6868)	−0.00000058 (0.6754)	−0.00000034 (0.8092)
Population density	0.00128 *** (0.0016)				
Coasts		0.449103 * (0.0948)			
Rivers			−0.131637 (0.6421)		
Latitude				0.033031 (0.1438)	
Longitude					0.034393 (0.2207)
Obs	360	360	360	360	360
R^2	0.324237	0.240256	0.235699	0.239488	0.244169

Note: The symbols *, **, and *** stand for 10%, 5%, and 1% significance, respectively. Other explanatory variables used in Table 2 are included but not reported here.

To summarize, the regression analysis with the control variables reveals robust correlations between per capita GDP growth rate and disasters, and education degree and disasters. From another perspective, to check the robustness of our empirical findings and further explore the relationship between MDs and economic growth in the second period, we examined the causal relationship among variables using the Granger causality test. We used the Akaike Info Criterion (AIC) to determine the optimal lag. Table 5 shows the results of the Granger causality test.

Table 5. Granger test in the second period (1999–2011).

Null Hypothesis:	Lags	Obs	*F*-Statistic	Prob.	Result
Meteorological disasters are not Granger cause of GDP growth rate	6	210	2.46543	0.0254 **	Reject the null hypothesis
Education is not Granger cause of GDP growth rate	6	210	4.06220	0.0007 ***	Reject the null hypothesis
GDP growth rate is not Granger cause of Education			11.7246	0.0000 ***	Reject the null hypothesis
Meteorological disasters are not Granger cause of education	4	270	2.20990	0.0683 *	Reject the null hypothesis

Note: The symbols *, **, and *** stand for 10%, 5%, and 1% significance, respectively.

From the above, the results demonstrate that MDs influence education and GDP growth rate, and education level and GDP growth rate influence each other. These regression analyses confirm that MDs influence economic growth mainly through human capital investments.

4. Conclusions and Discussion

Using a novel synthesized disaster indicator and the panel data of 30 Chinese provinces, this paper examined the impacts of NDs on China's economic growth during two periods. The empirical results show that in the first period, neither MDs nor GDs have significant relationship with GDP growth rate. In the second period, MDs have a marginally positive and causal relationship with GDP growth rate through human capital investment rather than physical capital investment. GDs have no impact on GDP growth rate.

Why could MDs promote economic growth through the accumulation of human capital while GDs are not related to economic growth since 1999? There are two possible explanations. First, the Chinese government places an increasing emphasis on the physical infrastructure, laws and regulations and public education of disaster prevention and reduction since 1999. The comprehensive observation system with a combination of ground-based, air-based and space-based observation has been established, which has greatly improved the accuracy and quality of predication and forecast of MDs. Since 1999, the Chinese government has successively promulgated about 40 laws, regulations and emergency response plans, such as "Disasters Mitigation Plan of the People's Republic of China" (1998–2010), *etc.* (Zhang *et al.* [65]), and carried out multi-level education of disaster prevention and reduction in different units, schools and communities. The public awareness of disaster prevention and reduction has been enhanced unceasingly and the quality of human capital has been improved accordingly. Second, MDs often occurred in richer coastal areas where the ocean–atmosphere and land–atmosphere interactions are relatively intensive. When they occurred, local governments put great emphasis on the post-disaster construction of infrastructures, raising public awareness of disaster prevention and reduction, and improving human capital investment. All would promote long-term economic growth.

The current economic, social, and cultural situations of China may support speedy recovery after disasters through physical capital in the short term. However, accumulating human capital in the long term is the vital key to the adoption of new technology and improvement of productivity toward economic recovery and sustained economic development after disasters. The counterpart-support policies with Chinese characteristics can ensure the speed and scale of the reconstruction work after disasters. For example, 4121 counterpart-support projects from 19 provinces with RMB 84.38 billion financial support were implemented after the Wenchuan earthquake in 2008. The funds from counterpart-support together with central government funds were invested mainly in the physical capital, which promotes economic growth. Although the current system ensures a speedy recovery, we have found that maintaining long-term sustainable economic growth is a challenge. Thus, we propose that the local officials or decision makers should not only pay attention to the immediate recovery of the economy, but also to the improvement of the investments in human capital and the quality of human resources. Shang [66] has concluded that the lack of knowledge of farmers to counter drought following disasters increases the damage of drought. Zhao [67] has determined that the education levels of family members have a significant influence on the improvement of the economy after a disaster in the long term. Reducing the vulnerability of future disasters, human factor plays an essential role in the recovery and the long-term economic development after each natural disaster.

Three caveats need to be mentioned. First, the EM-DAT does not have complete economic damage data. Tremendous effort was exerted to calculate or estimate some of the missing data to carry out the study. Second, some EM-DAT data and economic statistics data have different scopes in relation to the spatial and temporal dimensions. The duration of most NDs, such as earthquakes, is short, while the time span of economic statistics data is long. Moreover, NDs often occur in one location, whereas the related reconstruction involves the support of other provinces. Crowding-out may occur. Therefore, obtaining the uniform standard data is difficult. Finally, death toll and economic losses are related to the income level. With advanced medical treatment, strict building laws, and other safety rules, the impact of NDs may be reduced in the developed regions. Thus, future research should search for a better indicator that can reflect both income levels and the actual level of disasters.

Acknowledgments: This research was supported by: The Natural Science Foundation of China (71373131, 71140014, 11371292, and 91546117); National Social and Scientific Fund Program (14BJY043 and 15BTJ019); National Soft Scientific Fund Program (2011GXQ4B025); National Industry-specific Topics (GYHY200806017 and GYHY201506051); Scientific Research Foundation for the Returned Overseas Chinese Scholars of MOE (No.2013-693, Ji Guo); and Tianjin Higher Education Innovation Team Program (TD12-5063). This research was also supported by the Priority Academic Program Development of Jiangsu Higher Education Institutions and the Flagship Major Development of Jiangsu Higher Education Institutions. Sincere thanks should be given to the editor, three anonymous reviewers and Yehua Dennis Wei for their constructive comments and suggestions that greatly helped to improve the quality of this article.

Author Contributions: Ji Guo and Xianhua Wu originated the study and led the writing of the article. Hui Liu and Jiong Gu collected data and conducted the qualitative analysis. Shunfeng Song and Yinshan Tang analyzed data and contributed to the writing of the article.

Conflicts of Interest: The authors declare no conflict of interest.

Appendix

The data on the Wenchuan earthquake in EM-DAT are listed in Table A1.

Table A1. Decomposed damage results of Wenchuan earthquake on 12 May 2008.

Location	EM-DAT		Previous Literature (Yuan, 2008 [68])		Computed Value	
	Death Toll	Economic Losses (1000 US$)	Death Toll	Economic Losses (100,000,000 Yuan)	Death Toll	Economic Losses (1000 US$)
Sichuan(Chongqing)			68678	6231.52	86,854	765,512.0186
Gansu			365	442.8	462	54,395.83309
Shaanxi			122	228.14	154	28,025.89287
Yunnan			1	16.82	1	2066.255449
Shanxi	87,476	850,000	0	0	0	0
Guizhou			0	0	0	0
Hubei			1	0	1	0
Hunan			1	0	1	0
Henan			2	0	3	0

According to the data on the earthquake from relevant literature, the loss of each province is calculated through the following formula:

$$L_{cij} = \frac{L_{pij}}{\sum_{i=1}^{30} L_{pij}} \times L_{Ej}, (i = 1, 2, ..., 30; j = 1, 2) \tag{13}$$

where i refers to 30 different provinces; j refers to the death toll and economic loss; L_{cij} refers to the computed value of j of Province i; L_{pij} refers to the value of j of Province i retrieved from previous literature; and L_{Ej} refers to the total value of j of all provinces recorded in EM-DAT.

Table A2 contains all variables used in Equations (6)–(9).

Table A2. Unit root tests.

Variable	Test for unit root in	Method	in the First Period (1985–1998)		in the Second Period (1999–2011)	
			Statistic (Prob.)	Result	Statistic (Prob.)	Result
Per capita GDP growth rate G	Level	ADF-Choi Z-stat	−5.30819 (0.0000)	Smooth	−2.20509 (0.0000)	Smooth
Degree of natural disasters ND	Level	ADF-Choi Z-stat	−6.29742 (0.0000)	Smooth	-4.97697 (0.0000)	Smooth
Degree of meteorological disasters MD	Level	ADF-Choi Z-stat	−6.30789 (0.0000)	Smooth	−5.04097 (0.0008)	Smooth
Degree of geological disasters GD	Level	ADF-Choi Z-stat	−6.57734 (0.0000)	Smooth	−5.64198 (0.0000)	Smooth
Investment growth rate I	Level	PP-Choi Z-stat	−8.68497 (0.0000)	Smooth	−3.31127 (0.0005)	Smooth
Openness O	Level	ADF-Choi Z-stat	−4.35787 (0.0000)	Smooth	1.83879 (0.8520)	Rough
D (Openness) ΔO	1st Difference	ADF-Choi Z-stat			−6.81695 (0.0000)	Smooth
SOE S	Level	ADF-Choi Z-stat	−4.12426 (0.0000)	Smooth	−11.9920 (0.0000)	Smooth
Birth rate B	Level	ADF-Choi Z-stat	−4.19320 (0.0000)	Smooth	−2.15761 (0.0155)	Smooth
Income level IN	Level	ADF-Choi Z-stat		Smooth	9.89938 (1.0000)	Rough
D (Income level) ΔIN	1st Difference	ADF-Choi Z-stat		Smooth	-3.00439 (0.0013)	Smooth
Degree of education E	Level	PP-Choi Z-stat		Smooth	-5.32522 (0.0000)	Smooth

References and Notes

1. Benson, C.; Clay, E. *The Impact of Drought on Sub-Saharan African Economies: A Preliminary Examination, Geneva and New York*; World Bank Technical Paper; The World Bank: Washington, DC, USA, 1998.
2. Xiao, Y. Local economic impacts of natural disasters. *J. Reg. Sci.* **2011**, *51*, 804–820.
3. Gall, M.; Borden, K.A.; Emrich, C.T.; Cutter, S.L. The unsustainable trend of natural hazard losses in the United States. *Sustainability* **2011**, *3*, 2157–2181.

4. Coffman, M.; Noy, I. Hurricane Iniki: Measuring the long-term economic impact of a natural disaster using synthetic control. *Environ. Dev. Econ.* **2011**, *17*, 187–205.
5. Okuyama, Y. Disaster and economic structural change: Case study on the 1995 Kobe earthquake. *Econ. Syst. Resear.* **2014**, *26*, 98–117.
6. Dercon, S. Growth and shocks: Evidence from rural Ethiopia. *J. Dev. Econ.* **2004**, *74*, 309–329.
7. Mechler, R.; Hochrainer, S.; Asjorn, A. A risk management approach for assessing adaptation to changing flood and drought risks in Europe. In *Making Climate Change Work for Us: European Perspectives on Adaptation and Mitigation Strategies*; Hulme, M., Neufeld, H., Eds.; Cambridge University Press: Cambridge, UK, 2009; pp. 200–229.
8. Raddatz, C. *The Wrath of God: Macroeconomic Costs of Natural Disasters*; World Bank Policy Research Working Paper No. 5039; Central Bank of Chile: Santiago, Chile; The World Bank: Washington, DC, USA, 2009.
9. Arndt, C.; Tarp, F.; Thurlow, J. The economic costs of climate change: A multi-sector impact assessment for Vietnam. *Sustainability* **2015**, *7*, 4131–4145.
10. Arndt, C.; Asante, F.; Thurlow, J. Implications of climate change for Ghana's economy. *Sustainability* **2015**, *7*, 7214–7231.
11. Porfiriev, B. Economic issues of disaster and disaster risk reduction policies: International *vs.* Russian perspectives. *Int. J. Disaster Risk Reduct.* **2012**, *1*, 55–61.
12. Hallegatte, S.; Dumas, P. Can natural disasters have positive consequences? Investigating the role of embodied technical change. *Ecol. Econ.* **2009**, *68*, 777–786.
13. Gassebner, M.; Keck, A.; Shaken, R. Not stirred: The impact of disasters on international trade. *Rev. Int. Econ.* **2010**, *18*, 351–368.
14. Strobl, E. The economic growth impact of natural disasters in developing countries: Evidence from hurricane strikes in the central American and Caribbean regions. *J. Dev. Econ.* **2012**, *97*, 130–141.
15. Rodriguez-Oreggia, E.; de la Fuente, A.; de la Torre, R.; Moreno, H.A. Natural disasters, human development and poverty at the municipal level in Mexico. *J. Dev. Studies* **2013**, *3*, 442–455.
16. Cuaresma, J.C.; Hlouskova, J.; Obersteiner, M. Natural disasters as creative destruction? *Evidence from developing countries. Econ. Inquiry* **2008**, *46*, 214–226.
17. Skidmore, M.; Toya, H. Do natural disasters promote long-run growth? *Econ. Inquiry* **2002**, *40*, 664–687.
18. Albala-Bertrand, J.M. *Political Economy of Large Natural Disasters: With Special Reference to Developing Countries*; Oxford University Press: Oxford, UK, 1993.
19. Albala-Bertrand, J.M. *Disasters and the Networked Economy*; Routledge: Oxford, UK, 2013.
20. Dacy, D.C.; Kunreuther, H. *The Economics of Natural Disasters-Implications for Federal Policy*; The Free Press: New York, NY, USA, 1969.

21. Otero, R.; Marti, R. The impacts of natural disasters on national economies and the implications for the international development and disaster community. In *Disaster Prevention for Sustainable Development: Economic and Policy Issues*; Munasinghe, M., Clarke, C., Eds.; The International Bank for Reconstruction and Development: Geneva, Switzerland; The World Bank: Washington, DC, USA, 1995; pp. 11–40.
22. Tol, R.; Leek, F. Economic analysis of natural disasters. In *Climate, Change and Risk*; Downing, T.E., Olsthoorn, A.J., Tol, R.S., Eds.; Routledge: London, UK, 1999.
23. Wang, Y.M.; Chen, M.L.; Wang, X. The impacts of natural disasters on long-term growth: Theoretical and empirical study. *Econ. Manage.* **2008**, *19*, 144–150.
24. Kim, C. The effects of natural disasters on long-run economic growth. *Michi. J. Bus.* **2010**, *41*, 15–49.
25. Bradley, T.E.; Jamie, B.K.; Mark, A.T. Twister! Employment responses to the 3 May 1999 Oklahoma city tornado. *Appl. Econ.* **2009**, *41*, 691–702.
26. Lucas, R. On the mechanics of economic development. *J. Monetary Econ.* **1988**, *22*, 3–42.
27. Okuyama, Y. Economics of natural disasters: A critical review. *Resear. Paper* **2003**, *12*, 20–22.
28. Okuyama, Y.; Hewings, G.J.D.; Sonis, M. Measuring economic impacts of disasters: Interregional input-output analysis using sequential interindustry model. In *Modeling the Spatial and Economic Impacts of Disasters*; Okuyama, Y., Chang, S.E., Eds.; Springer: New York, NY, USA, 2004; pp. 77–101.
29. Horwich, G. Economic lessons of the Kobe earthquake. *Econ. Dev. Cult. Change* **2000**, *48*, 521–542.
30. Zou, X.H. Research on the Influence of Natural Disasters on Economic Growth in China. Master's Thesis, University of Xiamen, Xiamen, China, 2009.
31. Leiter, A.M.; Oberhofer, H.; Raschky, P.A. Creative disasters? Flooding effects on capital, labor and productivity within European firms. *Environ. Resour. Econ.* **2009**, *43*, 333–350.
32. Loayza, V.N.; Olaberria, E.; Rigolini, J. Natural disasters and growth: Going beyond the averages. *World Dev.* **2012**, *40*, 1317–1336.
33. Shabnam, N. Natural disasters and economic growth: A review. *Int. J. Disaster Risk Sci.* **2014**, *5*, 157–163.
34. Guha-Sapir, D.; Below, R. *The Quality and Accuracy of Disaster Data: A Comparative Analyses of Three Global Data Sets, Technical Document*; The ProVention Consortium: Geneva, Switzerland; Disaster Management Facility: Washongton, DC, USA; The World Bank: Washongton, DC, USA; WHO Centre for Research on the Epidemiology of Disasters: Brussels, Belgium; University of Louvain School of Medicine: Brussels, Belgium, 2002.
35. Wen, K.G. *China Meteorological Disasters Book*, 1st ed.; China Meteorological Press: Beijing, China, 2008.
36. Disaster Information Department of China national Disaster Reduction Center. *The National Disaster Monthly Report 2008–2011*; China Disaster Reduction: Beijing, China, 2008–2011.

37. The State Flood Control and Drought Relief Headquarters Office. *The National Flood Disasters 2001–2011*; China Flood Control and Drought Relief: Beijing, China, 2001–2011.
38. China's National Bureau of Statistics. *China Statistical Yearbook 1986–2012*; China Statistics Press: Beijing, China, 1986–2012.
39. The National Economy Comprehensive Statistics Division of The National Bureau of Statistics. *China Compendium of Statistics 1949–2008*; China Statistics Press: Beijing, China, 2009.
40. Kellenberg, D.K.; Mobarak, A.M. Does rising income increase or decrease damage risk from natural disasters? *J. Urban Econ.* **2008**, *63*, 788–802.
41. Li, H. An empirical study on relationship between natural disasters and economic growth—based on the 14 countries panel data in 1970–2008. *Prices Mon.* **2011**, *1*, 47–51.
42. Noy, I.; Nualsri, A. *What do Exogenous Shocks Tell Us about Growth Theories*; SCCIE Working Paper Series; Santa Cruz Center for International Economics: Santa Cruz, CA, USA, 2007.
43. Zhao, A.X.; Ma, Z.J. Appraising study for the loss evaluation system of natural disasters. *J. Nat. Disasters* **1993**, *3*, 1–7.
44. Tang, W.B.; Han, Z.J. A comprehensive evaluation method of financial position based on information entropy. *J. Nanjing Univ. Sci. Tech.* **2001**, *25*, 650–653.
45. Zhang, S.; Zhang, M.; Chi, G.T. The science and technology evaluation model based on entropy weight and empirical research during the 10th Five-Year of China. *Chinese J. Manag.* **2010**, *7*, 34–42.
46. Burg, J.P. Maximum Entropy Analysis. Ph.D. Thesis, Stanford University, Stanford, CA, USA, 1975.
47. Darbellay, G.A.; Wuertz, D. The entropy as a tool for analysing statistical dependences in financial time series. *Phys. A: Sta. Mech. Its Appl.* **2000**, *287*, 429–439.
48. Li, L.; Liu, F.; Li, C. Customer satisfaction evaluation method for customized product development using entropy weight and analytic hierarchy process original research article. *Comput. Ind. Eng.* **2014**, *77*, 80–87.
49. Kunreuther, H.; Kleffner, A. E. Should earthquake mitigation measures be voluntary or required? *J. Regul. Econ.* **1992**, *4*, 321–333.
50. Wisner, B. Risk and the neoliberal state: Why post-mitch lessons didn't reduce EI Salvador's earthquake losses. *Disaster* **2001**, *25*, 251–268.
51. Jiang, J.W. Disaster Relief and Reduction Work in China since 1978. Ph.D. Thesis, Central Party School of China, Beijing, China, 2009.
52. Guo, J. Human capital, the birth rate and the narrowing of the urban-rural income gap. *Soc. Sci. China* **2005**, *3*, 27–38.
53. Liu, R.M.; Shi, L. The dual efficiency loss of state-owned enterprises and economic growth. *Econ. Resear. J.* **2010**, *1*, 127–137.
54. Barro, R.; Lee, J. *Data Set for a Panel of 138 Countries*; Harvard University: Cambridge, MA, USA, 1994.
55. Lin, Y.F.; Li, Z.Y. Policy burden, moral hazard and soft budget constraint. *Econ. Resear. J.* **2004**, *2*, 17–27.
56. Kornai, J. *Economics of Shortage*; North-Holland: Amsterdam, The Netherlands, 1980.

57. Becker, G.S.; Barro, R.J. A reformulation of the economic theory of fertility. *Quart. J. Econ.* **1988**, *103*, 1–25.
58. Notice by the PRC State Council on "Disasters Mitigation Plan of the People's Republic of China (1998–2010)". The State Council Bulletin of the People's Republic of China, 1998, 12, 526–539.
59. Earthquake Disaster Mitigation Law of the People's Republic of China. The State Council Bulletin of the People's Republic of China, 1997, 39, 1680–1688.
60. National Disaster Relief Emergency Plan. China Flood Control and Drought Relief, 2006, 01, 29–34.
61. Lu, M.; Chen, Z.; Wan, G. Equality for the sake of growth: The nexus of inequality, investment, education and growth in China. *Econ. Resear. J.* **2005**, *12*, 4–14.
62. Gao, W.X.; Gao, Q.H.; Ma, Z.J. Mitigating natural disasters will be an industries to develop in the 21st century. *Optim. Cap. Constr.* **1998**, *19*, 5–7.
63. Hong, W.T. Study on flood Disasters Risk Management System in China. Master's Thesis, Wuhan University, Wuhan, China, 2012.
64. Wei, Q.; Wang, A.; Wang, J. The Rise of China's coastal areas: Power of market. *Econ. Resear.* **2014**, *8*, 170–183.
65. Zhang, P.; Li, N.; Fan, B.H.; Liu, X.Q.; Wen, Y.T. The evolution of quantity and enact time of Disaster laws and Regulations in Recent 30 years in China. *Disaster* **2011**, *3*, 109–114.
66. Shang, Y.R. The analysis of drought, agricultural drought disaster and the farmhouses' vulnerability: Taking the typical farmhouse of Xingtai county as an example. *J. Nat. Disasters* **2000**, *9*, 55–61.
67. Zhao, Y.D. Social capital and post-disaster recovery: A sociological study of natural disaster. *Soc. Studies* **2007**, *5*, 164–187.
68. Yuan, F. Loss assessment of Wenchuan earthquake. *J. Earthq. Eng. Eng. Vib.* **2008**, *5*, 10–19.

Do Urban Rail Transit Facilities Affect Housing Prices? Evidence from China

Xu Zhang, Xiaoxing Liu, Jianqin Hang, Dengbao Yao and Guangping Shi

Abstract: Urban rail transit facilities play a critical role in citizen's social activities (e.g., residence, work and education). Using panel data on housing prices and urban rail transit facilities for 35 Chinese cities for 2002 to 2013, this study constructs a panel data model to evaluate the effect of rail transit facilities on housing prices quantitatively. A correlation test reveals significant correlations between housing prices and rail transit facilities. Empirical results demonstrate that rail transit facilities can markedly elevate real estate prices. Quantitatively, a 1% increase in rail transit mileage improves housing prices by 0.0233%. The results highlight the importance of other factors (e.g., per capita GDP, land price, investment in real estate and population density) in determining housing prices. We also assess the effects of expectations of new rail transit lines on housing prices, and the results show that expectation effects are insignificant. These findings encourage Chinese policy makers to take rail transit facilities into account in achieving sustainable development of real estate markets.

Reprinted from *Sustainability*. Cite as: Zhang, X.; Liu, X.; Hang, J.; Yao, D.; Shi, G. Do Urban Rail Transit Facilities Affect Housing Prices? Evidence from China. *Sustainability* **2016**, *8*, 380.

1. Introduction

In recent decades, residential conditions have been a critical factor of the formation of sustainable cities [1]. Vast price bubbles in real estate markets increase the risk of bubbles bursting, which may damage the sustainable urban development. Meanwhile, urban rail transit facilities play an important role in alleviating traffic pressures [2] and promoting sustainable urban development [3], which may expand city boundaries. Accordingly, it is of enormous interest to ask whether urban rapid rail transit affects housing prices.

Starting in the early 2000s, a housing boom coupled with strong economic growth emerged in China. However, affected by the global financial crisis of 2008, the Chinese housing market dipped into a short recession. Following the institution of a series of economic stimuli (e.g., expansionary fiscal and monetary policies), China's real estate market quickly recovered. Recently, whether a bubble in China's housing market exists has been discussed by Shen [4], Ren *et al.* [5], and Feng *et al.* [6]. Economic fundamentals [7,8], regional productivity [9], monetary [10], demographic [11,12], and behavioral factors [13] are recognized as major determinants of housing values. More recently, impacts of transit factors on real estate values have attracted broad interest from transportation researchers and policymakers around the world.

At present, however, whether urban rail transit plays a critical role in determining housing values still remains controversial, though some case studies [14] provide supporting evidence. Zhang *et al.* [14] examine the capitalization of transit access in Beijing and find that the premium of metro rail transit is 248.31 yuan/m^2 for every 100 m closer to a metro station. Similar results are presented by Hewitt and Hewitt [15], Jun [16], Kim and Lahr [17] and Chatman *et al.* [18]. However, Seo *et al.* [19] claim that proximity to a light rail transit station has both positive (*i.e.*, accessibility to a rail transit station) and negative effects (*i.e.*, air pollution and noise).

Urban rail transit in China has entered a phase of rapid expansion since the start of this century, during which the real estate market has experienced a golden era. Although some case studies provide supporting evidence for impacts of urban rail transit on housing prices, this issue has not been researched adequately, especially for China, where urban rail transit facilities and economic growth have boomed over the past decades. As of today, however, little panel data evidence exists on the relationship between rail transit facilities and housing prices at the city level. In this paper, we propose a theoretical framework and employ a panel model to examine the impacts of urban rail transit on housing prices. Our work contributes to the existing literature in several ways. First, we collect annual rail transit mileage records for operations of each sampled city, and we conduct a correlational analysis of rail transit facilities and housing prices in China. Second, we extend existing research on this relationship using panel data at the city level (Rail transit mileage refers to the length of rail transit that has been in service.). More specifically, we employ a panel model to assess the significance of the effects of rail transit on real estate prices. Third, we study the influence of rail transit expectations on housing booms. Fourth, we compare the effects of various factors (e.g., economic fundamentals, demographic factors, educational resources and urban rapid transit) on housing prices.

The remainder of the paper is organized as follows. Section 2 reviews existing literature related to the determinants of urban housing prices. Section 3 presents a brief description of the development of urban rail transit systems in China and a correlational analysis of housing prices and rail transit facilities. Section 4 describes model and data used. Section 5 estimates the panel data model and discusses the results. Section 6 provides the study's conclusions and implications.

2. Literature Review

Our study touches on two strands of literature: micro- and macro-level studies on determinants of housing prices. Micro-level studies are typically conducted by case studies, which are usually based on a hedonic house modelling and estimated by cross-sectional micro-level data. In contrast, macro-level studies, in most cases, focus on economic determinants of average urban housing prices, which are usually conducted by panel data model or time series model.

2.1. Micro-Level Studies on Housing Prices

In regard to the first strand of literature, a number of studies have focused on whether and how transportation, amenity, education, building structure and distance to the central business district (CBD) affect housing markets. For the literature on transportation, some studies assess the effects of subway, light rail transit and other transportation facilities on housing prices. For example, Andersson *et al.* [20] assess the implicit housing price of a new high-speed railway line (HSR) that connects seven metropolitan areas in Taiwan using a hedonic price function. Their results show that HSR accessibility has at most a minor effect on house prices. Zhang *et al.* [14] study the capitalization of transit access in Beijing. In their study, they divide urban mass transit into three categories: bus rail transit (BRT), light rail transit (LRT) and metro rail transit (MRT). These researchers' results demonstrate that MRT has the highest premium on housing prices, that LRT has a moderate premium, and that BRT has the lowest premium. Similar studies include Lewis–Workman and Brod [21], Zheng and Liu [22], Zhang *et al.* [23], Zhang [24], Hewitt and Hewitt [15] and Kim and Lahr [17].

Using an approach similar to De Vany [25], Seo *et al.* [19] construct a theoretical model on how amenity and disamenity should decay differently from links and nodes of rail and road networks. The authors thus examine the effects of nodes and links on housing prices. Their study shows that proximity to transport nodes is significantly and positively associated with housing prices. As a function of distance from highway exits and LRT stations, distance-band coefficients are found to form an inverted-U pattern consistent with positive longer-range distance-decay accessibility effects minus smaller, shorter range distance-decay disamenity effects.

For the case of China, educational resources, among the various factors that influence real estate prices, are a key consideration for home buyers, which differs from the case of other countries due to the "school district" policy for primary school students (In China, primary school students are limited to attend school near their residence, and the district owns high-quality primary schools known as a "school district house"). As a result, "school district house" has become a focal social issue and has conjured heated discussions between economics and sociological scholars. Wen and Jia [26] construct a hedonic model that included 15 housing feature variables. Their results show that the proxy variables of school and kindergarten have insignificant effects on housing prices. Feng and Lu [27] conduct a case study on Shanghai using panel data on 52 regional distributions of high schools. Their findings provide sufficient evidence that housing prices to some extent depend on one's distance from a high school and on the quality of nearby schools. Wen *et al.* [28] conduct a more detailed case study on Hangzhou, China and find that educational facilities have positive capitalization effects on housing prices. In addition, some studies have revealed significant links between housing prices and structural characteristics of real estate markets (e.g., vacancy rates, construction permits, and building ages) [3,28,29].

2.2. Macro-Level Studies on Housing Prices

Macro-level studies on determinants of housing prices are quite extensive. According to the existing literature, the major macro determinants of housing price can be divided into two categories: economic fundamentals and demographic factors. Related empirical studies initially focused on economic factors. Fortura and Kushner [30] find a significant positive correlation between income and housing prices. Quigley [29] assesses the effects of per capita income, employment, and other variables on housing prices. The model shows that these factors can explain 10%–40% of housing price variations. The model also suggests that the lagged housing price can be served as an important factor in determining housing price. Zou and Chau [31] claim that the consumer price index (CPI) is a long-term determinant of house prices. At the national level, business cycles, industrial production levels and employment rates are also the main factors which contribute to the change in housing prices [32–34]. More recently, monetary variables such as monetary supplies and interest rates have been found to be associated with housing prices [10,35].

As a component of housing costs, land prices are considered to be significantly associated with housing prices. Yang [36] and Bao [37] claim that the rise in land prices has increased housing prices. Wen and Goodman [38] construct a simultaneous-equations model to examine the interplay between housing and land prices. Their model was estimated using sample data on 21 provincial cities in China from 2000 to 2005, and their results show that housing and land prices are endogenously interrelated. Du *et al.* [39] employ a Granger causality approach to test this relation. Their results suggest the presence of unidirectional Granger causality between housing and land prices.

Demographic factors are recognized as major determinants of urban housing prices [11,40,41]. Parker [11] and Jud and Winkler [40] claim that population growth is significantly correlated with the housing market. Takáts [41] notes that that demographic dividend serves as an important driving force behind housing prices. Zhang *et al.* [12] use a panel vector autoregression (PVAR) model to assess direct and indirect mechanisms of the demographic dividend that cause real estate price escalation. Their results show that the demographic dividend can increase housing prices though both direct and indirect channels.

In summary, we can examine determinants of housing prices from two perspectives: micro and macro perspectives. Micro-level studies focus on determinants of housing prices in a certain community or a district, while macro-level studies focus on determinants of average housing prices in the city or province levels. From both micro- and macro-level studies, we can conclude that factors that affect housing prices are quite extensive and micro-level factors such as building structure, distance to CBD, *etc.*, are less important in macro-level studies. Hence, it is necessary for us to control as many factors as possible to assess the effect of rail transit on housing prices. However, most of the existing studies on the effect of transportation on housing prices have been conducted by case study on the base of micro-level data. To our knowledge, no studies have based on panel data at the city level. We thus conduct a study using city-level panel data and select an available proxy for rail transit facilities. To accurately assess impacts of urban rail transit services on housing prices, we consider possible factors based on previous studies. More specifically, we use selected factors as "control" variables in our panel data model.

3. Urban Rapid Rail Transit in China

3.1. Rapid Rail Transit Development in China

Beijing is home to the oldest metro rail transit system in China (established in 1969). It was initially developed as the capital's defence system for transporting troops when necessary. Management rights over the line were then transferred to a local transit company in 1977 and thereafter for public transit purposes. In China, rail transit line development has always constituted a major objective for city administrators. Through Chinese rail transit construction history; however, two setbacks occurred. Due to economic conditions and construction techniques employed at the time, the State Council suspended the approval of new rapid rail transit projects by publicizing the Notice of suspension for the approval of rapid rail transit projects in 1995. Prior to October of 2002, a number of cities published rail transit network plans and prepared for urban rail construction. To curb this new wave of rail transit construction, the State Council decided to freeze metro projects in certain cities (Hangzhou, Qingdao, and Xi'an, among others).

After improvements were made to the rail transit project approval process, the urban rail transit system in China entered a phase of rapid expansion. As shown in Table 1, more and more cities opened rail transit lines after 2004. By the end of 2013, 23 cities had created rail transit lines, and more than 35 cities had published rail transit plans. Notably, cities that own rail transit networks are concentrated in eastern regions of China, implying that economic levels are a major determinant of rail transit construction. Along with the rapid development of urbanization in China, mass transit, and urban rail transit in particular, will certainly enter another golden era.

Table 1. Overview of urban rapid rail transit systems in China (by the end of 2013).

City	Length (km)	Lines	Opening Year	City Location	City	Length (km)	Lines	Opening Year	City Location
Beijing	465.0	17	1969	Eastern	Wuhan	72.1	4	2004	Central
Tianjin	142.6	4	2004	Eastern	Guangzhou	263.9	9	1997	Eastern
Shenyang	55.1	2	2010	Eastern	Shenzhen	173.1	5	2004	Eastern
Dalian	87.0	3	2002	Eastern	Chongqing	169.9	5	2004	Western
Ha'erbin	27.3	1	2013	Eastern	Chengdu	50.5	2	2010	Western
Shanghai	567.4	14	1993	Eastern	Kunming	38.5	3	2012	Western
Nanjing	108.6	2	2005	Eastern	Xi'an	45.5	2	2011	Western
Hangzhou	48.0	1	2013	Eastern	Changchun	47.9	5	2002	Eastern

Source: Royal Flush iFind Database and Baidu Encyclopedia.

3.2. The Correlation between Rail Transit Facilities and Housing Prices

To achieve a basic understanding of the relationship between rail transit facilities and housing prices, we present temporal trends for the two variables (see Figure 1). Both total rail transit mileage levels and average housing prices have rapidly increased from 2002 to 2013. In regards to housing prices, the year 2008 can be considered the start of more significantly increasing housing price trends. In contrast, the turning point for rail transit mileage levels occurred in 2009. Similar characteristics for the two variables imply that there may be a high correlation between them.

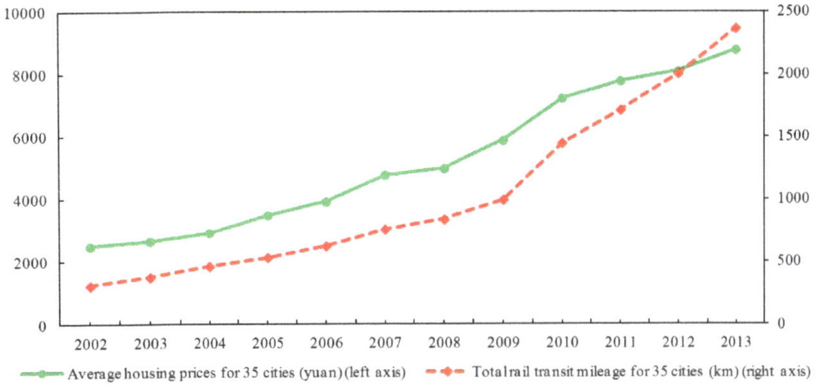

Figure 1. Total rail transit mileage records and average housing prices for 35 cities. Source: Royal Flush iFind Database [42] and *China Real Estate Statistical Yearbook* [43].

Table 2 presents Spearman correlational coefficients between rail transit mileage levels and housing prices for the cities that own metro lines. Overall, correlational coefficients in most of the cities are close to 1, and especially for larger cities, such as Beijing, Shanghai and Guangzhou, denoting a high correlation between rail transit facilities and housing prices. Next, we conduct a further analysis using an econometric model.

Table 2. Correlational coefficients for large cities.

City	Spearman	City	Spearman	City	Spearman
Beijing	0.9296 *** (7.9759)	Wuhan	0.8005 *** (4.2231)	Nanjing	0.9497 *** (9.5991)
Tianjin	0.8503 *** (5.1098)	Guangzhou	0.9736 *** (13.5017)	Changchun	0.8753 *** (5.7254)
Dalian	0.8849 *** (6.007)	Shenzhen	0.9406 *** (8.7651)		
Shanghai	0.9555 *** (10.2445)	Chongqing	0.9841 *** (17.5579)		

Spearman correlation coefficients were calculated using Eviews 6.0. *** denotes significance at the 1% level. *t*-statistics are reported in parentheses below correlational coefficients.

4. Methodology and Data

4.1. Econometric Model Specification

In light of the highly possible positive effect of rail transit facilities on housing prices, we use rail transit mileage as the main independent variable and economic fundamentals, demographic factors and educational resources as "control" variables. These variables are listed in Table 3. Our empirical model is described by the following equations:

$$\begin{aligned} \ln housingp_{it} = \; & c + \beta_1 \ln gdpp_{it} + \beta_2 unemploy_{it} + \beta_3 popgrowth_{it} + \beta_4 \ln popdensity_{it} \\ & + \beta_5 \ln landp_{it} + \beta_6 \ln invest_{it} + \beta_7 \ln college_{it} + \beta_8 \ln primary_{it} \\ & + \beta_9 \ln book_{it} + \beta_{10} \ln hospital_{it} + \beta_{11} \ln rail_{it} + \alpha_i + \varepsilon_{it} \end{aligned} \quad (1)$$

where the subscript i indexes cities, α_i denotes an individual specific effect term and ε_{it} the error term. Panel data are also called cross-sectional time-series data. A panel data set contains n entities or subjects, each of which includes T observations measured at 1 through t time period. Thus, the total number of observations in the panel data is $n \times T$. Cross-sectional data set is a type of one-dimensional data set and refers to a data set collected by observing subjects (such as individuals, firms or countries/regions at the same point of time. Panel data model and cross-sectional data model are econometric models estimated by various methods based on panel data set and cross-sectional data set, respectively. As noted above, the local government typically publishes rail transit network plans and construction periods a few years in advance. Thus, rail transit expectations may have unique effects on housing prices. We thus apply special specifications to the sample data by using the ln*rail* variable for the n (n = 1, 2, 3) period when estimating the models.

4.2. Data and Descriptive Statistics

Data for this study were gathered from the Royal Flush iFind Database [42], *China Real Estate Statistical Yearbook* [43], the *China City Statistical Yearbook* [44] and the *China Statistical Yearbook* [45], which includes data on 35 major Chinese cities for 2002 to 2013. To eliminate impacts of inflation on the economic variables, we treat 2000 as a base year and make price adjustments to the variables. We also use natural logarithms for the variables with the exception of those for unemployment and popgrowth. If one city does not own a metro line, variable values remain at zero.

Table 3. Definitions of variables.

Catalogue	Variable	Description
Dependent variable	Housing price (*housingp*)	This variable represents an average price for all commodity house types
Economic factors	GDP per capita (*gdpp*)	This variable measures per capita GDP
	Unemployment rate (*unemploy*)	This variable presents the unemployed population as a proportion of labourers
	Land price (*landp*)	This variable represents the urban land price
Demographic factors	Population growth rate (*popgrowth*)	This variable represents the annual ratio of natural demographic increase to the current average population
	Population density (*popdensity*)	This variable measures population density within a area of one square kilometre
Housing market factors	Completed investment in real estate market (*invest*)	This variable measures the annual investment spending that has completed in the real estate market, which differs from the planned investment spending
Educational factors	Colleges per capita (*college*)	This variable measures the number of colleges and universities per capita
	Primary schools per capita (*primary*)	This variable measures the number of primary schools per capita
	Books per capita (*book*)	This variable is a proxy of scientific resource and indirectly measure the level of science, technology and education in a certain city
Health facility	Hospitals per capita (*hospital*)	This variable measures the number of hospitals per capita
Rapid transit factor	Rail transit (*rail*)	The variable measures the operation mileage of rail transit systems

This study models housing prices as a function of eleven explanatory variables (listed in Table 3). Table 4 shows the descriptive statistics for the variables used in this paper. The average housing price is 5,238.803 yuan/m^2 for the sample period. The highest and lowest housing prices are 24,402 and 1464 yuan/m^2, respectively, denoting the existence of major housing price variations in different cities for the sample period. The per capita GDP level is 43,184.61 yuan. To compare the difference in the development of urban rapid rail transit between cities, we report the descriptive statistics for rail transit for the 35 cities and for the cities that have a rail transit system respectively. The means and standard deviations of rail transit for different groups are significantly distinguishing, which indicates a huge imbalance in the development of urban rail transit in China. On average, land prices are lower than housing prices and are more volatile than housing prices. It can be concluded that there are significant variations between the variables for the sample period.

Table 4. Descriptive statistics.

Variable	Obs	Mean	Std. Dev.	Min	Max
Housing price	420	5238.80	3544.25	1464.00	24,402.00
GDP per capita	420	43,184.61	25,982.08	6027.00	153,206.00
Unemployment rate	420	2.9816	1.3016	0.4295	8.1423
Population growth rate	420	5.0762	3.5636	−3.31	19.80
Population density	420	701.82	599.21	123.95	3809.08
Land price	420	3685.84	3922.92	492.00	21,395.00
Completed investment in real estate market	420	530.82	584.20	14.83	3483.41
Colleges per capita	420	5.83×10^{-6}	2.28×10^{-6}	9.31×10^{-7}	1.20×10^{-5}
Primary schools per capita	420	1.74×10^{-4}	8.46×10^{-5}	4.32×10^{-5}	4.37×10^{-4}
Books per capita	420	1.2631	1.2189	0.1252	9.3724
Hospitals per capita	420	5.99×10^{-5}	7.21×10^{-5}	2.26×10^{-5}	7.28×10^{-4}
Rail transit for the 35 cities	420	29.52	76.31	0	567.40
Rail transit for the cities that have a rail transit	180	68.73	104.50	10.00	567.40

To avoid estimation errors caused by a multicollinearity issue, we conduct a VIF (variance inflation factor) test and a correlational coefficient test. Tables 5 and 6 report the test results of VIF and correlational coefficient based on the pooled sample, respectively. From Table 5, we can observe that all VIF values are lower than 10 and the mean VIF is as low as 2.05, suggesting there doesn't exist a serious collinearity issue among the variables. In addition, the test results of correlational coefficients support the conclusion of the VIF test.

Table 5. Test results of variance inflation factor.

	VIF	1/VIF		VIF	1/VIF
ln*gdpp*	3.26	0.306333	*unemploy*	1.32	0.757158
ln*landp*	2.32	0.430743	ln*hospital*	1.39	0.721666
ln*invest*	3.9	0.256533	ln*primary*	1.74	0.573883
ln*rail*	2.05	0.488405	ln*college*	1.41	0.709692
popgrowth	1.38	0.725234	ln*book*	1.99	0.502968
ln*popdensity*	1.81	0.553413	Mean VIF	2.05	

Table 6. Correlational coefficients between variables.

	ln*housingp*1	ln*gdpp*	ln*landp*	ln*invest*	ln*rail*	*popgrowth*	ln*popdensity*	*unemploy*	ln*hospital*	ln*primary*	ln*college*	ln*book*
ln*housingp*1	1											
ln*gdpp*	0.89	1										
ln*landp*	0.70	0.56	1									
ln*invest*	0.77	0.73	0.66	1								
ln*rail*	0.50	0.49	0.38	0.61	1							
popgrowth	−0.10	−0.15	−0.24	−0.32	−0.33	1						
ln*popdensity*	0.53	0.48	0.57	0.50	0.43	−0.15	1					
unemploy	−0.44	−0.32	−0.37	−0.26	−0.06	−0.15	−0.13	1				
ln*hospital*	−0.38	−0.34	−0.35	−0.36	−0.28	0.04	−0.45	0.13	1			
ln*primary*	−0.34	−0.41	−0.21	−0.53	−0.17	0.30	−0.15	0.11	0.35	1		
ln*college*	0.15	0.23	−0.10	−0.02	−0.09	0.28	0.09	−0.01	0.12	−0.01	1	
ln*book*	0.52	0.58	0.33	0.34	0.39	−0.14	0.30	−0.13	−0.08	−0.06	0.30	1

5. Results and Discussion

5.1. *Effects of Urban Rail Transit Facilities on Housing Prices*

The empirical analysis consists of two parts: the direct effects and the expectation effects of rail transit facilities on housing prices. First, we use a panel data model to assess the direct effect of rail transit facilities on housing prices. We use the average commercial housing prices (*housingp1*) as the explained variable. The coefficient on rail transit facilities is estimated when the effects of other variables listed in Table 3 are controlled. To check the robustness, we also use the residential housing prices (*housingp2*) as the explained variable instead of the average commercial housing prices. In China, commercial house refers to buildings used to reside or to generate a profit, either from capital gain or rental income. Residential house refers to commercial house with living space less than 144 square meters. Thus, the term commercial house has a larger scope than a residential house.

Table 7 reports the regression results. Models 1 and 2 are estimated by an ordinary least squares (OLS) estimator using a pooled sample. Random-effect models are shown in the fourth and fifth columns, respectively. Fixed-effect estimates are listed in the last two columns. For the case of this study, the functional form of fixed-effect model is $\ln housingp_{it} = (c + \alpha_i) + X'_{it}\beta + \varepsilon_{it}$, while the functional form of random-effect model is $\ln housingp_{it} = c + X'_{it}\beta + (\alpha_i + \varepsilon_{it})$, where X'_{it} denotes the explanatory variable vector, and α_i is a fixed or random effect specific to individual (group) or time period. Error variances in fixed-effect model are constant, and intercepts are varying across group or time. In contrast, error variances in random-effect model are randomly distributed across group or time, and intercepts are constant. If we put all the panel data together and do not make any distinction between cross section and time series, we can of course run a regression over all the data using ordinary least squares. This is called a pooled OLS regression. Pooled OLS is often used as a rough and ready means of analyzing the data. It is a simple and quick benchmark to which more sophisticated regressions can be compared. The Hausman test for models with ln*housingp1* as the explained variable suggests that fixed-effect models exhibit better performance than the random-effect models. In contrast, the Hausman test for models with ln*housingp2* as the explained variable suggests that random-effect models exhibit better performance than fixed-effect models. According to the *F* statistic, we fail to reject the null hypothesis of coefficients being zero at the 1% level. The adjusted R^2 of each model are greater than 0.85. In particular, the adjusted R^2 for the suggested model (Model 3a) by Hausman test is as high as 0.8636, which indicates that all the independent variables can explain the housing prices variation by 86.36%. Overall, most of the coefficients on the "control" variables in the fifth column are significant at the 5% level or better and have the expected signs.

We shift focus to the impact of urban rail transit facilities on housing prices. The significance of the coefficient on ln*rail* in Model 3a precisely verifies our previous deduction. The sign on the coefficient, as expected, is positive, which is consistent with the findings in the case study documented by Seo *et al.* [19]. The magnitude of the coefficient is 0.0233, which means that each one-degree increase in rail transit mileage improves housing prices by 0.0233%. The coefficient, as one would expect, is much smaller in value than most of the other variables, such as per capita GDP, land price, completed investment and population growth, indicating that urban rapid trail transit generates relatively small but significant impact on housing prices. These findings suggest that economic factors such as income, population growth and employment are major determinants of housing prices.

The significant impact of urban rail transit facilities on housing prices suggests that the development of urban rail transit network plays an important role in house price increases in China during the past decade. Consequently, how the rail transit facilities influence the housing price is what we are interested in. There exist two possible channels (direct and scale effects) through which urban rail transit facilities affect housing prices. As previous case studies have shown, one's proximity to transportation nodes is significantly and positively associated with housing prices, and we refer to this as a direct effect. Compared to other transportation instruments (e.g., bus rail transit systems), metro rail transit systems have an entirely different impact on human activities and even on economic growth patterns. That is, metro rail transit systems extend the boundaries of urban functions due to their speed advantages relative to other mass transit systems, resulting in the convergence of suburban housing prices to central areas. From a microeconomics perspective, metro line system expansions may result in scale effects on housing prices. First, rail transit network developments create extra amenities that may lead to the development of additional accessibility premiums. Second, in terms of ripple effects, rail transit network expansion causes housing prices farther from stations to rise.

Some other findings are noteworthy. First, economic fundamentals such as per capita GDP and unemployment rate exhibit strong effects on the property values. Meanwhile, land price, as the cost proxy, shows strong impact on housing prices, which supports the view that cost drives housing prices [9,29,30]. Theses findings are consistent with the view that economic fundamentals are the major driving factors in determining housing prices [31–34]. Second, the coefficient on population is significant and presents the expected sign, which indicates that the demographic factors play an important role in forming housing price bubbles.

Table 7. Regression results of the basic models.

Model	Pooled-OLS		Random-Effects		Fixed-Effect	
	1a	1b	2a	2b	3a	3b
explained variable	ln*housingp*1	ln*housingp*2	ln*housingp*1	ln*housingp*2	ln*housingp*1	ln*housingp*2
ln*gdpp*	0.4733 *** (−16.6)	0.4890 *** (−16.39)	0.3689 *** (−8.99)	0.3699 *** (−8.71)	0.2486 *** (−5)	0.2481 *** (−4.77)
ln*landp*	0.1644 *** (−9.64)	0.1779 *** (−9.96)	0.2320 *** (−8.6)	0.2386 *** (−8.55)	0.3110 *** (−9.17)	0.3179 *** (−8.95)
ln*invest*	0.0641 *** (−4.05)	0.0588 *** (−3.55)	0.0907 *** (−4.02)	0.1070 *** (−4.58)	0.1182 *** (−4.15)	0.1392 *** (−4.67)
ln*rail*	0.0298 *** (−4.57)	0.0261 *** (−3.82)	0.0216 *** (−2.69)	0.0147 * (−1.76)	0.0233 *** (−2.65)	0.0163 * (−1.77)
popgrowth	0.0193 *** (−6.61)	0.0201 *** (−6.58)	0.0151 *** (−4.73)	0.0165 *** (−4.96)	0.0079 ** (−2.26)	0.0097 *** (−2.64)
ln*popdensity*	0.0356 ** (−2.08)	0.0523 *** (−2.92)	0.0282 (−0.9)	0.0326 (−1.01)	0.0798 * (−1.8)	0.0727 (−1.57)
unemploy	−0.0424 *** (−5.41)	−0.0452 *** (−5.51)	−0.0341 *** (−4.26)	−0.0373 *** (−4.47)	−0.0303 *** (−3.66)	−0.0334 *** (−3.86)
ln*college*	−0.0147 (−0.63)	−0.0132 (−0.54)	0.00529 (−0.16)	0.0217 (−0.62)	−0.0141 (−0.34)	0.00778 (−0.18)
ln*primary*	0.0794 (−1.65)	0.0434 (−0.86)	−0.202 ** (−2.51)	−0.182 ** (−2.20)	−0.444 *** (−4.31)	−0.394 *** (−3.66)
ln*hospital*	0.0104 (−0.49)	−0.0102 (−0.46)	0.0207 (−0.92)	0.0029 (−0.12)	0.0337 (−1.45)	0.0163 (−0.67)
ln*book*	0.0442 ** (−2.52)	0.0470 ** (−2.56)	0.0891 *** (−4.62)	0.103 *** (−5.12)	0.103 *** (−5.06)	0.121 *** (−5.68)
constant	1.260 *** (−4.32)	0.953 *** (−3.12)	2.060 *** (−4.82)	1.768 *** (−4)	2.641 *** (−4.96)	2.334 *** (−4.19)
Obs.	420	420	420	420	420	420
F/Wald-statistic	281.65 ***	281.59 ***	2439.09 ***	2487.53 ***	215.31 ***	217.06 ***
Adjusted-R^2	0.8805	0.8805	0.8575	0.8595	0.8636	0.8646
Hausman test	34.12 ***	6.36	34.12 ***	6.36	34.12 ***	6.36

t-statistics are reported in parentheses below the coefficient estimates. *, ** and *** denote significance at the 10%, 5% and 1% levels, respectively. Models a and b refer to the models with ln*housingp*1 and ln*housingp*2 as the explained variable, respectively.

Third, housing market factors show strong effects on housing prices. Completed investment in housing markets demonstrates a notable effect, which, to some extent, reflects the developers' expectations of housing prices. Fourth, scientific resource measured by books per capita generates a significantly positive impact on housing prices. In contrast, the effect of health facility per capita is insignificant. Fifth, the variable primary school per capita generates a significantly negative effect on housing prices, while the effect of the variable colleges per capita is insignificant. These findings imply that the scarcity of primary educational resource may increase housing prices, which is remarkably different from the effects of health facility per capita and scientific resource per capita. From the data set, we observe that the variable primary school per capita shows a diminishing trend with the population growth, which is opposite to health facility per capita and scientific resource per capita. Indeed, if the primary educational resource per capita is scarcer, the houses near the primary school are more expensive in China, which is more obvious in large cities.

5.2. *Rail Transit Expectation Effects on Housing Prices*

Real estate developers in China are keen to broadly advertise metro housing facilities from the planning of a rail transit line to its opening. While a metro line generally takes approximately 3–4 years to construct, overall transit network plans may be designed a few years earlier. As a result, the public is well informed of plans and of expected opening dates, resulting in differences between this variable and other endogenous variables. Thus, it is of enormous interest to investigate premiums of rail transit facilities on housing prices from the perspective of expectation.

To assess the impact of the expectation of new metro lines on housing price, we estimate different models with different expectation specification. Table 8 reports the estimates with fixed-effect specification. From the estimates of Model 5, the coefficient on ln*rail* is not significant at any significant levels, indicating that expectations of new metro lines 1 year prior to opening cannot generate significant effects on housing prices. Similar situations can be found in Models 6 and 7. Furthermore, coefficients on ln*rail* present a diminishing trend form Model 4 to Model 7, which seems to show that, with the planning of new lines, rail transit facilities generate more and more effects on housing prices, although most of the coefficients are insignificant.

Table 8. Regression results of the models that consider the expectation of new rail lines.

Model	4	5	6	7
ln*gdpp*	0.2486 ***	0.276 ***	0.277 ***	0.264 ***
	(−5)	(−5.19)	(−4.87)	(−3.04)
ln*landp*	0.3110 ***	0.389 ***	0.446 ***	0.535 ***
	(−9.17)	(−9.89)	(−9.38)	(−4.6)
ln*invest*	0.1182 ***	0.117 ***	0.103 ***	0.0768
	(−4.15)	(−4.01)	(−3.24)	(−1.16)
ln*rail*	0.0233 ***	0.0133	0.0108	0.00875
	(−2.65)	(−1.41)	(−1)	(−0.42)
popgrowth	0.0079 **	0.00930 **	0.00868 *	0.00574
	(−2.26)	(−2.5)	(−1.97)	(−1.1)

Table 8. *Cont.*

Model	4	5	6	7
ln*popdensity*	0.0798 *	0.0648	0.0309	−0.024
	(−1.8)	(−1.26)	(−0.56)	(−0.33)
unemploy	−0.0303 ***	−0.0283 ***	−0.0257 ***	−0.0197
	(−3.66)	(−3.31)	(−2.82)	(−1.24)
ln*college*	−0.0141	−0.0245	−0.0212	−2.8E-05
	(−0.34)	(−0.59)	(−0.50)	(−0.00)
ln*primary*	−0.444 ***	−0.384 ***	−0.414 ***	−0.475 **
	(−4.31)	(−3.56)	(−3.52)	(−2.13)
ln*hospital*	0.0337	0.0322	0.0292	0.0331
	(−1.45)	(−1.37)	(−1.17)	(−1.2)
ln*book*	0.103 ***	0.135 ***	0.168 ***	0.195 ***
	(−5.06)	(−5.89)	(−5.77)	(−3.13)
constant	2.641 ***	1.701 ***	1.524 **	1.462
	(−4.96)	(−2.98)	(−2.47)	(−1.65)
Obs.	420	385	350	315
F-statistic	215.31 ***	190.74 ***	151.36 ***	72.82 ***
Adjusted-R^2	0.8636	0.8609	0.8456	0.8296

t-statistics are reported in parentheses below the coefficient estimates. *, ** and *** denote significance at the 10%, 5% and 1% levels, respectively. Models 4, 5, 6, and 7 present the models based on current-period, one-period-ahead, two-period-ahead and three-period-ahead rail transit data, respectively.

6. Conclusions

The purpose of this study is to measure effects of urban rail infrastructure on housing values using a panel data model on 35 Chinese cities for 2002 to 2013. Previous studies have focused on relationships between urban transit systems and housing prices (e.g., between housing prices and high speed railway systems [20], road and rail systems [20], rail accessibility levels [46–48], and highway and light rail accessibility levels [49]. However, nearly all of the studies presented above are case studies. To our knowledge, no city-level panel data have been used to assess urban rail transit system impacts on housing prices. During estimation, numerous independent variables selected based on the existing literature were included to control and measure rail transit effects.

The main results of the study show that the rail transit facilities have significantly positive effects on housing prices. Quantitatively, each 1% increase in rail transit mileage improves housing prices by 0.0233%. As one would expect, such effects are smaller than those of some other variables such as per capita GDP, land price, real estate investment and population growth, which are recognized as fundamental determinants of housing price. We also determine whether rail transit expectations have significant effects on housing prices and use various sample specifications to test expectation effects. The expectations of new rail transit lines cannot generate statistically significant effects on housing price. It can be concluded that urban rail transit systems play an indispensable role in determining housing prices.

This study is of practical significance, as it provides relative information for policy makers to use when applying regulatory powers to improve urban functions while stabilizing real estate prices. First, the central government should take utmost precautions to mitigate economic fluctuations. Second, local governments should increase or reduce the supply of urban land to affect land prices and the cost of development according to changes in real estate market conditions. Third, in order to mitigate real estate bubbles in big cities with highly-concentrated resources like Beijing, Shanghai, Guangzhou and Shenzhen, the central government should allocate more infrastructure spending, education and human resources to small cities. Fourth, the urban governor should take measures to control demographic growth, which contributes to stabilize housing prices and to increase the infrastructure per capita. The results of this study may also help home buyers select reasonable times and locations in making beneficial housing purchases. As China's real estate bubble is at risk of bursting, it is necessary for policy makers to take the endogenous factors such as urban rail transit into account and to diminish such threats to the housing market. The findings of this study can help home buyers and investors reasonably evaluate the value of rail accessibility.

Acknowledgments: We gratefully acknowledge the financial support from the National Natural Science Foundation of China (No. 71473036). The authors are grateful to three anonymous reviewers whose comments have helped to improve the manuscript.

Author Contributions: In this paper, Xu Zhang designed the research programs, analyzed the results, and completed the writing of the results and discussion; Xiaoxing Liu committed to the data collection and completed the writing of the introduction; Jianqin Hang carried out the data process and completed the writing work of corresponding parts; Dengbao Yao and Guangping Shi completed the writing of Sections 2 and 3. Both authors have read and approved the final manuscript.

Conflicts of Interest: The authors declare no conflict of interest.

References

1. Yigitcanlar, T.; Kamruzzaman, M.; Teriman, S. Neighborhood sustainability assessment: Evaluating residential development sustainability in a developing country context. *Sustainability* **2015**, *7*, 2570–2602.
2. Wei, Y.; Huang, C.; Lam, P.T.; Sha, Y.; Feng, Y. Using urban-carrying capacity as a benchmark for sustainable urban development: An empirical study of Beijing. *Sustainability* **2015**, *7*, 3244–3268.

3. Shimizu, C. Estimation of Hedonic single-family house price function considering neighborhood effect variables. *Sustainability* **2014**, *6*, 2946–2960.
4. Shen, L. Are house prices too high in China? *China Econ. Rev.* **2012**, *23*, 1206–1210.
5. Ren, Y.; Xiong, C.; Yuan, Y. House price bubbles in China. *China Econ. Rev.* **2012**, *23*, 786–800.
6. Feng, Q.; Wu, G. Bubble or riddle? An asset-pricing approach evaluation on China's housing market. *Econ. Modell.* **2015**, *46*, 376–383.
7. Quigley, J. A simple hybrid model for estimating real estate price indexes. *J. Hous. Econ.* **1995**, *4*, 1–12.
8. Leung, C.K.Y. Economic growth and increasing housing prices. *Pac. Econ. Rev.* **2003**, *8*, 183–190.
9. Glaeser, E.L.; Gyourko, J.; Saks, R.E. Urban growth and housing supply. *J. Econ. Geo.* **2006**, *6*, 71–89.
10. Tan, Z.; Wu, S. A comparison of two housing markets. *Appl. Econ. Lett.* **2014**, *21*, 118–124.
11. Parker, J.A. Comments and discussion on: Case KE. *Brook. Pap. Econ. Act.* **2000**, *2*, 146–159.
12. Zhang, Y.; Liu, X.; Ding, Y.; Su, A. Assessing the impact of the demographic dividend on real estate prices: Empirical evidence from China. *Appl. Econ. Lett.* **2015**, *22*, 1450–1456.
13. Huston, J.H.; Spencer, R.W. Housing and behavioral factors. *Appl. Econ. Lett.* **2014**, *21*, 215–219.
14. Zhang, M.; Meng, X.; Wang, L.; Xu, T. Transit development shaping urbanization: Evidence from the housing market in Beijing. *Habitat Int.* **2014**, *44*, 545–554.
15. Hewitt, C.M.; Hewitt, W.E. The effect of proximity to urban rail on housing prices in Ottawa. *J. Publ. Trans.* **2012**, *15*, 43–65.
16. Jun, M. Redistributive effects of bus rapid transit (BRT) on development patterns and property values in Seoul, Korea. *Transp. Policy* **2012**, *19*, 85–92.
17. Kim, K.; Lahr, M. The impact of Hudsone-Bergen light rail on residential property appreciation. *Pap. Reg. Sci.* **2013**.
18. Chatman, D.; Tulach, N.; Kim, K. Evaluating the economic impacts of light rail by measuring home appreciation: A first look at New Jersey's River Line. *Urban Stud.* **2012**, *49*, 467–487.
19. Seo, K.; Golub, A.; Kuby, M. Combined impacts of highways and light rail transit on residential property values: A spatial hedonic price model for Phoenix, Arizona. *J. Trans. Geo.* **2014**, *41*, 53–62.
20. Andersson, D.E.; Shyr, O.F.; Fu, J. Does high-speed rail accessibility influence residential property prices? Hedonic estimates from southern Taiwan. *J. Trans. Geo.* **2010**, *18*, 166–174.
21. Lewis-Workman, S.; Brod, D. Measuring the neighbourhood benefits of rail transit accessibility. *J. Transp. Res. Board* **1997**, *1576*, 147–153.
22. Zheng, J.; Liu, H. Impacts of Shenzhen metro rail on housing price. *J. Chn. Rail. Soc.* **2005**, *27*, 11–18. (In Chinese)

23. Zhang, H.; Ma, J.; Zhu, H. Price influence of urban mass transit on housing projects along the line. *J. Beijing Jiaotong Univ.* **2007**, *31*, 10–13. (In Chinese)
24. Zhang, M. Bus versus rail: Meta-analysis of cost characteristics, carrying capacities, and land use impacts. *Transp. Res. Rec. J. Transp. Res. Board Natl. Acad.* **2009**, *2110*, 87–95.
25. De Vany, A.S. An economic model of airport noise pollution in an urban environment. In *Theory and Measurement of Economic Externalities*; Lin, S.A.Y., Ed.; Academic Press: New York, NY, USA, 1976; pp. 205–216.
26. Wen, H.Z.; Jia, S.H. Housing characteristics and hedonic price: Based on hedonic price model. *J. Zhejiang Unive. (Eng. Sci.)* **2004**, *38*, 38–42. (In Chinese)
27. Feng, H.; Lu, M. Choose a school through buying a house: Evidences and policy implications of how education effect the housing price. *World Econ.* **2010**, *12*, 89–104.
28. Wen, H.; Zhang, Y.; Zhang, L. Do educational facilities affect housing price? An empirical study in Hangzhou, China. *Habitat Int.* **2014**, *42*, 155–163.
29. Quigley, J.M. Real estate prices and economic cycles. *Int. Re. Est. Rev.* **1999**, *2*, 1–20.
30. Fortura, P.; Kushner, J. Canadian inter-city house price differentials. *Am. Real Estate Urban Econ. Assoc. J.* **1986**, *14*, 525–536.
31. Zou, G.L.; Chau, K.W. Determinants and sustainability of house prices: The case of Shanghai, China. *Sustainability* **2015**, *7*, 4524–4548.
32. Hwang, M.; Quigley, J. Economic fundamentals in local housing markets: Evidence from US metropolitan regions. *J. Reg. Sci.* **2006**, *46*, 425–453.
33. Ceron, J.A.; Suarez, J. *Hot and Cold Housing Markets: International Evidence*; Centre for Economic Policy Research: Washington, DC, USA, 2006; Volume 5411.
34. Agnello, L.; Schuknecht, L. Booms and busts in housing markets: Determinants and implications. *J. Hous. Econ.* **2011**, *20*, 171–190.
35. Kasparova, D.; White, M. The responsiveness of house prices to macroeconomic forces: A cross-country comparison. *Eur. J. Hous. Poli.* **2001**, *1*, 385–416.
36. Yang, S. Objectively look on the problem of rising housing price. *China Re. Est. Inf.* **2003**, *2*, 4–5.
37. Bao, Z.H. How to look on housing price in our country. *China Re. Est.* **2004**, *1*, 18–19.
38. Wen, H.; Goodman, A.C. Relationship between urban land price and housing price: Evidence from 21 provincial capitals in China. *Habitat Int.* **2003**, *40*, 9–17.
39. Du, H.; Ma, Y.; An, Y. The impact of land policy on the relation between housing and land prices: Evidence from China. *Q. Rev. Econ. Financ.* **2011**, *51*, 19–27.
40. Jud, D.; Winkler, D. The dynamics of metropolitan housing prices. *J. Real Est. Res.* **2002**, *23*, 29–45.
41. Takáts, E. Aging and house prices. *J. Hous. Econ.* **2012**, *21*, 131–141.
42. Economic database. Royal Flush iFind Database. Available online: http://www.51ifind.com (accessed on 18 April 2016).
43. National Bureau of Statistics of China. *China Real Estate Statistical Yearbook, 2003–2014*; China Statistics Press: Beijing, China, 2003–2014.
44. National Bureau of Statistics of China. *China City Statistical Yearbook, 2003–2014*; China Statistics Press: Beijing, China, 2003–2014.

45. National Bureau of Statistics of China. *China Statistical Yearbook, 2003–2014*; China Statistics Press: Beijing, China, 2003–2014.
46. Cervero, R. Transit-based housing in the San Francisco bay area: Market profiles and rent premiums. *Trans. Q.* **1996**, *50*, 33–47.
47. Hess, D.B.; Almeida, T.M. Impact of proximity to light rail rapid transit on station-area property values in Buffalo, New York. *Urban Study* **2007**, *44*, 1041–1068.
48. Golub, A.; Guhathakurta, S.; Sollapuram, B. Spatial and temporal capitalization effects of light rail in Phoenix: From conception, planning, and construction to operation. *J. Plan. Edu. Res.* **2012**, *32*, 415–429.
49. Ryan, S. The value of access to highways and light rail transit: Evidence for industrial and office firms. *Urban Study* **2005**, *42*, 751–764.

MDPI AG
St. Alban-Anlage 66
4052 Basel, Switzerland
Tel. +41 61 683 77 34
Fax +41 61 302 89 18
http://www.mdpi.com

Sustainability Editorial Office
E-mail: sustainability@mdpi.com
http://www.mdpi.com/journal/sustainability

www.ingramcontent.com/pod-product-compliance
Lightning Source LLC
Chambersburg PA
CBHW061553220926
51850CB00005B/328